Diagnosis, Classification, and Monitoring of Pulmonary Diseases

Diagnosis, Classification, and Monitoring of Pulmonary Diseases

Guest Editors

Barbara Ruaro
Paola Confalonieri
Francesco Salton

MDPI

Basel • Beijing • Wuhan • Barcelona • Belgrade • Novi Sad • Cluj • Manchester

Guest Editors
Barbara Ruaro
Department of Medical
Surgical and Health Sciences
University of Trieste
Trieste
Italy

Paola Confalonieri
Department of Medical
Surgical and Health Sciences
University of Trieste
Trieste
Italy

Francesco Salton
Department of Medical
Surgical and Health Sciences
University of Trieste
Trieste
Italy

Editorial Office
MDPI AG
Grosspeteranlage 5
4052 Basel, Switzerland

This is a reprint of the Special Issue, published open access by the journal *Diagnostics* (ISSN 2075-4418), freely accessible at: https://www.mdpi.com/journal/diagnostics/special_issues/PRX3VQ75A5.

For citation purposes, cite each article independently as indicated on the article page online and as indicated below:

Lastname, A.A.; Lastname, B.B. Article Title. *Journal Name* **Year**, *Volume Number*, Page Range.

ISBN 978-3-7258-6550-5 (Hbk)
ISBN 978-3-7258-6551-2 (PDF)
https://doi.org/10.3390/books978-3-7258-6551-2

Contents

About the Editors

Barbara Ruaro

Barbara Ruaro, MD, European PhD, is an Associate Professor of Respiratory Diseases at the University of Trieste and a Senior Medical Consultant in the Pulmonology Department, University Hospital of Cattinara, Trieste, Italy. She is a licensed physician in Italy (University of Turin). She completed her specialization in Genoa (Italy) and pursued fellowships in Ghent (Belgium) and Chicago (USA). She later also earned her European PhD. She then served as a Postdoctoral Research Fellow at the University of Genoa and was awarded an Investigator Starting Grant at a National Congress of Medicine. Barbara is a frequent speaker at national and international events. She currently coordinates the Specialization Course on Respiratory Diseases at the University of Trieste. She leads multiple research projects and has authored numerous papers, posters, and book chapters. Her research portfolio includes autoimmune diseases with lung involvement, telemedicine, imaging in pulmonary and rheumatic diseases, and COVID-19. Her scholarly output comprises more than 160 publications (h-index 44) and more than 100 invited lectures at international and national congresses.

Paola Confalonieri

Paola Confalonieri is a medical doctor at the Pulmonary Department of the University Hospital of Trieste (Italy), specialized in Respiratory Diseases at the University of Trieste with a focus on interstitial lung diseases, pulmonary fibrosis, and vascular lung diseases. She is earning a PhD at the University of Trieste in Personalized Medicine and Innovative Therapies, contributing to studies on molecular therapies and innovative diagnostic approaches for progressive pulmonary fibrosis. She is a member of AIPO and ERS and serves as coordinator of the AIPO scientific group on Interstitial and Vascular Lung Diseases, with numerous international publications in the field.

Francesco Salton

Francesco Salton is a Medical Consultant at the Pulmonology Department of the University Hospital of Trieste (Italy). He graduated in Medicine and Surgery and specialized in Respiratory Diseases at the University of Trieste, with previous experience in translational and clinical research. He is an author and co-author of over 100 international scientific publications. His main interests include interstitial lung diseases, community-acquired pneumonia, ARDS, innovative therapies, and the application of artificial intelligence to clinical pulmonology. Since 2024, he has coordinated the Young Pulmonologists section of the Italian Thoracic Society (AIPO/ITS).

Preface

This Special Issue is primarily intended to inform, inspire, and assist researchers engaged in the study of lung diseases by highlighting recent advancements, emerging perspectives, and evolving methodologies within this dynamic field. Authored by the Guest Editors, Professor Barbara Ruaro, Dr Paola Confalonieri, and Dr Francesco Salton, it aims to present a comprehensive overview of innovative imaging techniques, novel diagnostic approaches, and strategies for identifying biological pathways and biomarkers relevant to lung health. The scope encompasses recent scientific progress, particularly developments from the past year, reflecting the heightened global interest in respiratory health driven by the COVID-19 pandemic. The purpose of this work is to showcase technological and conceptual innovations that have shaped current research paradigms, fostering a deeper understanding of disease mechanisms, triggers, and progression. This Reprint is addressed to researchers, clinicians, and academics dedicated to advancing lung disease research and improving patient outcomes.

Barbara Ruaro, Paola Confalonieri, and Francesco Salton
Guest Editors

diagnostics MDPI

Editorial

Editorial for "Diagnosis, Classification, and Monitoring of Pulmonary Diseases"

Paola Confalonieri, Francesco Salton and Barbara Ruaro *

Pulmonology Unit, Department of Medical Surgical and Health Sciences, Hospital of Cattinara, University of Trieste, 34149 Trieste, Italy
* Correspondence: barbara.ruaro@yahoo.it

This Special Issue offers a comprehensive overview of recent advances and innovative approaches in the field of lung disease research, emphasizing the importance of technological and conceptual innovations that are shaping our understanding of respiratory conditions. The collected studies highlight the development of novel diagnostic tools, biomarkers, and risk stratification strategies that aim to improve early detection, personalized treatment, and patient management.

Several articles focus on enhancing prognostic accuracy through the identification of biomarkers—whether in pulmonary hypertension, COPD, or lung cancer—that can predict disease progression and outcomes more effectively [1–4]. Others explore cutting-edge imaging techniques, such as advanced microscopy, ultrasound, and oscillometry, which provide deeper insights into lung pathology while offering less invasive options for diagnosis and monitoring [4–8].

In addition, the research demonstrates progress in procedural safety and efficacy, exemplified by studies on minimally invasive pleuroscopy and real-world treatment outcomes. Emerging diagnostic modalities, like the thermal imaging of exhaled CO_2 and microvascular assessments via non-invasive imaging, open new avenues for non-contact, rapid detection of respiratory abnormalities [4–8].

Importantly, the collection underscores a trend toward integrating molecular, imaging, and clinical data to foster personalized medicine approaches, ultimately aiming to improve patient outcomes across a spectrum of lung diseases [1–4]. This body of work not only advances scientific understanding but also highlights the critical role of research and innovation in addressing the complex challenges faced in respiratory healthcare.

The importance of ongoing research and studies cannot be overstated, as they serve as the foundation for the development of new diagnostic approaches that could revolutionize how we detect and manage pulmonary diseases. By leveraging innovative technologies and methodologies, researchers are paving the way for more accurate, early diagnoses that are crucial for effective intervention [4–8]. This progress is vital, considering the often insidious progression of many lung conditions, which can lead to delayed treatment and poorer outcomes.

Furthermore, continuous research efforts are essential for translating scientific discoveries into clinical practice, ensuring that patients benefit from the latest advancements. The integration of novel diagnostic strategies not only enhances our ability to identify diseases at earlier stages but also enables more personalized and targeted therapies, reducing side effects and increasing treatment efficacy.

Ultimately, sustained investment in research fuels innovation, leading to the discovery of new biomarkers, imaging techniques, and therapeutic options. This, in turn, can

https://doi.org/10.3390/diagnostics15192520

significantly transform patient care, improve quality of life, and reduce the burden of lung diseases on healthcare systems worldwide. The promising advances documented here exemplify how dedication to scientific inquiry and technological development can propel the field forward, opening new horizons for future breakthroughs and more effective management of pulmonary pathologies.

Funding: This research received no external funding.

Conflicts of Interest: The authors declare no conflicts of interest.

References

1. Celli, B.; Fabbri, L.; Criner, G.; Martinez, F.J.; Mannino, D.; Vogelmeier, C.; Montes de Oca, M.; Papi, A.; Sin, D.D.; Han, M.K.; et al. Definition and Nomenclature of Chronic Obstructive Pulmonary Disease: Time for Its Revision. *Am. J. Respir. Crit. Care Med.* **2022**, *206*, 1317–1325. [CrossRef] [PubMed]
2. Screm, G.; Mondini, L.; Salton, F.; Confalonieri, P.; Trotta, L.; Barbieri, M.; Romallo, A.; Galantino, A.; Hughes, M.; Lerda, S.; et al. Vascular Endothelial Damage in COPD: Where Are We Now, Where Will We Go? *Diagnostics* **2024**, *14*, 950. [CrossRef] [PubMed]
3. Desai, U.; Joshi, J.M. Impulse oscillometry. *Adv. Respir. Med.* **2019**, *87*, 235–238. [CrossRef] [PubMed]
4. Ruaro, B.; Baratella, E.; Caforio, G.; Confalonieri, P.; Wade, B.; Marrocchio, C.; Geri, P.; Pozzan, R.; Andrisano, A.G.; Cova, M.A.; et al. Chronic Thromboembolic Pulmonary Hypertension: An Update. *Diagnostics* **2022**, *12*, 235. [CrossRef] [PubMed]
5. Monteleone, G.; Cameli, P.; Bonella, F. The role of heat shock protein 90 in idiopathic pulmonary fibrosis: State of the art. *Eur. Respir. Rev.* **2025**, *34*, 240147. [CrossRef] [PubMed]
6. Baratella, E.; Cernic, S.; Minelli, P.; Furlan, G.; Crimì, F.; Rocco, S.; Ruaro, B.; Cova, M.A. Accuracy of CT-Guided Core-Needle Biopsy in Diagnosis of Thoracic Lesions Suspicious for Primitive Malignancy of the Lung: A Five-Year Retrospective Analysis. *Tomography* **2022**, *8*, 2828–2838. [CrossRef] [PubMed]
7. Rotondo, C.; Busto, G.; Rella, V.; Barile, R.; Cacciapaglia, F.; Fornaro, M.; Iannone, F.; Lacedonia, D.; Quarato, C.M.I.; Trotta, A.; et al. Transthoracic Lung Ultrasound in Systemic Sclerosis-Associated Interstitial Lung Disease: Capacity to Differentiate Chest Computed-Tomographic Characteristic Patterns. *Diagnostics* **2025**, *15*, 488. [CrossRef] [PubMed]
8. Connolly, C.M.; Gupta, L.; Fujimoto, M.; Machado, P.M.; Paik, J.J. Idiopathic inflammatory myopathies: Current insights and future frontiers. *Lancet Rheumatol.* **2024**, *6*, e115–e127. [CrossRef] [PubMed]

diagnostics

MDPI

Article

Evaluating NT-proBNP-to-Albumin (NTAR) and RDW-to-eGFR (RGR) Ratios as Biomarkers for Predicting Hospitalization Duration and Mortality in Pulmonary Arterial Hypertension (PAH) and Chronic Thromboembolic Pulmonary Hypertension (CTEPH)

Dragos Gabriel Iancu [1,2], Liviu Cristescu [1,3,*], Razvan Gheorghita Mares [2], Andreea Varga [3] and Ioan Tilea [2]

1 Doctoral School, George Emil Palade University of Medicine, Pharmacy, Science, and Technology of Targu Mures, 540142 Targu Mures, Romania; dragos-gabriel.iancu@umfst.ro

2 Faculty of Medicine, George Emil Palade University of Medicine, Pharmacy, Science, and Technology of Targu Mures, 540142 Targu Mures, Romania; razvan.mares@umfst.ro (R.G.M.); ioan.tilea@umfst.ro (I.T.)

3 Faculty of Medicine in English, George Emil Palade University of Medicine, Pharmacy, Science, and Technology of Targu Mures, 540142 Targu Mures, Romania; andreea.varga@umfst.ro

* Correspondence: liviu.cristescu@umfst.ro

Abstract: Background/Objectives: Prognostic biomarkers are essential for guiding the clinical management of pulmonary hypertension (PH). This study aimed to assess both established and novel biomarkers—specifically, the red cell distribution width-to-estimated glomerular filtration rate ratio (RGR) and the NT-proBNP-to-albumin ratio (NTAR)—for their ability to predict length of hospital stay (LOS), prolonged LOS (ELOS), in-hospital mortality, and 3-month all-cause mortality in patients with pulmonary arterial hypertension (PAH) and chronic thromboembolic pulmonary hypertension (CTEPH). **Methods**: A retrospective analysis was conducted on 275 PH-related hospital regular admissions (148 PAH; 127 CTEPH). Established biomarkers—including serum albumin, neutrophil-to-lymphocyte ratio (NLR), Log NT-proBNP, red cell distribution width (RDW), and estimated glomerular filtration rate (eGFR)—as well as novel indices (RGR, and NTAR) were examined for their relationships with LOS, ELOS, in-hospital mortality, and 3-month all-cause mortality. Spearman correlation, univariate logistic regression, and ROC analyses evaluated biomarker relationships and predictive performance. **Results**: Serum albumin independently predicted in-hospital and 3-month mortality in PAH, while in CTEPH, it inversely correlated with LOS and strongly predicted prolonged hospitalization and mortality (AUC = 0.833). NLR had limited correlation with LOS but predicted mortality across both groups. RDW correlated weakly with LOS, significantly predicting prolonged hospitalization (threshold > 52.1 fL) in PAH but not in CTEPH. Preserved renal function (eGFR > 60 mL/min/1.73 m^2) was inversely associated with LOS in CTEPH patients, suggesting a protective effect. Additionally, reduced eGFR significantly predicted mortality in both PAH (AUC = 0.701; optimal cut-off $\leq$ 97.4 mL/min/1.73 m^2) and CTEPH (AUC = 0.793; optimal cut-off $\leq$ 59.2 mL/min/1.73 m^2) groups. NTAR (AUC = 0.817) outperformed Log NT-proBNP alone in predicting extended hospitalization and mortality, whereas RGR correlated with LOS and predicted in-hospital mortality. Phenotype-specific analysis demonstrated that inflammatory and renal biomarkers had a stronger prognostic impact in CTEPH. **Conclusions**: Stratification by PH phenotype highlighted the greater prognostic significance of inflammatory and renal indices, particularly in patients with CTEPH. Incorporating NTAR and RGR into clinical workflows may enhance risk stratification and enable more precisely targeted interventions to improve outcomes in pulmonary hypertension.

https://doi.org/10.3390/diagnostics15172126

Keywords: pulmonary arterial hypertension; chronic thromboembolic pulmonary hypertension; prognostic; biomarkers; length of hospital stay; mortality

1. Introduction

Pulmonary hypertension (PH) is a complex and multifaceted disease characterized by elevated pulmonary arterial pressure and progressive right ventricular dysfunction, ultimately leading to right heart failure and increased mortality [1].

The estimated global prevalence of PH varies depending on the underlying etiology, affecting approximately 1% of the general population and up to 10% of individuals aged over 65 years [2,3]. Despite therapeutic advancements, PH remains a significant clinical challenge due to its heterogeneous pathophysiology, late diagnosis, and limited therapeutic options [4].

In PH management, length of hospital stay (LOS) serves as a surrogate marker for disease severity, therapeutic efficacy, and healthcare resource utilization. Prolonged LOS (ELOS) is often indicative of complications or suboptimal care strategies [5,6]. Data from the REVEAL Registry indicate a mean LOS of 6.5 days for patients with Group 1 PH (PAH), while analyses of PH admissions via the emergency department report an average LOS of 6.9 days [7,8].

Hypoalbuminemia is a frequent manifestation in PH, reflecting systemic inflammation and endothelial dysfunction that increase capillary permeability and promote fluid extravasation into the lungs, myocardium, and splanchnic vasculature. Reduced serum albumin also alters the pharmacokinetics of loop diuretics, contributing to diuretic resistance and worsening hemodynamic status. Beyond its pathophysiological roles, hypoalbuminemia often signals underlying malnutrition or chronic inflammatory states—both established predictors of prolonged hospitalization and higher mortality. In a retrospective cohort of 163 PAH patients, low albumin levels independently predicted longer hospital stays, higher morbidity, and increased risk of death [9–13].

The neutrophil-to-lymphocyte ratio (NLR)—a simple, cost-effective marker of systemic inflammation—has demonstrated robust prognostic value in pulmonary hypertension [14]. Elevated NLR is consistently linked to longer hospital stays, higher rates of morbidity, and increased all-cause mortality [15,16]. In a large multicenter study among PH cohort patients, an NLR above the cohort median independently predicted both early and long-term mortality [17]. These findings support the incorporation of NLR into routine risk stratification algorithms, facilitating early identification of high-risk patients.

The 2022 ESC/ERS Guidelines for the Diagnosis and Treatment of Pulmonary Hypertension highlight NT-proBNP as a pivotal biomarker in evaluating right ventricular function and overall prognosis in PH patients [1]. Due to its longer plasma half-life, NT-proBNP is a stable indicator of cardiac dysfunction [18]. It provides valuable insights into disease severity and progression and is influenced by age and sex, showing higher concentrations in older individuals and females [19,20].

Red cell distribution width (RDW)—an index of erythrocyte size variability (anisocytosis)—has been linked to adverse outcomes across cardiovascular and pulmonary diseases. In PH, elevated RDW is associated with increased mortality and prolonged hospital stays. Similar associations have been noted in chronic obstructive pulmonary disease (COPD) patients with concomitant PH [21–23]. Importantly, Hampole et al. demonstrated that RDW independently predicts mortality in PH patients and outperforms NT-proBNP in prognostic accuracy [24].

Renal function, as measured by the estimated glomerular filtration rate (eGFR), is a critical determinant of clinical outcomes in patients with PAH [25]. Data from the REVEAL Registry demonstrated that a decline in eGFR of 10% or more over a year independently predicted poorer survival in PAH patients [26]. Additionally, a study conducted by Shah et al. found that serum creatinine and eGFR were independent risk predictors of mortality in PAH patients [27].

The aim of this study was to evaluate the prognostic performance of two novel laboratory-based biomarker ratios—the red cell distribution width-to-eGFR ratio (RGR) and the NT-proBNP-to-albumin ratio (NTAR)—for predicting LOS, ELOS, in-hospital mortality, and 3-month all-cause mortality in patients with pulmonary arterial hypertension and chronic thromboembolic pulmonary hypertension. Their discriminative accuracy was benchmarked against five established biomarkers (albumin, NLR, NT-proBNP, RDW, and eGFR). Multivariable logistic regression analyses were then conducted to identify independent prognostic indicators.

2. Materials and Methods

2.1. Study Design and Population

This retrospective observational study was carried out in the Department of Internal Medicine II—Cardiology at the County Emergency Clinical Hospital Targu Mures, Romania, from 1 September 2015 to 31 October 2024. To the best of our knowledge, it is the first European investigation to evaluate the prognostic utility of two novel biomarker ratios in pulmonary hypertension—NTAR and RGR—in relation to hospital LOS, ELOS, in-hospital mortality, and 3-month all-cause mortality. These exploratory indices were benchmarked against established prognostic markers to determine their incremental predictive value. The analytic cohort comprised 468 consecutive PH-related admissions, identified through the hospital's electronic information system. As a complete census of our registry, no sample size calculation was performed [28].

2.2. Inclusion and Exclusion Criteria

All consecutive hospital admissions of patients who met the ESC/ERS 2015 or 2022 criteria for PAH or CTEPH and remained hospitalized for at least 48 h were eligible for inclusion [1,29].

Admissions complicated by active inflammation or infection (e.g., sepsis), solid or hematologic malignancies (including myelodysplastic syndromes), autoimmune disorders, primary or infectious liver disease, or end-stage renal disease (CKD stage V) were ineligible, as were records with incomplete clinical data. After applying these criteria, the study cohort consisted of 275 admissions, of which 148 were PAH and 127 were CTEPH.

2.3. Treatment, Management, and Disease Severity Parameters

All patients received guideline-directed medical therapy in line with the current ESC/ERS recommendations for PAH and CTEPH, ensuring consistent treatment exposure across the cohort. Disease severity was quantified by WHO-FC and circulating NT-proBNP concentrations to capture both functional impairment and hemodynamic stress. Extended length of stay (ELOS) was defined as a hospitalization exceeding seven days.

2.4. Laboratory Assessments

Venous blood was collected after an 8-hour fast and within two hours of hospital admission. Complete blood counts—including total white blood cell (WBC) count with neutrophil and lymphocyte differentials—were performed on a Sysmex XN-550 analyzer (Sysmex Corporation, Kobe, Japan). Serum biochemistry was measured using the Konelab Prime 60i system (Thermo Fisher Scientific Inc., Waltham, MA, USA), and NT-proBNP concentrations were quantified with the Nano-Checker™ 710 Reader (Nano-Ditech Corporation, Cranbury, NJ, USA). All assays were carried out in an ISO 15189-accredited laboratory.

2.5. Comorbidities

Comorbidity conditions were systematically recorded and their prevalence calculated separately for the PAH and CTEPH cohorts.

Documented cardiovascular disorders were systemic hypertension, coronary artery disease, atrial fibrillation, and prior deep vein thrombosis (DVT). Metabolic conditions such as type II diabetes mellitus (T2DM) and thyroid dysfunction were noted, while respiratory diagnoses included obstructive sleep apnea, asthma, COPD, and other chronic lung diseases. A documented history of SARS-CoV-2 infection was also considered.

2.6. LOS, ELOS, and Mortality Assessment

Based on the REVEAL Registry, which reports a mean LOS of 6.9 days, we defined ELOS in our study as a hospitalization duration exceeding 7 days [7,8].

Mortality outcomes were evaluated for both in-hospital mortality and post-discharge (3 months). In-hospital mortality was defined as any death occurring during the index hospital admission. Post-discharge mortality was assessed through electronic medical records and structured phone calls with patients' families or caregivers, in compliance with the General Data Protection Regulation (GDPR).

2.7. Proposed New Biomarkers

In addition to established laboratory parameters, two novel, unit-independent indices were introduced to enhance prognostic stratification—the RDW-to-eGFR ratio (RGR) and the NT-proBNP-to-albumin ratio (NTAR). Both indices are calculated as simple quotients of routinely measured biomarkers, yielding dimensionless metrics that obviate unit conversion challenges and support seamless incorporation into clinical risk models. Detailed calculation methods and analytical performance for RGR and NTAR are outlined in the subsequent sections.

2.7.1. RDW-SD-to-eGFR Ratio (RGR)

Renal impairment reduces erythropoietin production, leading to increased anisocytosis and elevated RDW. The estimated glomerular filtration rate—calculated using the 2021 CKD-EPI equation—served as the measure of renal function.

The RGR reflects both anisocytosis and ineffective erythropoiesis—often driven by chronic inflammation, hypoxemia, or nutritional deficiencies—and renal function, as indicated by the eGFR. Renal impairment is common in PH and is strongly associated with adverse outcomes, primarily due to reduced renal perfusion and venous congestion resulting from right heart failure. Elevated RDW has been independently linked to poor prognosis in PH, while reduced eGFR signifies systemic end-organ dysfunction. Therefore, RGR captures the interplay between hematologic abnormalities and renal dysfunction—two key systems involved in the progression of PH. RGR was derived by dividing the RDW standard deviation (measured in femtoliters, fl) by eGFR (mL/min/1.73 m^2):

$$RGR = \frac{RDW\ standard\ deviation\ (\mathrm{fl})}{eGFR\ (\mathrm{mL/min/1.73\ m^2})}$$

2.7.2. NT-proBNP-to-Albumin Ratio (NTAR)

The NT-proBNP-to-albumin ratio (NTAR) is an emerging biomarker that reflects both myocardial wall stress and systemic congestion in PH. Its underlying mechanism is rooted in the complex interaction between cardiac overload and hepatic dysfunction. Elevated right ventricular strain stimulates the release of natriuretic peptides, particularly NT-proBNP, while progressive PH leads to hepatic congestion, disrupting cellular homeostasis and impairing albumin synthesis. NTAR was determined as the base-10 logarithm of the NT-proBNP-to-albumin ratio:

$$\mathrm{NTAR} = \log(10)\frac{\mathrm{NT-proBNP}\ \left(\frac{\mathrm{pg}}{\mathrm{mL}}\right)}{\mathrm{Albumin}\ \left(\frac{\mathrm{g}}{\mathrm{dL}}\right)}$$

2.8. Statistical Analysis

Descriptive and analytical statistics for the final dataset was performed using MedCalc® Statistical Software (version 23.1.6, MedCalc Software Ltd., Ostend, Belgium; https://www.medcalc.org; accessed on 30 March 2025). A p-value of <0.05 was considered the threshold for statistical significance.

Continuous variables were tested for normality using the Kolmogorov–Smirnov test. Parametric continuous variables are presented as the mean ± standard deviation (SD), while non-parametric continuous variables are reported as the median and interquartile range (IQR). Categorical variables are expressed as counts (percentages).

Between-group comparisons of continuous variables were performed using Student's t-test when data followed a normal distribution and the Mann–Whitney U test for non-normally distributed data. Relationships between continuous measures were quantified using Spearman's rank correlation coefficient for non-parametric variables.

Logistic regression was applied to classify data into binary variables (defined as 1 for ELOS, in-hospital, and 3-month all-cause mortality). The odds ratio (OR) along with the 95% confidence interval (CI) was computed to evaluate the relationship between predictors and outcomes. Model goodness-of-fit was determined using the Hosmer–Lemeshow test, indicated by a p-value greater than 0.05.

To assess predictive accuracy, a receiver operating characteristic (ROC) curve analysis was conducted, and the area under the curve (AUC) was determined with a 95% CI. The AUC was interpreted according to the classification proposed by Çorbacıoğlu and Aksel [30]. Comparisons between ROC curves were carried out using DeLong's test. Optimal cut-off values were identified based on the Youden index.

Analyzed data are presented raw and unadjusted for confounders.

3. Results

3.1. Characteristics of the Study Cohort

Of the 468 pulmonary hypertension-related hospital admissions initially identified, 275 met the predefined inclusion and exclusion criteria and were included in the analysis. Among these, 148 (53.8%) admissions were identified as PAH and 127 (46.2%) as CTEPH. The PAH cohort demonstrated a higher proportion of female patients, whereas sex distribution in the CTEPH group was more balanced. Functional limitation was primarily moderate to severe, with most participants being assigned WHO-FC II or III. Within the PAH subgroup, congenital heart disease-associated PAH (CHD-PAH) was the most common etiology, accounting for 102 patients (68.92%), followed by idiopathic PAH (IPAH) with 21 patients (14.19%), connective tissue disease-associated PAH (CTD-PAH) with 20 patients (13.51%), and portopulmonary hypertension (PoPH) with 5 patients (3.38%). Key demographic and clinical characteristics are presented in Table 1.

Table 1. Baseline characteristics of the PAH and CTEPH admissions.

Parameter	PAH (n = 148)	CTEPH (n = 127)
Age (years, median, IQR)	51 (42.00–67.00)	67 (57.50–71.00)
Sex (female, n, %)	84 (56.75)	57 (44.88)
BMI (kg/m^2, median, IQR)	28.16 (23.20–32.3)	26.7 (24.64–31.14)
WHO-FC (n, %)		
I	8 (5.41)	8 (6.30)
II	50 (33.78)	53 (41.73)
III	72 (48.65)	46 (36.22)
IV	18 (12.16)	20 (15.75)
Comorbidities (n, %)		
Cardiovascular Diseases		
Systemic hypertension	47 (31.76)	51 (40.16)
Coronary artery disease	11 (22.45)	19 (14.96)
Atrial fibrillation	49 (33.11)	24 (18.90)
Previous deep vein thrombosis	-	116 (91.34)
Metabolic Disorders		
Excess body weight	97 (65.54)	89 (70.08)
T2DM	50 (33.78)	38 (29.92)
Thyroid disease	40 (27.03)	23 (18.11)
Respiratory Diseases		
Obstructive sleep apnea	14 (9.46)	13 (10.24)
Asthma	11 (7.43)	7 (5.51)
COPD	10 (6.76)	20 (15.75)
Lung disease—other than COPD and asthma	52 (35.14)	28 (22.05)
History of SARS-CoV-2 infection	62 (41.90)	54 (42.52)
Length of hospital stay		
LOS (days, median, IQR)	7 (5.00–9.00)	7 (5.00–9.00)
ELOS (n, %)	61 (41.22)	58 (45.67)
ELOS (days, median, IQR)	10 (8.00–15.00)	14 (8.00–15.75)
Mortality (n, %)		
Overall mortality	15 (10.14)	7 (5.51)
In-hospital mortality	8 (5.41)	4 (3.15)
3-month all-cause mortality	7 (4.73)	3 (2.36)

BMI, body mass index; COPD, chronic obstructive pulmonary disease; CTEPH, chronic thromboembolic pulmonary hypertension; ELOS, extended length of hospital stay; IQR, interquartile range; LOS, length of hospital stay; n, number of admissions; PAH, pulmonary arterial hypertension; T2DM, type 2 diabetes mellitus; WHO-FC, World Health Organization functional classification.

3.2. Comorbidities

Comorbidity profiles differed substantially between the PAH and CTEPH groups. Excess body weight was common in both cohorts. Systemic hypertension and COPD were more prevalent in the CTEPH group. A history of DVT was observed almost exclusively in CTEPH patients (91.34%), consistent with its known pathophysiological basis. In contrast, atrial fibrillation and thyroid disease were more frequently encountered in the PAH group. The prevalence of T2DM and a prior SARS-CoV-2 infection was comparable between the two groups.

3.3. Length of Hospital Stay and Extended Length of Hospital Stay

The median length of hospital stay (LOS) was 7 days in both groups. Extended length of stay (ELOS) was observed in 41.22% of PAH patients and 45.67% of those with CTEPH, reflecting a substantial burden of hospitalization. Notably, patients requiring ELOS were significantly more likely to present with advanced functional limitation (WHO-FC III–IV) compared with those with shorter admissions ($\leq$7 days; $p = 0.001$ in both groups), suggesting an association between clinical severity and prolonged hospitalization.

3.4. Laboratory Data

A total of seven biomarkers were evaluated across the study population, encompassing parameters derived from hematological indices and routine biochemical measurements. These markers were selected based on their potential relevance in pulmonary hypertension pathophysiology. An overview of the biomarker values distribution in PAH and CTEPH groups is provided in Table 2.

Table 2. Distribution of established and novel biomarker indices in PAH and CTEPH groups.

Parameter	PAH (n = 148)	CTEPH (n = 127)
Laboratory data		
Neutrophils ($\times 10^3/\mu L$, mean ± SD)	4.92 ± 2.44	4.88 ± 1.62
Lymphocytes ($\times 10^3/\mu L$, median, IQR)	1.56 (1.13–2.13)	1.39 (0.97–1.77)
Creatinine (mg/dL, mean ± SD)	0.98 ± 0.38	1.23 ± 0.52
NT-proBNP (pg/mL, mean ± SD)	3120.26 ± 3702.10	4534.54 ± 4926.93
Prior studied biomarkers		
Albumin (g/dL, median, IQR)	4.14 (3.82–4.48)	4.15 (3.83–4.38)
NLR (mean ± SD)	3.57 ± 2.25	4.07 ± 2.47
Log NT-proBNP (median, IQR)	3.22 (2.60–3.68)	3.4 (2.81–3.84)
RDW (fl, mean ± SD)	51.31 ± 9.75	51.88 ± 8.20
eGFR (ml/min/1.73 m^2, median, IQR)	85.1 (63.62–110.32)	63.19 (47.29–90.01)
Proposed biomarkers		
RGR (mean ± SD)	0.85 ± 2.36	0.91 ± 0.50
NTAR (median, IQR)	2.62 (1.95–3.08)	2.79 (2.21–3.23)

CTEPH, chronic thromboembolic pulmonary hypertension; eGFR, estimated glomerular filtration rate; IQR, interquartile range; n, number; NTAR, NT-proBNP-to-albumin ratio; PAH, pulmonary arterial hypertension; RDW, red cell distribution width; RGR, red cell distribution width-to-eGFR ratio.

3.5. Biomarkers and LOS

Analysis by PH subtype (Spearman correlations) revealed distinct biomarker associations with LOS. In the CTEPH cohort, serum albumin and eGFR showed negative correlations with LOS, indicating that hypoalbuminemia and reduced renal function were linked to prolonged hospitalization in this subgroup. These associations were not statistically significant in the PAH group. In contrast, positive correlations with LOS were consistently observed across both groups for Log NT-proBNP, NTAR, NLR, RGR, and RDW, reflecting their potential role as markers of disease severity and hospitalization burden. These correlations were consistently stronger in the CTEPH group (Table 3).

Table 3. Phenotype-specific Spearman correlations of biomarkers with length of hospitalization.

Parameters	Length of Hospital Stay (LOS)			
	PAH		CTEPH	
	r	*p*	*r*	*p*
Prior studied biomarkers				
Albumin	−0.13	0.122	−0.24	0.007
NLR	0.23	0.004	0.33	<0.001
Log NT-proBNP	0.23	0.005	0.48	<0.001
RDW	0.21	0.009	0.19	0.028
eGFR	−0.10	0.216	−0.31	<0.001
Proposed biomarkers				
RGR	0.20	0.016	0.33	<0.001
NTAR	0.23	0.005	0.46	<0.001

CTEPH, chronic thromboembolic pulmonary hypertension; eGFR, estimated glomerular filtration rate; NLR, neutrophil-to-lymphocyte ratio; NTAR, NT-proBNP-to-albumin ratio; PAH, pulmonary arterial hypertension; RDW, red cell distribution width; RGR, red cell distribution width-to-eGFR ratio.

3.6. Biomarkers and ELOS

To identify predictors of ELOS within each phenotype group, univariate logistic regression analysis was performed (Table 4). In the PAH group, NLR, RDW, RGR, NTAR, and Log NT-proBNP showed weak but statistically significant associations with ELOS, based on AUC values and satisfactory model calibration. Among these, NLR and RDW exhibited the highest AUCs (0.647 and 0.642, respectively). In the CTEPH group, Log NT-proBNP, NTAR, RGR, NLR, and eGFR emerged as significant predictors of ELOS, with Log NT-proBNP and NTAR again demonstrating the strongest discriminative performance (AUCs: 0.748 and 0.743, respectively).

These findings support the potential clinical utility of cardiac stress markers (NT-proBNP, NTAR), inflammatory indices (NLR, RGR), and red cell distribution measures (RDW) in identifying patients at increased risk of prolonged hospitalization. In patients with CTEPH, reduced eGFR also emerged as an independent predictor of ELOS.

Table 4. Univariate logistic regression analysis of predictors for ELOS in PAH and CTEPH groups.

Independent Variables	Extended Length of Hospital Stay (ELOS)					
	AUC (95% CI)	*p*	OR (95% CI)	*p*	Hosmer–Lemeshow Test, *p*	Associated Criterion to Youden Index J
PAH group						
Prior studied biomarkers						
Albumin	0.579 (0.495–0.659)	0.102	0.587 (0.301–1.146)	0.118	0.449	≤4.18
NLR	0.647 (0.564–0.723)	0.001	1.150 (0.989–1.338)	0.068	0.095	>3.11
Log NT–proBNP	0.630 (0.547–0.708)	0.006	1.753 (1.098–2.800)	0.018	0.506	>3.00
RDW	0.642 (0.559–0.719)	0.002	1.049 (1.011–1.089)	0.010	0.659	>52.10
eGFR	0.580 (0.496–0.660)	0.096	0.990 (0.980–1.000)	0.061	0.896	≤84.11
Proposed biomarkers						
RGR	0.637 (0.554–0.715)	0.002	3.722 (1.187–11.671)	0.024	0.084	>0.60
NTAR	0.627 (0.544–0.705)	0.007	1.773 (1.116–2.817)	0.015	0.800	>2.37

Table 4. *Cont.*

Independent Variables	Extended Length of Hospital Stay (ELOS)					
	AUC (95% CI)	*p*	OR (95% CI)	*p*	Hosmer–Lemeshow Test, *p*	Associated Criterion to Youden Index J
CTEPH group						
Prior studied biomarkers						
Albumin	0.640 (0.550–0.723)	0.006	1.031 (0.890–1.195)	0.680	0.012	≤3.95
NLR	0.645 (0.555–0.727)	0.003	1.208 (1.035–1.410)	0.016	0.667	>2.77
Log NT-proBNP	0.748 (0.664–0.821)	<0.001	3.947 (2.085–7.471)	<0.001	0.510	>3.47
RDW	0.551 (0.460–0.639)	0.333	1.030 (0.986–1.075)	0.180	0.139	>52.20
eGFR	0.645 (0.555–0.728)	0.003	0.984 (0.972–0.997)	0.014	0.766	≤60.00
Proposed biomarkers						
RGR	0.648 (0.559–0.731)	0.003	3.766 (1.582–8.960)	0.002	0.309	>0.58
NTAR	0.743 (0.658–0.817)	<0.001	3.641 (1.992–6.654)	<0.001	0.979	>3.09

AUC, area under the curve; CI, confidence interval; CTEPH, chronic thromboembolic pulmonary hypertension; eGFR, estimated glomerular filtration rate; ELOS, extended length of hospital stay; NLR, neutrophil-to-lymphocyte ratio; NTAR, NT-proBNP-to-albumin ratio; PAH, pulmonary arterial hypertension; RDW, red cell distribution width; RGR, red cell distribution width-to-eGFR ratio.

An ROC curve analysis was conducted to compare the relative performance of the biomarkers in predicting ELOS (Figure 1). In PAH, no significant differences were observed between the markers in terms of discriminatory power. In contrast, in CTEPH, Log NT-proBNP and NTAR demonstrated significantly greater AUCs compared with albumin, NLR, eGFR, and RGR ($p < 0.001$), although no significant difference was found between Log NT-proBNP and NTAR themselves.

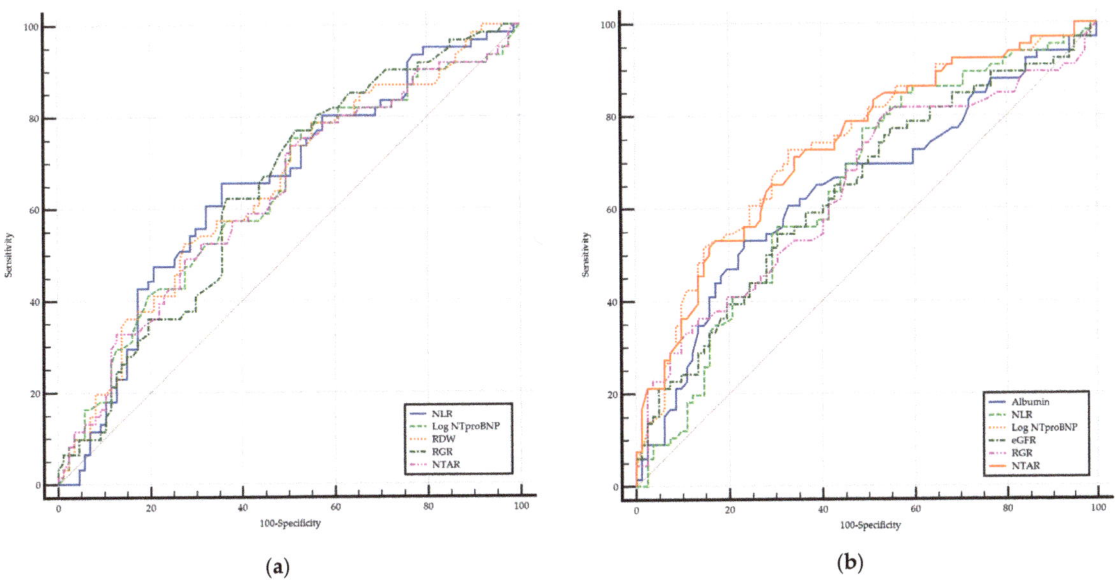

Figure 1. Comparison of the ROC curves for ELOS as the dependent value expected in PAH (**a**) and CTEPH (**b**) groups. eGFR, estimated glomerular filtration rate; Log NT-proBNP, logarithm of the NT-proBNP; NLR, neutrophil-to-lymphocyte ratio; NTAR, NT-proBNP-to-albumin ratio; RDW, red cell distribution width; RGR, red cell distribution width-to-eGFR ratio.

3.7. Biomarkers and In-Hospital Mortality

Twelve deaths occurred (eight PAH, four CTEPH), and the male-to-female death ratio was 1:1.2. All were WHO-FC III-IV patients at the time of admission. Univariate logistic regression identified multiple significant predictors of in-hospital mortality (Table 5).

Table 5. Predictors for in-hospital mortality in PAH and CTEPH groups.

Independent Variables	In-Hospital Mortality					
	AUC (95% CI)	***p***	**OR (95% CI)**	***p***	**Hosmer–Lemeshow Test, *p***	**Associated Criterion to Youden Index J**
PAH group						
Prior studied biomarkers						
Albumin	0.804 (0.731–0.865)	<0.001	0.152 (0.037–0.619)	0.008	0.490	≤4.16
NLR	0.858 (0.791–0.910)	<0.001	1.492 (1.163–1.913)	0.001	0.480	>3.28
Log NT-proBNP	0.784 (0.709–0.848)	<0.001	10.291 (1.490–71.065)	0.018	0.498	>3.65
RDW	0.700 (0.619–0.773)	0.054	1.050 (0.993–1.111)	0.082	0.018	>58.60
eGFR	0.701 (0.621–0.774)	0.031	0.974 (0.948–1.000)	0.058	0.411	≤97.39
Proposed biomarkers						
RGR	0.765 (0.688–0.830)	<0.001	1.018 (0.799–1.298)	0.881	0.458	>0.65
NTAR	0.815 (0.743–0.874)	<0.001	12.083 (1.743–83.748)	0.011	0.093	>3.08
CTEPH group						
Prior studied biomarkers						
Albumin	0.772 (0.690–0.842)	0.010	0.455 (0.179–1.156)	0.098	0.305	≤3.77
NLR	0.706 (0.619–0.784)	0.079	1.197 (0.869–1.649)	0.268	0.372	>2.96
Log NT-proBNP	0.799 (0.718–0.865)	0.005	13.824 (0.620–308.177)	0.097	0.738	>3.40
RDW	0.578 (0.487–0.665)	0.569	1.016 (0.903–1.144)	0.780	0.214	>55.1
eGFR	0.644 (0.555–0.727)	0.492	0.988 (0.952–1.026)	0.546	0.470	≤49.46
Proposed biomarkers						
RGR	0.637 (0.547–0.721)	0.511	4.229 (1.172–15.264)	0.027	0.434	>1.11
NTAR	0.817 (0.739–0.880)	0.005	7.779 (0.959–63.066)	0.054	0.577	>3.24

AUC, area under the curve; CI, confidence interval; CTEPH, chronic thromboembolic pulmonary hypertension; eGFR, estimated glomerular filtration rate; NLR, neutrophil-to-lymphocyte ratio; NTAR, NT-proBNP-to-albumin ratio; PAH, pulmonary arterial hypertension; RDW, red cell distribution width; RGR, red cell distribution width-to-eGFR ratio.

In the PAH group, several biomarkers demonstrated strong predictive value for in-hospital mortality. NLR, NTAR, and albumin exhibited good predictive capacity, while Log NT-proBNP, RGR, eGFR, and RDW showed acceptable discrimination. These findings were supported by a well-calibrated model, as indicated by the Hosmer–Lemeshow test. Notably, higher serum albumin levels emerged as a significant protective factor.

In CTEPH, NTAR demonstrated the highest predictive performance for in-hospital mortality (AUC = 0.817), indicating good discriminative ability. A very similar AUC, considered acceptable, was observed for Log NT-proBNP, followed by albumin and NLR. Poor discriminatory power was noted for eGFR and RGR, with RGR showing the lowest AUC.

In the ROC curve analysis comparison, both in Figure 2a,b, there were no significant differences in terms of AUC in the leading biomarkers.

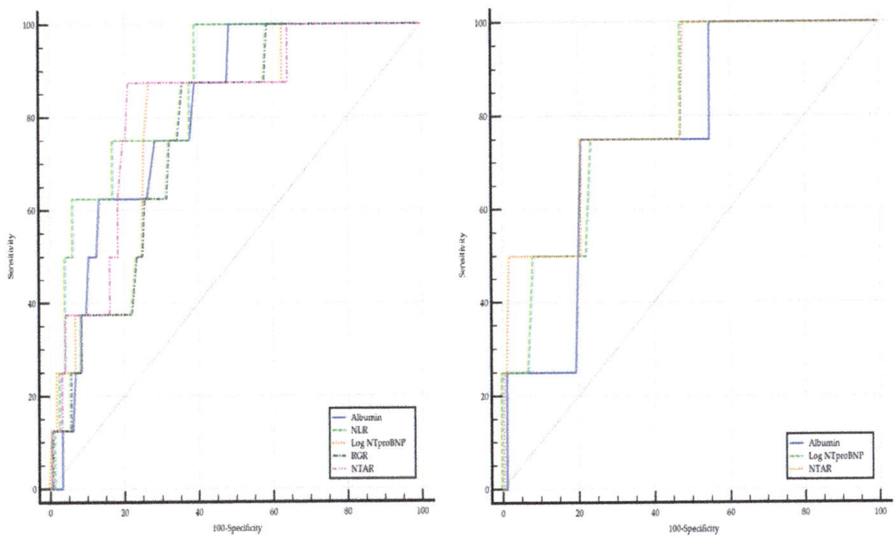

(a) (b)

Figure 2. Comparison of the ROC curves for in-hospital mortality as the dependent value expected in PAH (**a**) and CTEPH (**b**) groups. Log NT-proBNP; NLR, neutrophil-to-lymphocyte ratio; NTAR, NT-proBNP-to-albumin ratio; RGR, red cell distribution width-to-eGFR ratio.

3.8. Biomarkers and 3-Month All-Cause Mortality

In the PAH cohort, univariate logistic regression analysis identified serum albumin as the only statistically significant predictor of 3-month all-cause mortality (AUC = 0.782), with lower concentrations being independently associated with increased mortality risk. In the CTEPH group, both NLR and serum albumin demonstrated very good prognostic performance for predicting 3-month all-cause mortality, with AUCs of 0.875 and 0.833, respectively.

These findings suggest a phenotype-specific prognostic pattern for 3-month all-cause mortality, where hypoalbuminemia is most relevant in the PAH group, while elevated NLR and low albumin emerge as the dominant predictors in the CTEPH group (Table 6). Additionally, reduced eGFR demonstrated good predictive capacity in this subgroup.

Table 6. Predictors for 3-month all-cause mortality in PAH and CTEPH groups.

Independent Variables	3-Month All-Cause Mortality					
	AUC (95% CI)	*p*	OR (95% CI)	*p*	Hosmer–Lemeshow Test, *p*	Associated Criterion to Youden Index J
PAH group						
Prior studied biomarkers						
Albumin	0.782 (0.707–0.846)	0.006	0.135 (0.030–0.604)	0.008	0.686	≤3.88
NLR	0.608 (0.525–0.688)	0.342	1.060 (0.784–1.435)	0.701	0.068	>3.47
Log NT-proBNP	0.564 (0.480–0.646)	0.601	1.340 (0.453–3.958)	0.596	0.324	>3.70
RDW	0.547 (0.463–0.629)	0.680	0.983 (0.901–1.072)	0.703	0.428	≤47.10
eGFR	0.599 (0.515–0.678)	0.399	1.009 (0.988–1.030)	0.382	0.923	>74.63
Proposed biomarkers						
RGR	0.598 (0.514–0.677)	0.396	1.578 (0.009–5.800)	0.374	0.387	≤0.63
NTAR	0.578 (0.494–0.658)	0.508	1.506 (0.501–4.527)	0.465	0.897	>2.36

Table 6. *Cont.*

Independent Variables	3-Month All-Cause Mortality					
	AUC (95% CI)	***p***	**OR (95% CI)**	***p***	**Hosmer–Lemeshow Test, *p***	**Associated Criterion to Youden Index J**
CTEPH group						
Prior studied biomarkers						
Albumin	0.833 (0.757–0.894)	<0.001	0.487 (0.169–1.405)	0.183	0.799	≤3.98
NLR	0.875 (0.805–0.927)	<0.001	1.491 (1.047–2.123)	0.026	0.925	>4.83
Log NT-proBNP	0.743 (0.658–0.817)	0.253	4.929 (0.360–67.448)	0.232	0.117	>4.16
RDW	0.581 (0.490–0.668)	0.654	1.033 (0.905–1.180)	0.625	0.512	>46.8
eGFR	0.793 (0.712–0.860)	0.018	0.945 (0.879–1.017)	0.136	0.822	≤59.21
Proposed biomarkers						
RGR	0.741 (0.655–0.814)	0.081	3.347 (0.807–13.882)	0.095	0.578	>0.78
NTAR	0.748 (0.670–0.816)	0.142	4.232 (0.525–34.083)	0.175	0.277	>3.52

AUC, area under the curve; CI, confidence interval; CTEPH, chronic thromboembolic pulmonary hypertension; eGFR, estimated glomerular filtration rate; NLR, neutrophil-to-lymphocyte ratio; NTAR, NT-proBNP-to-albumin ratio; PAH, pulmonary arterial hypertension; RDW, red cell distribution width; RGR, red cell distribution width-to-eGFR ratio.

For the prediction of 3-month all-cause mortality, serum albumin was the only biomarker in the PAH group that demonstrated statistically significant discriminative ability and was thus selected for ROC analysis (Figure 3a). In the CTEPH group, albumin, NLR, and eGFR were included in the comparison; however, no statistically significant differences were observed between their respective AUCs (Figure 3b).

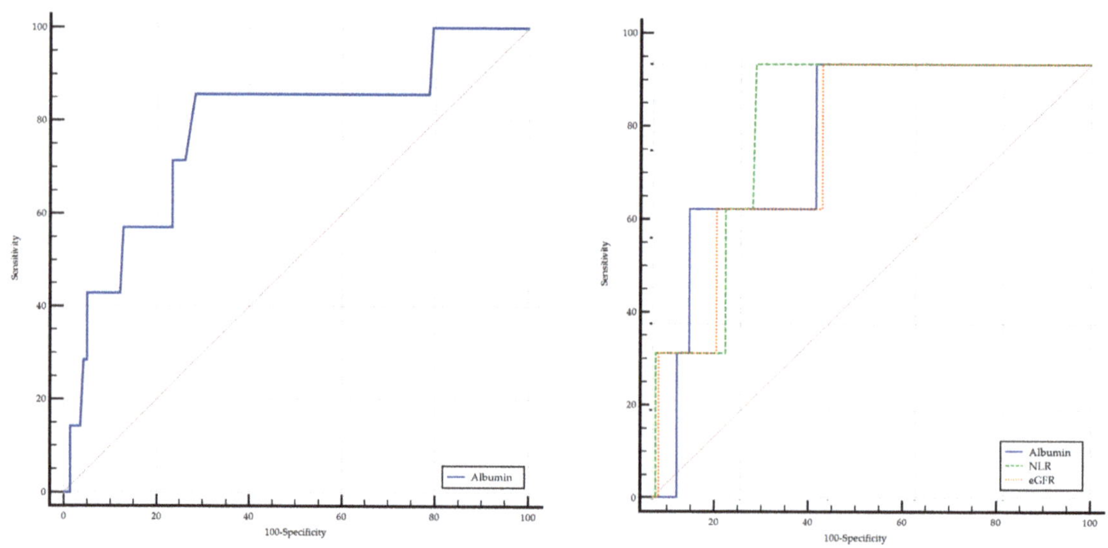

(a) (b)

Figure 3. Comparison of ROC curves for 3-month all-cause mortality as the dependent value expected in PAH (**a**) and CTEPH (**b**) groups. eGFR, estimated glomerular filtration rate; NLR, neutrophil-to-lymphocyte ratio.

4. Discussion

In this study, we evaluated the prognostic significance of both established and novel biomarkers in relation to hospitalization outcomes and mortality across 275 PH-related hospital admissions, stratified by clinical phenotype (PAH and CTEPH). Our findings underscore the utility of a composite biomarker strategy—encompassing inflammatory, hematologic, cardiac, and renal indicators—tailored to the underlying disease profile. Notably, this is the first investigation to propose and validate RGR and the NTAR as prognostic markers in this context. NTAR demonstrated particularly strong discriminative performance in CTEPH patients, especially for ELOS and in-hospital mortality, whereas RGR showed weaker standalone performance but may offer added prognostic value in future multimodal assessment models.

4.1. Prognostic Utility of Conventional Biomarkers and Cut-Off Determination

In our analysis, serum albumin exhibited distinct phenotype-specific prognostic patterns in the PAH and CTEPH groups. In PAH patients, lower albumin concentrations were significantly associated with increased in-hospital and 3-month all-cause mortality, respectively, but showed no meaningful association with LOS or ELOS. Conversely, in CTEPH patients, albumin demonstrated a moderate yet significant negative correlation with LOS and emerged as a predictor of ELOS. It also showed good and very good discriminative performance for in-hospital and 3-month mortality, with AUCs of 0.772 and 0.833, respectively. These results underscore the prognostic relevance of hypoalbuminemia as phenotype-dependent: primarily mortality-driven in PAH, and more closely associated with hospitalization burden in CTEPH. Our findings are consistent with prior evidence linking hypoalbuminemia to both adverse clinical outcomes and prolonged hospitalizations [31–34].

NLR, a widely studied marker of systemic inflammation, demonstrated a weak yet statistically significant correlation with LOS exclusively in the CTEPH group. It did not reliably predict ELOS in PAH patients, although it reached statistical significance in the CTEPH group, albeit with limited discriminative capacity. Its prognostic utility was more robust in relation to mortality outcomes. In PAH patients, an NLR value > 3.28 was strongly associated with increased in-hospital mortality risk. In the CTEPH group, NLR thresholds >2.96 and >4.83 were predictive of in-hospital and 3-month all-cause mortality, respectively. These findings are in line with previous studies: Özpelit et al. identified an NLR cut-off of 3.0 for predicting all-cause mortality in PAH patients, while Yanartaş et al. reported that an NLR > 2.54 was significantly associated with increased postoperative mortality in CTEPH [35,36].

Log NT-proBNP demonstrated a moderate positive correlation with LOS in both PAH and CTEPH groups, with a stronger association observed in the CTEPH group. It emerged as a significant predictor of ELOS in both phenotypes, exhibiting the highest discriminative performance in CTEPH. Regarding mortality, elevated Log NT-proBNP values were associated with increased in-hospital mortality in PAH and showed similarly good predictive performance in CTEPH. For 3-month all-cause mortality, its prognostic utility was limited in PAH but good in CTEPH. These findings are supported by previous studies: Maurer et al. identified a threshold of 2.73 as predictive of all-cause mortality in PH patients, while Januzzi et al. reported that Log NT-proBNP levels exceeding 2.99 were associated with a 2.88-fold increase in one-year mortality among individuals with heart failure [37,38]. Clinically, each one-unit increase in Log NT-proBNP—reflecting a tenfold rise in NT-proBNP concentration—was associated with prolonged hospitalization and greater short-term mortality risk, particularly in patients with CTEPH.

RDW exhibited a weak positive correlation with LOS in both PAH and CTEPH patients, suggesting its potential utility as a marker of disease severity. In the PAH group, RDW was significantly associated with an increased likelihood of ELOS. Conversely, RDW demonstrated limited prognostic value in the CTEPH group, showing no significant association with either ELOS or mortality. These findings align closely with prior research, including a large-scale study of 167,292 admissions, which reported that elevated RDW at hospital admission was associated with longer stays compared with admissions with lower RDW values [39]. Elevated RDW has also been widely recognized as a predictor of in-hospital mortality and all-cause death across multiple patient populations [40,41]. Nevertheless, in our cohort, the association between RDW and in-hospital mortality was marginal, suggesting the need for further validation in larger, prospective studies.

eGFR demonstrated a significant negative correlation with LOS in the CTEPH group and independently predicted ELOS, with values > 60 mL/min/1.73 m^2 conferring a protective effect. In the PAH group, lower eGFR was associated with higher in-hospital mortality risk, while in the CTEPH group, it exhibited good discriminative performance for 3-month all-cause mortality. These results underscore the prognostic importance of renal dysfunction across PH phenotypes, particularly in CTEPH. Prior investigations have similarly linked reduced eGFR to prolonged hospitalizations and increased ICU admissions; reported a 15% rise in 30-day mortality per 10 mL/min/1.73 m^2 decline in heart failure cohorts; and identified eGFR < 56 mL/min/1.73 m^2 as an independent predictor of in-hospital death in patients with pneumonia [42–44]. Together, these data highlight the critical role of preserving renal function to mitigate mortality and reduce hospitalization duration.

4.2. RGR and NTAR in Focus: New Predictive Tools for Risk Stratification

RGR, a novel composite biomarker, demonstrated consistent prognostic utility across PAH and CTEPH phenotypes. RGR exhibited a positive correlation with LOS in both cohorts, with a stronger effect size in CTEPH. As a predictor of ELOS, RGR achieved acceptable discrimination in PAH and CTEPH. For in-hospital mortality, RGR showed good discriminative performance in PAH and maintained acceptable accuracy in CTEPH. Elevated RGR likely reflects combined disturbances in erythropoiesis and renal function—manifested as increased anisocytosis alongside reduced glomerular filtration—underscoring the interplay between hematologic variability and renal impairment in PH prognostication [45]. These results corroborate earlier studies linking RDW variability to adverse PH outcomes and highlight the integral role of cardiorenal dysfunction in shaping patient prognosis [46–48].

The NTAR demonstrated robust prognostic value in both the PAH and CTEPH cohorts. NTAR correlated moderately with LOS—more strongly in the CTEPH group—and was significantly predictive of ELOS in both phenotypes. Its discriminative performance for in-hospital mortality was particularly high, with NTAR emerging as an independent risk factor in PAH patients and showing borderline significance in the CTEPH group. These data indicate that NTAR serves as a dual indicator of hospitalization burden and short-term mortality in PAH and CTEPH. Prior studies have linked hypoalbuminemia—an important marker of nutritional status and a marker of illness—to poorer outcomes in hospitalized patients and identified low serum albumin as a marker of higher mortality risk in PAH [13,49,50].

While NTAR and RGR have demonstrated strong prognostic associations in PH, a more in-depth discussion of the pathophysiological mechanisms through which these biomarkers influence disease progression would be beneficial. Elevated NTAR likely reflects a convergence of increased myocardial wall stress—driving NT-proBNP release—and impaired albumin synthesis due to hepatic congestion or chronic systemic inflammation, thereby compounding right ventricular dysfunction. Conversely, an elevated RGR embodies the dual impact of erythropoietic disruption (manifested as anisocytosis) and renal impairment, conditions that potentiate hypoxia-driven pulmonary vascular remodeling and endothelial dysfunction, ultimately escalating pulmonary vascular resistance [51]. A deeper exploration of these mechanisms could refine risk stratification models and identify novel targets for therapeutic intervention in PH.

These analyses highlight the complex pathophysiological interdependence among cardiorenal dysfunction, systemic inflammation, and metabolic dysregulation in the prognostication of prolonged hospitalizations and mortality, underscoring the imperative for a precision medicine approach integrating multimodal biomarker profiling to enhance risk stratification and therapeutic optimization across the continuum of pulmonary arterial hypertension and chronic thromboembolic pulmonary hypertension progression.

4.3. Limitations and Future Directions

While this study provides valuable preliminary insights, its retrospective, single-center design introduces inherent limitations. The reliance on a single institution's electronic records may engender selection bias and constrain the generalizability of our findings to broaden PH populations. Moreover, the absence of invasive hemodynamic measurements—such as right heart catheterization-derived pulmonary artery pressures and pulmonary vascular resistance—precludes direct mechanistic correlations with NTAR and RGR. Biomarker sampling was limited to admission values, preventing the assessment of temporal dynamics, and the cohort size (275 admissions) lacked sufficient power for granular subgroup analyses by etiology, comorbidity burden, or specific therapeutic regimens.

Future research should emphasize large-scale, prospective multicenter studies to validate and expand upon these findings. Inclusion of geographically diverse pulmonary hypertension centers would help reduce selection bias and enhance external validity. Longitudinal biomarker assessments—ideally at baseline and at 1, 3, and 6 months of follow-up—could clarify how NTAR and RGR trajectories correlate with clinical outcomes over time. Incorporating invasive hemodynamic measurements and advanced imaging techniques (e.g., speckle-tracking echocardiography, three-dimensional echocardiography, cardiac magnetic resonance imaging, computed tomography, and hybrid imaging) would provide valuable insights into the pathophysiological mechanisms linking these biomarker ratios to right ventricular function and pulmonary vascular remodeling.

Our findings also underscore the need to translate NTAR and RGR from research metrics into practical clinical tools. Embedding automated alerts for high-risk thresholds (e.g., NTAR > 3.0; RGR > 0.6) within electronic health records could prompt timely multidisciplinary interventions, such as intensified decongestive therapy, nutritional support, or early nephrology referral. Additionally, targeted analyses of how specific PH treatments (prostacyclin analogs, endothelin receptor antagonists, PDE-5 inhibitors) modulate these biomarkers will refine their interpretive context. Finally, health economic evaluations comparing standard management to NTAR/RGR-guided pathways are essential to determine cost-effectiveness, resource utilization, and potential improvements in patient-centered outcomes.

5. Conclusions

The NT-proBNP-to-albumin ratio (NTAR) and the red cell distribution width-to-eGFR ratio (RGR) consistently predicted prolonged hospitalization and in-hospital mortality, with especially strong performance in the CTEPH phenotype. By capturing the interplay between cardiac stress, systemic inflammation, and renal function in a single metric, these novel indices augment traditional risk models and facilitate more nuanced patient stratification. To transition from proof-of-concept to clinical practice, large-scale, prospective multicenter validation is essential. Such efforts should assess NTAR and RGR across diverse PH groups, explore integration with hemodynamic and imaging data, and evaluate their impact on management algorithms. Ultimately, embedding these biomarkers into routine care pathways holds the promise of earlier risk identification and individualized therapeutic interventions, with the goal of improving outcomes in pulmonary arterial hypertension and chronic thromboembolic pulmonary hypertension.

Author Contributions: Conceptualization, D.G.I., A.V., L.C., and I.T.; methodology, D.G.I., A.V., L.C., R.G.M., and I.T.; software, D.G.I., A.V., L.C., and R.G.M.; validation, D.G.I., A.V., L.C., and I.T.; formal analysis, D.G.I., A.V., L.C., R.G.M., and I.T.; investigation, D.G.I., A.V., L.C., R.G.M., and I.T.; resources, D.G.I., A.V., L.C., and I.T.; data curation, D.G.I., A.V., L.C., and I.T.; writing—original draft preparation, D.G.I., A.V., L.C., and I.T.; writing—review and editing, D.G.I., A.V., L.C., R.G.M., and I.T.; visualization, D.G.I., A.V., L.C., and I.T.; supervision, D.G.I., A.V., L.C., and I.T.; project administration, D.G.I., A.V., and I.T.; funding acquisition, A.V. and I.T. All authors have read and agreed to the published version of the manuscript.

Funding: This work was partially supported by the George Emil Palade University of Medicine, Pharmacy, Science, and Technology of Targu Mures, Romania, under contract number: 615/13/17.01.2019.

Institutional Review Board Statement: This study was conducted in accordance with the principles of the Declaration of Helsinki and approved by the Research Ethics Committee of the George Emil Palade University of Medicine, Pharmacy, Science, and Technology of Targu Mures, Romania (Approval Number: 1283, 25 February 2021). All study procedures complied with national and international ethical guidelines for human research.

Informed Consent Statement: Informed consent was obtained from all subjects involved in the study.

Data Availability Statement: The data presented in this study can be obtained from the corresponding author according to the local and national regulations.

Acknowledgments: We thank all the patients and collaborators involved in this study.

Conflicts of Interest: The authors declare no conflicts of interest. The funders had no role in the design of the study; in the collection, analyses, or interpretation of data; in the writing of the manuscript; or in the decision to publish the results.

Abbreviations

The following abbreviations are used in this manuscript:

AUC	Area under the curve
BMI	Body mass index
CHD-PAH	Pulmonary arterial hypertension associated with congenital heart disease
CI	Confidence interval
CKD	Chronic kidney disease
CKD-EPI	Chronic Kidney Disease Epidemiology Collaboration
COPD	Chronic obstructive pulmonary disease
CTD-PAH	Pulmonary arterial hypertension associated with connective tissue disease
CTEPH	Chronic thromboembolic pulmonary hypertension
DVT	Deep vein thrombosis
eGFR	Estimated glomerular filtration rate
ELOS	Extended length of hospital stay
ESC/ERS	European Society of Cardiology/European Respiratory Society
GDPR	General Data Protection Regulation
IQR	Interquartile range
IPAH	Idiopathic pulmonary arterial hypertension
LOS	Length of hospital stay
NLR	Neutrophil-to-lymphocyte ratio

NT-proBNP	N-terminal pro-brain natriuretic peptide
NTAR	NT-proBNP-to-albumin ratio
OR	Odds ratio
PAH	Pulmonary arterial hypertension
PH	Pulmonary hypertension
PoPH	Pulmonary arterial hypertension associated with portal hypertension
RDW	Red cell distribution width
REVEAL	Registry to Evaluate Early and Long-term Pulmonary Arterial Hypertension Disease Management
RGR	Red cell distribution width-to-estimated glomerular filtration rate ratio
ROC	Receiver operating characteristic
SD	Standard deviation
T2DM	Type 2 diabetes mellitus
WHO-FC	World Health Organization functional classification
WBC	White blood cell

References

1. Humbert, M.; Kovacs, G.; Hoeper, M.M.; Badagliacca, R.; Berger, R.M.F.; Brida, M.; Carlsen, J.; Coats, A.J.S.; Escribano-Subias, P.; Ferrari, P.; et al. 2022 ESC/ERS Guidelines for the diagnosis and treatment of pulmonary hypertension. *Eur. Heart J.* **2022**, *43*, 3618–3731. [CrossRef]
2. GBD 2021 Pulmonary Arterial Hypertension Collaborators. Global, Regional, and National Burden of Pulmonary Arterial Hypertension, 1990–2021: A Systematic Analysis for the Global Burden of Disease Study 2021. *Lancet Respir. Med.* **2025**, *13*, 69–79. [CrossRef] [PubMed]
3. Simonneau, G.; Montani, D.; Celermajer, D.S.; Denton, C.P.; Galiè, N.; Ghofrani, H.A.; Gomez Sanchez, M.A.; Kumar, R.K.; Langleben, D.; Lincoln, D.; et al. Haemodynamic definitions and updated clinical classification of pulmonary hypertension. *Eur. Respir. J.* **2019**, *53*, 1801913. [CrossRef]
4. Hoeper, M.M.; Humbert, M.; Souza, R.; Idrees, M.; Kawut, S.M.; Keogh, A.; Lam, C.S.P.; Mayosi, B.M.; Provencher, S.; Reinaque, J.M.; et al. A global view of pulmonary hypertension. *Lancet Respir. Med.* **2016**, *4*, 306–322. [CrossRef]
5. Studer, S.; Hull, M.; Pruett, J.; Koep, E.; Tsang, Y.; Drake, W., III. Treatment Patterns, Healthcare Resource Utilization, and Healthcare Costs Among Patients with Pulmonary Arterial Hypertension in a Real-World US Database. *Pulm. Circ.* **2019**, *9*, 2045894018816294. [CrossRef]
6. Dufour, R.; Pruett, J.; Hu, N.; Lickert, C.; Stemkowski, S.; Tsang, Y.; Lane, D.; Drake, W., III. Healthcare Resource Utilization and Costs for Patients with Pulmonary Arterial Hypertension: Real-World Documentation of Functional Class. *J. Med. Econ.* **2017**, *20*, 1178–1186. [CrossRef]
7. Burger, C.D.; Long, P.K.; Shah, M.R.; McGoon, M.D.; Miller, D.P.; Romero, A.J.; Benton, W.W.; Safford, R.E. Characterization of First-Time Hospitalizations in Patients with Newly Diagnosed Pulmonary Arterial Hypertension in the REVEAL Registry. *Chest* **2014**, *146*, 1263–1273. [CrossRef]
8. Bowman, J.K.; Dutta, S.; Zheng, H.; Kraus, C.; Wilcox, S.R. Patients with Pulmonary Hypertension Presenting to the Emergency Department. *Am. J. Emerg. Med.* **2020**, *38*, 2313–2317. [CrossRef] [PubMed]
9. McNeil, J.B.; Jackson, K.E.; Wang, C.; Siew, E.D.; Vincz, A.J.; Shaver, C.M.; Bastarache, J.A.; Ware, L.B. Linear Association Between Hypoalbuminemia and Increased Risk of Acute Respiratory Distress Syndrome in Critically Ill Adults. *Crit. Care Explor.* **2021**, *3*, e0527. [CrossRef] [PubMed]
10. Arques, S. Serum Albumin and Heart Failure: Recent Advances on a New Paradigm. *Ann. Cardiol. Angeiol.* **2011**, *60*, 272–278. [CrossRef]
11. Ebert, E.C.; Hagspiel, K.D. Gastrointestinal and Hepatic Manifestations of Systemic Lupus Erythematosus. *J. Clin. Gastroenterol.* **2011**, *45*, 436–441. [CrossRef]
12. Arques, S. Human Serum Albumin in Cardiovascular Diseases. *Eur. J. Intern. Med.* **2018**, *52*, 8–12. [CrossRef]
13. Snipelisky, D.; Jentzer, J.; Batal, O.; Dardari, Z.; Mathier, M. Serum Albumin Concentration as an Independent Prognostic Indicator in Patients with Pulmonary Arterial Hypertension. *Clin. Cardiol.* **2018**, *41*, 782–787. [CrossRef]
14. Jutras-Beaudoin, N.; Toro, V.; Lajoie, A.C.; Breuils-Bonnet, S.; Paulin, R.; Potus, F. Neutrophil-Lymphocyte Ratio as an Independent Predictor of Survival in Pulmonary Arterial Hypertension: An Exploratory Study. *CJC Open* **2021**, *4*, 357–363. [CrossRef]
15. Efros, O.; Beit Halevi, T.; Meisel, E.; Soffer, S.; Barda, N.; Cohen, O.; Kenet, G.; Lubetsky, A. The Prognostic Role of Neutrophil-to-Lymphocyte Ratio in Patients Hospitalized with Acute Pulmonary Embolism. *J. Clin. Med.* **2021**, *10*, 4058. [CrossRef] [PubMed]

16. Ge, H.; Zhang, L.; Zhang, W.; Yuan, Q.; Xiao, X. Neutrophil-to-Lymphocyte Ratio Predicts Poor Prognosis in Patients with Chronic Kidney Disease-Related Pulmonary Hypertension: A Retrospective Study. *Medicine* **2024**, *103*, e40161. [CrossRef] [PubMed]
17. Yogeswaran, A.; Tello, K.; Lund, J.; Klose, H.; Harbaum, L.; Sommer, N.; Oqueka, T.; Hennigs, J.K.; Grimminger, F.; Seeger, W.; et al. Risk Assessment in Pulmonary Hypertension Based on Routinely Measured Laboratory Parameters. *J. Heart Lung Transplant.* **2022**, *41*, 400–410. [CrossRef]
18. Mueller, T.; Gegenhuber, A.; Dieplinger, B.; Poelz, W.; Haltmayer, M. Long-Term Stability of Endogenous B-Type Natriuretic Peptide (BNP) and Amino Terminal proBNP (NT-proBNP) in Frozen Plasma Samples. *Clin. Chem. Lab. Med.* **2004**, *42*, 942–944. [CrossRef]
19. Fradley, M.G.; Larson, M.G.; Cheng, S.; McCabe, E.; Coglianese, E.; Shah, R.V.; Levy, D.; Vasan, R.S.; Wang, T.J. Reference Limits for N-Terminal-Pro-B-Type Natriuretic Peptide in Healthy Individuals (from the Framingham Heart Study). *Am. J. Cardiol.* **2011**, *108*, 1341–1345. [CrossRef] [PubMed]
20. Fu, S.; Ping, P.; Wang, F.; Luo, L. Synthesis, Secretion, Function, Metabolism and Application of Natriuretic Peptides in Heart Failure. *J. Biol. Eng.* **2018**, *12*, 2. [CrossRef]
21. Fava, C.; Cattazzo, F.; Hu, Z.D.; Lippi, G.; Montagnana, M. The Role of Red Blood Cell Distribution Width (RDW) in Cardiovascular Risk Assessment: Useful or Hype? *Ann. Transl. Med.* **2019**, *7*, 581. [CrossRef]
22. Petrauskas, L.A.; Saketkoo, L.A.; Kazecki, T.; Saito, S.; Jaligam, V.; deBoisblanc, B.P.; Lammi, M.R. Use of Red Cell Distribution Width in a Population at High Risk for Pulmonary Hypertension. *Respir. Med.* **2019**, *150*, 131–135. [CrossRef]
23. Yang, J.; Liu, C.; Li, L.; Tu, X.; Lu, Z. Red Blood Cell Distribution Width Predicts Pulmonary Hypertension Secondary to Chronic Obstructive Pulmonary Disease. *Can. Respir. J.* **2019**, *2019*, 3853454. [CrossRef]
24. Hampole, C.V.; Mehrotra, A.K.; Thenappan, T.; Gomberg-Maitland, M.; Shah, S.J. Usefulness of Red Cell Distribution Width as a Prognostic Marker in Pulmonary Hypertension. *Am. J. Cardiol.* **2009**, *104*, 868–872. [CrossRef] [PubMed]
25. Iancu, D.G.; Varga, A.; Cristescu, L.; Dumbrava, R.A.; Stoica, F.; Moldovan, D.A.; Suteu, R.A.; Tilea, I. Kidney Dysfunction, Hepatic Impairment, and Lipid Metabolism Abnormalities in Patients with Precapillary Pulmonary Hypertension. *Diagnostics* **2024**, *14*, 1824. [CrossRef]
26. Chakinala, M.M.; Coyne, D.W.; Benza, R.L.; Frost, A.E.; McGoon, M.D.; Hartline, B.K.; Frantz, R.P.; Selej, M.; Zhao, C.; Mink, D.R.; et al. Impact of Declining Renal Function on Outcomes in Pulmonary Arterial Hypertension: A REVEAL Registry Analysis. *J. Heart Lung Transplant.* **2018**, *37*, 696–705. [CrossRef]
27. Shah, S.J.; Thenappan, T.; Rich, S.; Tian, L.; Archer, S.L.; Gomberg-Maitland, M. Association of Serum Creatinine with Abnormal Hemodynamics and Mortality in Pulmonary Arterial Hypertension. *Circulation* **2008**, *117*, 2475–2483. [CrossRef]
28. Lenth, R.V. Some Practical Guidelines for Effective Sample Size Determination. *Am. Stat.* **2001**, *55*, 187–193. [CrossRef]
29. Galiè, N.; Humbert, M.; Vachiery, J.L.; Gibbs, S.; Lang, I.; Torbicki, A.; Simonneau, G.; Peacock, A.; Vonk Noordegraaf, A.; Beghetti, M.; et al. 2015 ESC/ERS Guidelines for the Diagnosis and Treatment of Pulmonary Hypertension: The Joint Task Force for the Diagnosis and Treatment of Pulmonary Hypertension of the European Society of Cardiology (ESC) and the European Respiratory Society (ERS): Endorsed by: Association for European Paediatric and Congenital Cardiology (AEPC), International Society for Heart and Lung Transplantation (ISHLT). *Eur. Heart J.* **2016**, *37*, 67–119. [CrossRef] [PubMed]
30. Çorbacıoğlu, Ş.K.; Aksel, G. Receiver operating characteristic curve analysis in diagnostic accuracy studies: A guide to interpreting the area under the curve value. *Turk. J. Emerg. Med.* **2023**, *23*, 195–198. [CrossRef] [PubMed]
31. Pan, D.; Chen, H. Relationship between Serum Albumin Level and Hospitalization Duration following Percutaneous Coronary Intervention for Acute Coronary Syndrome. *Sci. Rep.* **2024**, *14*, 23883. [CrossRef]
32. Pokhrel Bhattarai, S.; Dzikowicz, D.J.; Carey, M.G. Association Between Serum Albumin and the Length of Hospital Stay Among Patients With Acute Heart Failure. *Biol. Res. Nurs.* **2025**, *27*, 37–46. [CrossRef]
33. Ling, M.; Huiyin, L.; Shanglin, C.; Haiming, L.; Zhanyi, D.; Shuchun, W.; Meng, B.; Murong, L. Relationship between Human Serum Albumin and In-Hospital Mortality in Critical Care Patients with Chronic Obstructive Pulmonary Disease. *Front. Med.* **2023**, *10*, 1109910. [CrossRef]
34. Chao, P.; Cui, X.; Wang, S.; Zhang, L.; Ma, Q.; Zhang, X. Serum Albumin and the Short-Term Mortality in Individuals with Congestive Heart Failure in Intensive Care Unit: An Analysis of MIMIC. *Sci. Rep.* **2022**, *12*, 16251. [CrossRef]
35. Özpelit, E.; Akdeniz, B.; Özpelit, M.E.; Tas, S.; Bozkurt, S.; Tertemiz, K.C.; Sevinç, C.; Badak, Ö. Prognostic Value of Neutrophil-to-Lymphocyte Ratio in Pulmonary Arterial Hypertension. *J. Int. Med. Res.* **2015**, *43*, 661–671. [CrossRef] [PubMed]
36. Yanartas, M.; Kalkan, M.E.; Arslan, A.; Tas, S.G.; Koksal, C.; Bekiroglu, N.; Yildizeli, B. Neutrophil/Lymphocyte Ratio Can Predict Postoperative Mortality in Patients with Chronic Thromboembolic Pulmonary Hypertension. *Ann. Thorac. Cardiovasc. Surg.* **2015**, *21*, 229–235. [CrossRef] [PubMed]
37. Maurer, S.J.; Habdank, V.; Hörer, J.; Ewert, P.; Tutarel, O. NT-proBNP Is a Predictor of Mortality in Adults with Pulmonary Arterial Hypertension Associated with Congenital Heart Disease. *J. Clin. Med.* **2023**, *12*, 3101. [CrossRef]

38. Januzzi, J.L., Jr.; Sakhuja, R.; O'Donoghue, M.; Baggish, A.L.; Anwaruddin, S.; Chae, C.U.; Cameron, R.; Krauser, D.G.; Tung, R.; Camargo, C.A., Jr.; et al. Utility of Amino-Terminal Pro-Brain Natriuretic Peptide Testing for Prediction of 1-Year Mortality in Patients with Dyspnea Treated in the Emergency Department. *Arch. Intern. Med.* **2006**, *166*, 315–320. [CrossRef] [PubMed]
39. Kim, K.M.; Nerlekar, R.; Tranah, G.J.; Browner, W.S.; Cummings, S.R. Higher Red Cell Distribution Width and Poorer Hospitalization-Related Outcomes in Elderly Patients. *J. Am. Geriatr. Soc.* **2022**, *70*, 2354–2362. [CrossRef]
40. Huang, S.; Zhou, Q.; Guo, N.; Zhang, Z.; Luo, L.; Luo, Y.; Qin, Z.; Ge, L. Association between Red Blood Cell Distribution Width and In-Hospital Mortality in Acute Myocardial Infarction. *Medicine* **2021**, *100*, e25404. [CrossRef]
41. Tonelli, M.; Sacks, F.; Arnold, M.; Moye, L.; Davis, B.; Pfeffer, M.; for the Cholesterol and Recurrent Events (CARE) Trial Investigators. Relation Between Red Blood Cell Distribution Width and Cardiovascular Event Rate in People with Coronary Disease. *Circulation* **2008**, *117*, 163–168. [CrossRef]
42. Haas, L.; Eckart, A.; Haubitz, S.; Mueller, B.; Schuetz, P.; Segerer, S. Estimated Glomerular Filtration Rate Predicts 30-Day Mortality in Medical Emergency Departments: Results of a Prospective Multi-National Observational Study. *PLoS ONE* **2020**, *15*, e0230998. [CrossRef]
43. Verstreken, S.; Beles, M.; Oeste, C.L.; Moya, A.; Masuy, I.; Dierckx, R.; Heggermont, W.; Dauw, J.; Hens, D.; Bartunek, J.; et al. eGFR Slope as Predictor of Mortality in Heart Failure Patients. *ESC Heart Fail.* **2024**, *12*, 1217–1226. [CrossRef] [PubMed]
44. Suk, C.W.; Hsu, S.C.; Chen, C.Y.; Hsieh, H.L.; Kuo, H.T.; Hsu, Y.P.; Sue, Y.M.; Chen, T.H.; Lin, F.Y.; Shih, C.M.; et al. Point of Care eGFR and the Prediction of Outcomes in Pneumonia. *Sci. Rep.* **2019**, *9*, 8478. [CrossRef]
45. Deng, X.; Gao, B.; Wang, F.; Zhao, M.H.; Wang, J.; Zhang, L. Red Blood Cell Distribution Width Is Associated With Adverse Kidney Outcomes in Patients With Chronic Kidney Disease. *Front. Med.* **2022**, *9*, 877220. [CrossRef] [PubMed]
46. Ulrich, A.; Wharton, J.; Thayer, T.E.; Swietlik, E.M.; Assad, T.R.; Desai, A.A.; Gräf, S.; Harbaum, L.; Humbert, M.; Morrell, N.W.; et al. Mendelian Randomisation Analysis of Red Cell Distribution Width in Pulmonary Arterial Hypertension. *Eur. Respir. J.* **2020**, *55*, 1901486. [CrossRef]
47. Liu, J.; Yang, J.; Xu, S.; Zhu, Y.; Xu, S.; Wei, L.; Qian, P.; Lv, Y.; Zhang, C.; Xing, X.; et al. Prognostic Impact of Red Blood Cell Distribution Width in Pulmonary Hypertension Patients: A Systematic Review and Meta-Analysis. *Medicine* **2020**, *99*, e19089. [CrossRef]
48. Smukowska-Gorynia, A.; Tomaszewska, I.; Malaczynska-Rajpold, K.; Marcinkowska, J.; Komosa, A.; Janus, M.; Olasinska-Wisniewska, A.; Slawek, S.; Araszkiewicz, A.; Jankiewicz, S.; et al. Red Blood Cells Distribution Width as a Potential Prognostic Biomarker in Patients With Pulmonary Arterial Hypertension and Chronic Thromboembolic Pulmonary Hypertension. *Heart Lung Circ.* **2018**, *27*, 842–848. [CrossRef] [PubMed]
49. Thongprayoon, C.; Cheungpasitporn, W.; Chewcharat, A.; Mao, M.A.; Thirunavukkarasu, S.; Kashani, K.B. Impacts of Admission Serum Albumin Levels on Short-Term and Long-Term Mortality in Hospitalized Patients. *QJM* **2020**, *113*, 393–398. [CrossRef]
50. Zhang, S.; Li, S.; Gao, L.; Zhao, Q.; Yang, T.; Zeng, Q.; Huang, Z.; Li, X.; Duan, A.; Wang, Y.; et al. Effects of Malnutrition on Disease Severity and Adverse Outcomes in Idiopathic Pulmonary Arterial Hypertension: A Retrospective Cohort Study. *Respir. Res.* **2024**, *25*, 292. [CrossRef]
51. Wang, W.; Liu, J.; Yang, Y.H.; Zhai, Z.G.; Wang, C.; Wang, J. Red cell distribution width is increased in chronic thromboembolic pulmonary hypertension. *Clin. Respir. J.* **2016**, *10*, 54–60. [CrossRef] [PubMed]

diagnostics

MDPI

Article

Atrial Fibrillation Among ICU Patients with Type 2 Respiratory Failure: Who Is at Risk and What Are the Outcomes?

Oral Mentes [1], Deniz Celik [2], Murat Yıldız [3,*], Tarkan Özdemir [4], Maside Ari [3], Eda Nur Aksoy Güney [3], Emrah Ari [5], Fatma Canbay [6], Yusuf Taha Güllü [7], Abdullah Kahraman [8] and Mustafa Özgür Cırık [9]

1 Department of Intensive Care, Gulhane Training and Research Hospital, 06010 Ankara, Turkey; omentes@live.com
2 Department of Pulmonary Medicine, Faculty of Medicine, Alanya Alaaddin Keykubat University, 07425 Antalya, Turkey; drdenizcelik@hotmail.com
3 Department of Pulmonary Medicine, Atatürk Sanatorium Research Hospital, Faculty of Medicine, Turkey Health Sciences University, 06830 Ankara, Turkey; masidetuten@icloud.com (M.A.); aksoyedanur95@gmail.com (E.N.A.G.)
4 Department of Pulmonology, Konya Farabi Hospital, 42090 Konya, Turkey; tarkanozdemir78@gmail.com
5 Mamak Public Hospital, 06270 Ankara, Turkey; dremrahari25@gmail.com
6 Department of Pulmonary Medicine, Dr. Burhan Nalbantoğlu State Hospital, 99010 Lefkoşa, Cyprus; canbayfatma5@gmail.com
7 Anesthesia Program Mudany, Vocational School, Mudanya University, 16960 Bursa, Turkey
8 Department of Intensive Care, Etlik City Hospital, 06170 Ankara, Turkey; abdullahhero100@gmail.com
9 Department of Anesthesiology, Atatürk Sanatorium Research Hospital, Faculty of Medicine, Turkey Health Sciences University, 06100 Ankara, Turkey; dr.ozgurr@hotmail.com
* Correspondence: drmuratyildiz85@gmail.com; Tel.: +90-5548441305

Abstract: Background: Atrial fibrillation (AF) frequently occurs in individuals with hypercapnic type 2 respiratory failure and has the potential to adversely affect patient outcomes. This study sought to investigate the clinical features and prognostic significance of atrial fibrillation in patients admitted to the intensive care unit with hypercapnic type 2 respiratory failure. **Methods**: This retrospective, single-center study included 200 adult patients diagnosed with hypercapnic type 2 respiratory failure between May 2022 and May 2023. Patients were grouped according to whether atrial fibrillation was present or not. Demographic, laboratory, and echocardiographic findings, comorbidities, and outcomes were compared. Kaplan–Meier survival analysis and Cox regression were used to identify mortality predictors. **Results**: AF was present in 50.5% of patients. Those with AF were older, had higher Charlson Comorbidity Index scores, and a greater prevalence of heart failure ($p < 0.001$). No significant differences were found in arterial blood gas values. AF patients had higher urea, creatinine, and BNP levels, and lower hemoglobin, lymphocyte, eosinophil, and monocyte counts ($p < 0.05$). Echocardiography showed more severe tricuspid and mitral regurgitation, lower ejection fractions, and higher systolic pulmonary pressures in the AF group. About 20% of AF patients were not receiving anticoagulants at ICU admission. AF was associated with shorter survival (49.6 ± 4.07 vs. 61.4 ± 3.8 days, $p = 0.031$) and 1.6-fold higher mortality risk (HR: 1.60, 95% CI: 1.04–2.47). Advanced age and low hemoglobin were independent predictors of mortality. **Conclusions**: AF is frequent among patients with type 2 respiratory failure and is linked to increased mortality. Despite known complications, treatment remains underutilized. AF should be actively screened during ICU admissions for respiratory failure.

Keywords: hypercapnic respiratory failure; atrial fibrillation; mortality; COPD; survival analysis

 https://doi.org/10.3390/diagnostics15131612

1. Introduction

Hypercapnic respiratory failure is a systemic syndrome marked by elevated partial carbon dioxide (pCO_2) levels due to hypoventilation, commonly seen in chronic respiratory disorders such as chronic obstructive pulmonary disease (COPD). Respiratory failure can directly impact patient survival due to inadequate oxygenation and hemodynamic alterations resulting from hypercapnia [1,2]. Hypercapnic respiratory failure extends beyond the respiratory system and may result in cardiovascular complications. Atrial fibrillation (AF) is a frequent rhythmic condition observed in these patients and significantly impacts patient prognosis.

Analysis of the pathophysiology of atrial fibrillation revealed that hypoxia and hypercapnia are significant factors. In hypercapnic individuals, factors such as intrathoracic pressure fluctuations, alterations in autonomic nervous system activity, and atrial stretching and remodeling contribute to the onset of AF [3]. Moreover, hypercapnia and hypoxemia can induce right ventricular hypertension and right atrial dilation by increasing pulmonary arterial pressure. These alterations result in cardiac hemodynamic imbalances, promoting the onset of atrial fibrillation. This may not only increase the incidence of atrial fibrillation but also negatively impact clinical outcomes, including cardioversion efficacy, ablation results, and in-hospital mortality [4].

In several patients with hypercapnic respiratory failure monitored in pulmonary intensive care units, concurrent atrial fibrillation has been noted to be associated with a poor prognosis. AF may worsen respiratory function problems, impair cardiopulmonary balance, and make therapeutic procedures more difficult for this patient population. The frequency of AF in individuals with COPD has been reported to be twice as high as that in those without COPD. Moreover, atrial fibrillation is associated with an elevated risk of mortality in this patient population [5].

The presence of AF in patients with hypercapnic respiratory failure is often accompanied by additional clinical risks, such as heart failure. Furthermore, alterations in echocardiographic findings and the increased relevance of certain laboratory parameters are inevitable in this context. Identifying the clinical profiles of patients with type 2 respiratory failure who also have AF is of great importance. Recognizing these profiles can facilitate earlier diagnosis and help determine which patients should undergo more frequent ECG screening for timely detection of arrhythmias [6,7].

Considering the well-known severe complications of AF—such as acute coronary syndrome, heart failure, valvular diseases, and ischemic cerebrovascular events—demonstrating its potential additive negative impact on mortality in this patient population would provide a meaningful contribution to the medical literature.

This study investigated the impact of AF on survival in patients with hypercapnic type 2 respiratory failure in a multidimensional manner. This study specifically investigated how pulmonary hypertension, cardiac hemodynamics, and blood gas changes affect the onset of AF in this patient population. Additionally, the impact of AF on clinical outcomes such as diagnosis, treatment, and in-hospital mortality will be assessed. The findings of this study may aid in the formulation of clinical treatment plans for this diverse patient population.

2. Materials and Methods

The study was carried out in compliance with institutional and national ethical guidelines, as well as the principles of the 1964 Declaration of Helsinki and its subsequent revisions. Informed consent was obtained from all participants involved in the study. Consent to publish the study findings was also obtained from all participants. In our retrospective studies, during the process of obtaining informed consent regarding procedures to be performed in the intensive care unit, patients and/or their legal representatives are asked to sign a "Retrospective Data Use Consent Form." This form ensures that privacy regulations will be followed and authorizes the use of anonymized demographic data, laboratory results, clinical scores, ECGs, chest X-rays, and other relevant examinations for research purposes. Patients who explicitly refused permission by signing the form or who left the consent form unsigned were excluded from the study.

This study was approved by the Clinical Research Ethics Committee of the University of Health Sciences, Ankara Sanatorium Training and Research Hospital (Date: 12 June 2024, Decision No: 2024-BÇEK/81). The study group consisted of 512 patients who were followed up in the respiratory intensive care unit between 1 May 2022 and 1 May 2023. Among these, 200 patients diagnosed with type 2 respiratory failure meeting the inclusion criteria were included in the study, while patients under the age of 18, those with incomplete medical records, patients with type 1 respiratory failure, and those referred to other centers were excluded (Figure 1).

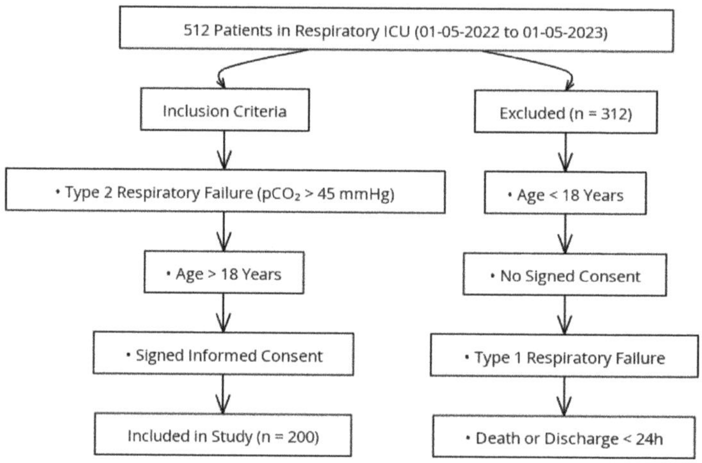

Figure 1. Flowchart of the study design.

All relevant data—such as age, sex, comorbidities, laboratory values, Charlson Comorbidity Index (CCI), and other demographic and clinical information—were retrieved from patient records and the hospital's information management system. The diagnosis of "respiratory failure" was confirmed through arterial blood gas (ABG) values obtained during hospitalization and follow-up. AF was diagnosed based on electrocardiographic (ECG) findings obtained at ICU admission, which is routinely performed for all patients. Patients were classified as having AF if the arrhythmia was present on admission ECG and persisted throughout the ICU stay. Individuals with paroxysmal AF—defined as AF episodes that spontaneously resolve or terminate with intervention within 7 days—were excluded. Accordingly, patients with chronic or persistent AF were included in the AF group. Due to the retrospective nature of the study, a more granular distinction between persistent and chronic AF was not possible. Echocardiographic findings within the last year and the use of antiarrhythmic and anticoagulant drugs were reviewed.

Patients were stratified into two groups based on the presence or absence of AF. Blood parameters routinely analyzed during intensive care follow-up were retrospectively examined. The blood parameters were analyzed comparatively between the two groups to assess their impact on in-hospital mortality and length of stay.

Statistical Analysis

Statistical analyses were conducted using IBM SPSS Statistics for Windows, Version 30.0 (IBM Corp., Armonk, NY, USA) and MedCalc version 23.2.1 (MedCalc Software Ltd., Ostend, Belgium). The distribution of continuous variables was evaluated through descriptive measures, including the Kolmogorov–Smirnov and Shapiro–Wilk tests, skewness–kurtosis values, histograms, and outlier inspection. Normally distributed data were expressed as means ± standard deviations, whereas non-normally distributed data were presented as medians with interquartile ranges (IQR). Categorical variables were reported as frequencies and percentages.

Comparisons between categorical variables were performed using Pearson's chi-square test, and Fisher's exact test was applied when expected cell counts were below 5. For continuous variables, Student's *t*-test was used when data met normality assumptions, while the Mann-Whitney U test was applied for non-normally distributed variables. Survival outcomes were assessed using Kaplan–Meier analysis, with the log-rank test employed to determine significance. Cox proportional hazards regression analysis was performed to identify independent predictors of mortality, and hazard ratios (HRs) with 95% confidence intervals (CIs) were reported. A two-tailed *p*-value < 0.05 was considered statistically significant.

3. Results

The study was conducted with 200 patients who met the eligibility criteria. Among the patients, 118 (59%) were male. The mean age of the patients was 72 ± 9 years. The most common comorbidity was chronic obstructive pulmonary disease (COPD). The demographic and clinical characteristics of the patients are presented in Table 1.

Table 1. Demographic and Clinical Characteristics of All Patients.

Variable	Value—*n* (%), Mean (±sd), Median (IQR)
Age, years	72 ± 9
Presence of atrial fibrillation	101 (50.5%)
Mortality status	85 (42.5%)
Gender	
Male	118 (59%)
Female	82 (41%)
Comorbidity	
Chronic obstructive pulmonary disease	169 (84.5%)
Hypertension	103 (51.5%)
Coronary artery disease	26 (13%)
Diabetes mellitus	77 (38.5%)
Chronic kidney disease	24 (12%)
Bronchiectasis	8 (4%)
Malignancy	8 (4%)
Obesity hypoventilation syndrome	6 (3%)
Past tuberculosis enfection	7 (3.5%)
Dementia	8 (4%)
Acute Kidney Failure	5 (2.5%)
Past cerebrovascular disease	11 (5.5%)
Pulmonary embolism	8 (4%)
Heart failure	68 (34%)
Charlson comorbidity index	5 (4–7)
Length of hospital stay	8.94 ± 5.24

When the demographic characteristics of patients were evaluated on the basis of the presence of AF, patients with atrial fibrillation were older ($p < 0.001$). An examination of comorbid conditions revealed that heart failure was more common and that the CCI was greater in these patients ($p < 0.001$). Additionally, mortality was found to be greater in patients with AF ($p = 0.043$) (Table 2).

Table 2. Demographic characteristics of patients based on the presence of atrial fibrillation.

Demographic Characteristics	Patients with AF (n = 101)	Patient Without AF (n = 99)	p Value
Age, years (Mean ± sd)	76 ± 8	69 ± 8	<0.001 *$^{\alpha}$
Gender			0.059 $^{\beta}$
Male n (%)	53 (%44.9)	65 (%55.1)	
Female n (%)	48 (%58.5)	34 (%41.5)	
Number of deceased patients n (%)	59 (%58.4)	45 (%45.4)	0.043 *$^{\beta}$
Comorbidity			
COPD n (%)	81 (%47.9)	88 (%52.1)	0.090 $^{\beta}$
HT n (%)	53 (%51.5)	50 (%48.5)	0.781 $^{\beta}$
DM n (%)	39 (%50.6)	38 (%49.4)	0.973 $^{\beta}$
HF n (%)	46 (%67.6)	22 (%32.4)	<0.001 *$^{\beta}$
CCI median (IQR)	6 (4–7)	4 (4–6)	<0.001 *$^{\gamma}$
ICU length of stay (Mean ± sd)	9.08 ± 6.29	8.79 ± 3.92	0.829 $^{\alpha}$

AF: Atrial Fibrillation; CCI: Charlson Comorbidity Index; COPD: chronic obstructive pulmonary disease; HT: hypertension; ICU: intensive care unit; $^{\alpha}$: Student's t test; $^{\beta}$: chi-square test; $^{\gamma}$: Mann-Whitney U test; *: statistically significant values.

When patients with type 2 respiratory failure with AF were compared to those without AF based on arterial blood gas values at ICU admission, no statistically significant difference was found. The admission blood gas values of patients who were followed up with hypercapnic respiratory failure are presented in Table 3.

Table 3. The admission blood gas values of patients.

Admission Blood Gas	Patients with AF (n = 101) Median (IQR)	Patients Without AF (n = 99) Median (IQR)	p Value
pH	7.33 (7.24–7.42)	7.33 (7.29–7.39)	0.754 $^{\gamma}$
paO_2	50 (32.2–72.7)	48 (33–62)	0.422 $^{\gamma}$
pCO_2	66 (56–83)	67 (59–78)	0.835 $^{\gamma}$
HCO_3	36 (32.2–42)	37 (32–42)	0.607 $^{\gamma}$

AF: Atrial fibrillation; $^{\gamma}$: Mann-Whitney U test.

AF was present in 50.5% of the patients included in the study. The blood test results of the patients based on the presence of AF are presented in Table 4. Urea and creatinine levels were found to be higher in patients with AF ($p < 0.001$). Among the complete blood count parameters, the hemoglobin, lymphocyte, eosinophil, and monocyte levels were lower in patients with AF ($p < 0.001$, $p = 0.004$, $p = 0.029$, $p = 0.047$, respectively). On the other hand, natriuretic peptide levels were greater in patients with AF ($p < 0.001$) (Table 4).

Table 4. Evaluation of Patient Admission Blood Tests.

Laboratory Findings	Patient with AF (n = 101) Median (IQR)	Patient Without AF (n = 99) Median (IQR)	p Value
Urea (mg/dL)	58 (41.5–82.2)	47.5 (29.2–64.5)	<0.001 *ϒ
Creatinine (mg/dL)	1.07 (0.89–1.37)	0.96 (0.77–1.29)	0.014 *ϒ
Hemoglobin (g/dL)	11.05 (9.6–13.5)	12.9 (11.2–14.6)	<0.001 *ϒ
Leukocyte ($\times 10^3/\mu L$)	9.8 (7.9–13.05)	9.6 (7.9–14.5)	0.301 ϒ
Lymphocyte ($\times 10^3/\mu L$)	0.86 (0.55–1.4)	1.09 (0.66–1.6)	0.004 *ϒ
Neutrophil ($\times 10^3/\mu L$)	8.03 (5.7–10.9)	7.75 (5.7–11.7)	0.967
Platelet ($\times 10^3/\mu L$)	235 (174–313)	245 (191–315)	0.537 ϒ
Eosinophil ($\times 10^3/\mu L$)	0.02 (0–0.1)	0.025 (0–0.11)	0.029 *ϒ
Monocyte ($\times 10^3/\mu L$)	0.57 (0.32–0.76)	0.62 (0.37–0.82)	0.047 *ϒ
Natriuretic Peptide	358 (150–910)	100 (41–280)	<0.001 *ϒ
D-dimer	1645 (617–3287)	1315 (592–2455)	0.466 ϒ
Troponin	17 (8–35)	13 (5.2–35.5)	0.156 ϒ

ϒ: Mann-Whitney U test, *: Statistically significant values.

Among the patients included in the study, 42.5% passed away within 30 days of follow-up. The admission blood test results of the patients were evaluated on the basis of overall averages and mortality status. In deceased patients, urea, brain natriuretic peptide (BNP), troponin, and D-dimer levels were significantly greater ($p = 0.046$, $p = 0.017$, $p = 0.02$, $p = 0.011$, respectively). Hemoglobin levels, on the other hand, were lower in deceased patients ($p = 0.06$) (Table 5). When we compared the categorical variables in our study in terms of mortality, we found that mortality was significantly higher in patients with heart failure and atrial fibrillation (Table 6).

A total of 125 patients underwent echocardiographic evaluation, of whom 65 had AF and 60 did not. Upon evaluating the echocardiographic findings of patients admitted to our intensive care unit with a diagnosis of type 2 respiratory failure, we identified certain statistically significant differences between those with and without AF. Patients with AF had more severe tricuspid and mitral regurgitation (p: 0.0047 and p: 0.0135, respectively). Ejection fraction values were also significantly lower in patients with AF (p: 0.001). Additionally, systolic pulmonary arterial pressures were significantly higher in patients with AF (p: 0.016). However, there was no statistically significant difference in the prevalence of aortic regurgitation between the two groups (Table 7).

To analyze the association of atrial fibrillation (AF) with other clinical conditions and comorbidities, the prevalence of additional diseases was examined in detail among patients with and without AF. Heart failure was the only comorbidity found to be significantly more common in patients with AF ($p < 0.001$). All other clinical conditions and comorbidities were statistically similar between the two groups (Table 8).

Table 5. Evaluation of Patient Admission Blood Tests on the basis of mortality.

Laboratory Findings	Surviving Patients (*n*: 115) Median (IQR)	Deceased Patients (*n*: 85) Median (IQR)	*p* Value
Urea (mg/dL)	51.00 (32.25–66.00)	57.50 (40.00–82.25)	0.046 *ϒ
Creatinine (mg/dL)	0.98 (0.80–1.20)	1.07 (0.84–1.39)	0.341 ϒ
Potassium (mEq/L)	4.50 (4.20–4.88)	4.60 (4.20–5.20)	0.115 ϒ
Hemoglobin (g/dL)	12.80 (11.03–14.55)	11.25 (9.58–13.50)	0.006 *ϒ
WBC ($\times 10^3/\mu$L)	9.8 (7.8–13.5)	9.6 (8–13.8)	0.651 ϒ
Lymphocyte ($\times 10^3/\mu$L)	0.94 (0.63–1.5)	0.92 (0.64–1.5)	0.670 ϒ
Platelet ($\times 10^3/\mu$L)	246.5 (198–311)	233 (167–328)	0.887 ϒ
Magnesium (mg/dL)	2.00 (1.80–2.20)	2.00(1.80–2.30)	0.103 ϒ
Natriuretic Peptide (ng/L)	158.50 (61.75–414.00)	279.50 (112.00–855.00)	0.017 *ϒ
D-dimer (ng/mL)	940 (552–2112)	1820 (940–3755)	0.002 *ϒ
Troponin I (ng/mL)	12 (5.23–27.25)	19.50 (10–36.5)	0.011 *ϒ
Age	70.2 ± 0.87	76.2 ± 0.93	<0.001
CCI	4.50 (4.0–6.0)	5.00 (4.0–7.0)	<0.001
pH	7.35 (7.27–7.41)	7.32 (7.26–7.39)	0.324
PO2	50.00 (33.8–72.0)	45.00 (31.5–66.0)	0.434
PCO2	66.00 (57.0–81.0)	66.00 (58.5–81.0)	0.641
HCO3	37.00 (33.0–42.0)	34.00 (31.0–42.5)	0.262
Monocyte count ($\times 10^3/\mu$L)	0.56 (0.34–0.76)	0.3 (0.11–0.61)	0.773
ICU Length of Stay (days)	90.00 (90.0–90.0)	8.00 (5.0–12.0)	<0.001

CCI: Charlson comorbidity index, ICU: Intensive care unit, WBC: White blood cell, ϒ: Mann-Whitney U test, *: Statistically significant values.

Table 6. Comparison of categorical variables between deceased and surviving patients using the Pearson chi-square test.

Variable	Deceased (*n*, %)	Survived (*n*, %)	*p*-Value
Gender (Male)	51 (43.2%)	67 (56.8%)	0.805
Gender (Female)	34 (41.5%)	48 (58.5%)	
NIMV Required	48 (41.4%)	68 (58.6%)	0.706
NIMV Not Required	37 (44.0%)	47 (56.0%)	
Anticoagulant Use (Yes)	46 (47.9%)	50 (52.1%)	0.137
Anticoagulant Use (No)	39 (37.5%)	65 (62.5%)	
Antiarrhythmic Use (Yes)	68 (44.2%)	86 (55.8%)	0.386
Antiarrhythmic Use (No)	17 (37.0%)	29 (63.0%)	
Atrial Fibrillation (Yes)	50 (49.5%)	51 (50.5%)	0.043 *
Atrial Fibrillation (No)	35 (35.4%)	64 (64.6%)	
Heart Failure (Yes)	36 (52.9%)	32 (47.1%)	0.032 *
Heart Failure (No)	49 (37.1%)	83 (62.9%)	
Diabetes Mellitus (Yes)	32 (41.6%)	45 (58.4%)	0.831
Diabetes Mellitus (No)	53 (43.1%)	70 (56.9%)	
Hypertension (Yes)	42 (40.8%)	61 (59.2%)	0.611
Hypertension (No)	43 (44.3%)	54 (55.7%)	
COPD (Yes)	72 (42.6%)	97 (57.4%)	0.945
COPD (No)	13 (41.9%)	18 (58.1%)	

COPD: Chronic obstructive pulmonary disease, NIMV: Noninvasive mechanic ventilation, * significant *p* values.

Table 7. Comparison of Echocardiographic Findings Based on Atrial Fibrillation Status.

Variable	AF Negative(−) Patients n (%), Median (IQR), n: 60	AF Positive (+) Patients n (%), Median (IQR), n: 65	p Value
Tricuspid Regurgitation	1: 31 (51.7%) 2: 26 (43.3%) 3: 3 (5%)	1: 23 (35.4%) 2: 25 (38.5%) 3: 17 (26.2%)	0.0047 *ᶿ
Mitral Regurgitation	1: 53 (88.3%) 2: 6 (10.0%) 3: 1 (1.7%)	1: 43 (66.2%) 2: 19 (29.2%) 3: 3 (4.6%)	0.0135 *ᶿ
Aortic Regurgitation	1: 58 (96.7%) 2: 2 (3.3%)	1: 61 (93.8%) 2: 4 (6.2%)	0.4629 ᶿ
SPAP (mmHg)	40.0 (33.25–47.75)	45.0 (35.0–52.5)	0.016 *ᵞ
Ejection fraction (%)	55 (55–60)	55 (45–60)	0.001

Regurgitation grades are defined as follows: 1 = mild, 2 = moderate, 3 = severe. SPAP: Systolic pulmonary arterial pressure. ᶿ Fisher's exact test, ᵞ Mann–Whitney U test, * significant p value.

Table 8. Comparison of clinical conditions between AF (+) and AF (−) patients.

Clinical Condition	AF (−) n: 99 n (%)	AF (+) n: 101 n (%)	p Value
Female Gender	34 (34.3%)	48 (47.5%)	0.058 ᵝ
Asthma	2 (2.0%)	5 (5.0%)	0.260 ᶿ
OSAS	3 (3.0%)	3 (3.0%)	0.980 ᶿ
Malignancy	2 (2.0%)	6 (5.9%)	0.157 ᶿ
COPD	88 (88.9%)	81 (80.2%)	0.090 ᵝ
Bronchiectasis	5 (5.1%)	3 (3.0%)	0.453 ᶿ
History of TB	2 (2.0%)	5 (5.0%)	0.260 ᶿ
Dementia/Alzheimer	2 (2.0%)	6 (5.9%)	0.157 ᶿ
Heart Failure	22 (22.2%)	46 (45.5%)	<0.001 *ᵝ
Chronic Kidney Disease	10 (10.1%)	14 (13.9%)	0.413 ᵝ
Pulmonary Thromboembolism	3 (3.0%)	5 (5.0%)	0.488 ᶿ
Coronary Artery Disease	13 (13.1%)	13 (12.9%)	0.956 ᵝ
Hypertension	50 (50.5%)	53 (52.5%)	0.780 ᵝ
Diabetes Mellitus	38 (38.4%)	39 (38.6%)	0.973 ᵝ
Acute Kidney Injury	1 (1.0%)	4 (4.0%)	0.181 ᶿ
History of Stroke	6 (6.1%)	5 (5.0%)	0.731 ᵝ

COPD: Chronic obstructive pulmonary disease, TB: Tuberculosis, ᵝ Pearson chi square test, ᶿ Fisher's exact test, * significant p value.

Among patients admitted to the intensive care unit with pre-existing atrial fibrillation (AF), 19.8% were not prescribed anticoagulant therapy and 6.9% were not receiving antiarrhythmic medications at the time of admission (Table 9).

Table 9. AF (+) patients without prescribed anticoagulants or antiarrhythmics at ICU admission.

Treatment Type	Not Prescribed (n)	Total AF (+) Patients (n)	Rate (%)
Anticoagulant	20	101	19.8%
Antiarrhythmic	7	101	6.9%

When comparing the survival times of patients with and without atrial fibrillation (AF), the mean survival time was found to be 49.6 (±4.07) days (95% CI: 41.6 to 57.6) for patients with AF and 61.4 (±3.8) days (95% CI: 53.8 to 69.05) for those without AF. The difference in survival times was statistically significant, favoring patients without AF (log-rank test: $p = 0.031$). Furthermore, according to this analysis, the mortality risk in patients with AF was calculated to be 1.6 times higher compared to those without AF (HR: 1.6084, 95% CI: 1.0437 to 2.4787) (Figure 2).

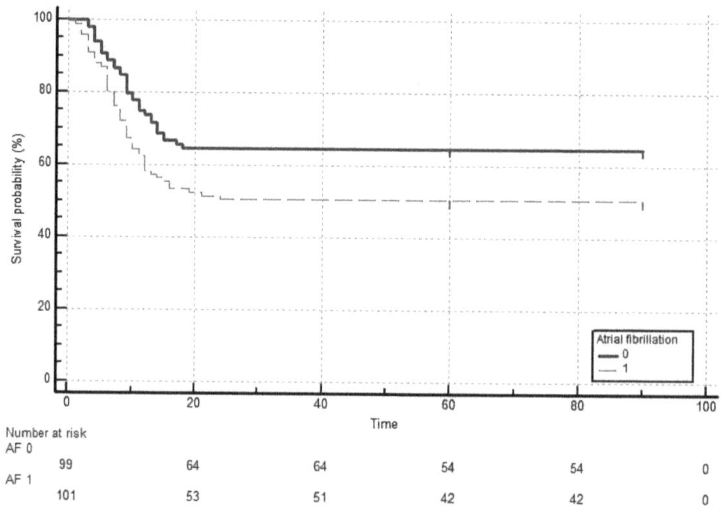

Figure 2. Kaplan-Meier survival curve for patients with and without atrial fibrillation.

A multivariable Cox regression analysis was performed using the enter method, including all variables that were significantly associated with mortality in univariate analyses. Among these, lower hemoglobin levels (HR: 0.901; 95% CI: 0.816–0.996; $p = 0.041$), older age (HR: 1.039; 95% CI: 1.007–1.073; $p = 0.018$), and higher Charlson Comorbidity Index scores (HR: 1.175; 95% CI: 1.004–1.375; $p = 0.045$) were found to be independent predictors of in-hospital mortality. Notably, the presence of atrial fibrillation (HR: 0.883; 95% CI: 0.532–1.467; $p = 0.631$) did not independently predict mortality in this model (Table 10).

Table 10. Enter method Cox regression analysis results of all patients.

Variable	*p* Value	HR (Exp (B))	95% CI Lower	95% CI Upper
Urea	0.435	0.997	0.991	1.004
Hemoglobin	0.041*	0.901	0.816	0.996
Natriuretic peptide	0.122	1.0	1.0	1.0
D-Dimer	0.661	1.0	1.0	1.0
Troponine	0.293	1.0	1.0	1.001
Presence of AF	0.631	0.883	0.532	1.467
Age	0.018 *	1.039	1.007	1.073
CCI	0.045 *	1.175	1.004	1.375

AF: Atrial fibrillation, CCI: Charlson comorbidity index, HR: Hazard ratio, CI: Confidence interval, * significant *p* value.

4. Discussion

4.1. Prevalence and Clinical Characteristics of AF in Type 2 Respiratory Failure

One of the most prominent findings of our study was the detection of AF in half of the patients admitted to our respiratory intensive care unit with a diagnosis of type 2 respiratory failure over a one-year period. In the literature, AF is recognized as the most common arrhythmia accompanying patients with type 2 respiratory failure and COPD [8]. When we compared the clinical characteristics of patients with AF to those without AF among individuals with type 2 respiratory failure, we observed that patients with AF were older, had higher CCI scores—indicating a greater burden of comorbidities—and had a notably higher prevalence of heart failure. In parallel with our findings, the study conducted by Chen and Liao also demonstrated that patients with AF were older and had a higher prevalence of cardiac comorbidities, including heart failure, compared to those without AF [9].

4.2. Laboratory Abnormalities and Their Pathophysiological Implications

When further examining the characteristics of patients with AF in our study through laboratory test results, we observed that urea and creatinine levels were higher in patients with AF compared to those without AF and that natriuretic peptide levels were also elevated, consistent with the higher prevalence of heart failure in this group. Additionally, hemoglobin and lymphocyte levels were found to be lower in patients with AF. Rodríguez-Manero et al. reported elevated natriuretic peptide levels in patients with AF, attributing this to the high prevalence of heart failure in this population. Similarly, Terzano et al. found that urea and creatinine levels were higher in AF patients, explaining this finding in the context of heart failure and associated systemic hypoperfusion. Chen and Liao not only noted that anemia was more common in patients with AF, but—consistent with our findings—also emphasized its association with poor prognosis in this group. Furthermore, Romiti et al. suggested that lower lymphocyte counts observed in patients with AF reflect systemic inflammation and a weakened immune response [8–11]. These laboratory abnormalities in our AF cohort may reflect a more advanced stage of cardio-respiratory compromise, particularly in the context of chronic hypercapnic respiratory failure. In diseases that typically cause type 2 respiratory failure—such as COPD and obesity hypoventilation syndrome—chronic hypoventilation and right heart strain can exacerbate renal hypoperfusion and systemic inflammation, amplifying neurohormonal activation. Elevated urea and BNP levels may therefore reflect not only cardiac dysfunction but also an adaptive response to chronic hypercapnia and fluid retention. Similarly, the observed anemia and lymphopenia may be consequences of systemic inflammation, reduced erythropoietin activity, and nutritional deficiencies common in chronic respiratory insufficiency. These findings suggest that the laboratory profile of AF patients in type 2 respiratory failure represents a complex interplay between cardiovascular strain, respiratory mechanics, and systemic inflammation.

4.3. Echocardiographic Features Suggestive of Cardiac Dysfunction

When comparing the echocardiographic findings of our AF patients to those without AF among individuals with hypercapnic respiratory failure, we observed that advanced mitral and tricuspid regurgitation were more common in the AF group. Additionally, systolic pulmonary arterial pressures were higher, and ejection fraction values were lower in patients with AF. These findings suggest that AF in the context of hypercapnic respiratory failure is associated with more severe underlying cardiac dysfunction. Similarly, Terzano et al. reported that patients with COPD exacerbations and AF had significantly higher pulmonary artery pressures and a greater incidence of valvular abnormalities, particularly mitral regurgitation. Reduced ejection fraction was also more common in the AF group, highlighting the interplay between AF and impaired cardiac performance in respiratory patients [5,11].

4.4. AF and Mortality: Interpretation of Survival Analyses

Based on our survival analysis, we observed that patients with AF had shorter survival times compared to those without AF among individuals admitted to the intensive care unit with type 2 respiratory failure. However, in the enter method Cox regression analysis—

which included the parameters that differed significantly between survivors and non-survivors—AF did not emerge as an independent predictor of mortality [12]. One possible explanation is that AF may function more as a surrogate marker of systemic burden rather than as a direct contributor to mortality. In critically ill patients, particularly those with advanced age and multiple comorbidities, AF often reflects underlying physiological stress, inflammation, or cardiac dysfunction. When variables such as age and Charlson Comorbidity Index are accounted for, the additive prognostic value of AF appears to diminish. Furthermore, in the context of acute-on-chronic respiratory failure, factors such as gas exchange impairment and baseline anemia may exert a more immediate impact on outcomes, overshadowing the contribution of arrhythmias. These findings suggest that while AF may signal increased clinical complexity, it may not independently drive mortality risk in this specific ICU population. This finding aligns with results from Rodríguez-Mañero et al., who also found that while AF was associated with worse unadjusted survival in patients with COPD, it did not independently predict mortality after adjusting for age and comorbidities, emphasizing the stronger prognostic weight of systemic factors such as age and overall disease burden [10]. Similarly, Xiao et al. reported that although AF prevalence was high in end-stage COPD patients, mortality was more strongly influenced by age, need for mechanical ventilation, and comorbidity burden rather than AF itself in multivariable analyses [13]. Considering the contribution of low hemoglobin levels to mortality in type 2 respiratory failure, future studies may focus on this issue to determine whether a new transfusion threshold should be established specifically for this patient population. Just as the presence of cardiac disease can raise the transfusion threshold in intensive care units, type 2 respiratory failure itself—independent of cardiac comorbidities—may warrant a reassessment of red blood cell replacement criteria [14].

4.5. Gas Exchange Parameters and AF

In our study, there were no statistically significant differences in pCO_2, pO_2, pH, or HCO_3 levels between patients with and without AF at the time of ICU admission. This suggests that the presence of AF may not be directly related to the severity of gas exchange abnormalities at presentation. Supporting this, Lahousse et al. found that while reduced lung function was associated with increased AF risk over time, cross-sectional arterial blood gas values—such as pCO_2 and pH—did not differ significantly at baseline between patients who developed AF and those who did not, emphasizing the role of chronic pulmonary and cardiovascular remodeling over acute respiratory derangement in AF pathophysiology [15].

4.6. Inadequate Pre-ICU AF Management and Therapy Gaps

In our study, we identified significant gaps in pre-ICU management by comparing AF diagnoses—based on ECGs obtained at ICU admission—with the patients' ongoing anticoagulant and antiarrhythmic therapies at the time of admission. These deficiencies were particularly notable in anticoagulant therapy, which is vital for preventing thrombotic complications. Despite having a diagnosis of AF, approximately 20% of patients were not receiving any anticoagulant treatment prior to ICU admission, while about 7% were not on antiarrhythmic therapy [16]. These findings are in line with the results of Wang et al., who reported substantial underutilization of anticoagulants in patients with AF, especially among those with chronic comorbidities such as COPD. Their study highlighted that up to one-quarter of eligible AF patients were not prescribed anticoagulants, often due to concerns about bleeding risk or a lack of cardiology follow-up, reflecting a broader issue of suboptimal adherence to evidence-based AF management in high-risk populations [17]. We share and support this perspective, as our findings similarly highlight a disconnect between AF diagnosis and the implementation of guideline-directed therapies. The underuse of anticoagulation observed in our cohort—despite a clear clinical indication—underscores the need for improved coordination between outpatient cardiology and ICU teams. It also points to the importance of early medication reconciliation and intervention planning at the point of ICU admission, particularly in patients with complex chronic diseases such as COPD.

4.7. Clinical Implications and Need for Multidisciplinary Care

Taken together, our findings highlight the multifactorial nature of AF in patients with type 2 respiratory failure, particularly within the intensive care setting. The coexistence of AF with elevated age, comorbidity burden, and cardiac dysfunction reflects a complex pathophysiological interaction rather than a direct effect of respiratory acidosis or gas exchange parameters. This complexity has been echoed in prior literature, where systemic inflammation, ventricular strain, and impaired myocardial oxygenation are increasingly recognized as key contributors to AF onset and progression in COPD and ICU cohorts. Furthermore, persistent gaps in the application of guideline-directed anticoagulation and rhythm control therapies suggest a real-world treatment inertia that may negatively impact outcomes. These findings underscore the need for a multidisciplinary approach to AF management in critically ill respiratory patients, incorporating early cardiology consultation, comprehensive risk stratification, and improved adherence to evidence-based therapies [18–20].

4.8. Limitations of the Study

This study presents several limitations. Primarily, its retrospective and single-center design may restrict the generalizability of the results to broader populations. Additionally, the diagnosis of atrial fibrillation was based solely on ECG recordings obtained at ICU admission, and paroxysmal AF cases may have been missed. Echocardiographic evaluations were not available for all patients and were conducted only in those with accessible records; therefore, cardiac functional data do not represent the entire study population. Furthermore, data on anticoagulant and antiarrhythmic therapy were extracted from hospital records, without access to detailed information regarding treatment adherence or reasons for discontinuation. Lastly, the absence of long-term follow-up data limits our ability to evaluate post-discharge outcomes and the long-term impact of atrial fibrillation on morbidity and mortality.

5. Conclusions

This study demonstrated that atrial fibrillation (AF) is highly prevalent among patients admitted to the intensive care unit with a diagnosis of type 2 respiratory failure. AF in this population was associated with older age, a greater burden of comorbidities, higher rates of heart failure, and abnormalities in both hematologic and cardiac biomarkers. Although AF appeared to be associated with increased mortality in univariate survival analyses, it did not emerge as an independent predictor in multivariable models. Instead, age and hemoglobin levels were identified as independent determinants of mortality. Our findings also revealed a noteworthy proportion of patients with AF who were not receiving appropriate anticoagulant therapy. Early recognition of AF, identification of cardiac dysfunction, and timely implementation of appropriate treatment strategies are crucial to improving prognosis in this patient group. In this context, multidisciplinary evaluation and future prospective studies with larger cohorts are warranted to enhance clinical management and outcomes.

Author Contributions: Conceptualization, O.M. and M.Y.; methodology, D.C.; software, F.C. and M.A.; validation, O.M., E.A., Y.T.G. and T.Ö.; formal analysis, M.Y., A.K. and O.M.; investigation, E.N.A.G. and M.Ö.C.; resources, O.M. and M.Y.; data curation, E.N.A.G.; writing—original draft preparation, O.M. and M.Y.; writing—review and editing, T.Ö., A.K. and M.Ö.C.; visualization, M.A. and E.A.; supervision, O.M. and E.A.; project administration, O.M. and D.C.; funding acquisition, M.Y. and Y.T.G. All authors have read and agreed to the published version of the manuscript.

Funding: This research received no external funding.

Institutional Review Board Statement: This research was approved by the Scientific Research Ethics Committee of the University of Health Sciences, Ankara Atatürk Training and Research Hospital (approval date: 12 June 2024, decision number: 2024-BÇEK/81).

Informed Consent Statement: Informed consent was obtained from all subjects involved in the study.

Data Availability Statement: The datasets used and/or analyzed during the current study are available from the corresponding author on reasonable request.

Conflicts of Interest: The authors declare no conflicts of interest.

References

1. Grippi, M.A. Respiratory failure: An overview. In *Fishman's Pulmonary Disease and Disorders*, 3rd ed.; Fishman, A.P., Ed.; Mc-Graw Hill: New York, NY, USA, 1998; pp. 2525–2535.
2. Vestbo, J.; Hurd, S.S.; Agustí, A.G.; Jones, P.W.; Vogelmeier, C.; Anzueto, A.; Barnes, P.J.; Fabbri, L.M.; Martinez, F.J.; Nishimura, M.; et al. Global strategy for the diagnosis, management, and prevention of chronic obstructive pulmonary disease: GOLD executive summary. *Am. J. Respir. Crit. Care Med.* **2013**, *187*, 347–365. [CrossRef] [PubMed]
3. Bemis, C.E.; Serur, J.R.; Borkenhagen, D.; Sonnenblick, E.H.; Urschel, C.W. Influence of right ventricular filling pressure on left ventricular pressure and dimension. *Circ. Res.* **1974**, *34*, 498–504. [CrossRef] [PubMed]
4. Terzano, C.; Conti, V.; Di Stefano, F.; Petroianni, A.; Ceccarelli, D.; Graziani, E.; Mariotta, S.; Ricci, A.; Vitarelli, A.; Puglisi, G.; et al. Comorbidity, hospitalization, and mortality in COPD: Results from a longitudinal study. *Lung* **2010**, *188*, 321–329. [CrossRef] [PubMed]
5. Konecny, T.; Park, J.Y.; Somers, K.R.; Konecny, D.; Orban, M.; Soucek, F.; Parker, K.O.; Scanlon, P.D.; Asirvatham, S.J.; Brady, P.A.; et al. Relation of chronic obstructive pulmonary disease to atrial and ventricular arrhythmias. *Am. J. Cardiol.* **2014**, *114*, 272–277. [CrossRef] [PubMed]
6. Ari, M.; Ari, E. Efficacy of Age-Adjusted Dyspnea, Eosinopenia, Consolidation, Acidemia and Atrial Fibrillation Score in Predicting Long-Term Survival in COPD-Related Persistent Hypercapnic Respiratory Failure. *Life* **2025**, *15*, 533. [CrossRef] [PubMed]
7. de Oliveira, C.W.M.; Silva, J.C.; Bezerra, A.L.Q.; del Corsso, C.; de Sousa, M.B.V.; Castro, B.; Fedatto, M.C.; Gutierrez, M.; Fausto, A.T. *Atrial Fibrillation: Diagnosis, Therapeutic Management and Prevention of Thromboembolic Complications in Cardiac Patients*; Seven Editora: São José dos Pinhais, Brazil, 2025. [CrossRef]
8. Romiti, G.F.; Corica, B.; Pipitone, E.; Vitolo, M.; Raparelli, V.; Basili, S.; Boriani, G.; Harari, S.; Lip, G.Y.H.; Proietti, M.; et al. Prevalence, management and impact of chronic obstructive pulmonary disease in atrial fibrillation: A systematic review and meta-analysis of 4,200,000 patients. *Eur. Heart J.* **2021**, *42*, 3541–3554. [CrossRef] [PubMed]
9. Chen, C.Y.; Liao, K.M. The impact of atrial fibrillation in patients with COPD during hospitalization. *Int. J. Chronic Obstr. Pulm. Dis.* **2018**, *13*, 2105–2112. [CrossRef] [PubMed]
10. Rodríguez-Mañero, M.; López-Pardo, E.; Cordero, A.; Ruano-Ravina, A.; Novo-Platas, J.; Pereira-Vázquez, M.; Martínez-Gómez, Á.; García-Seara, J.; Martínez-Sande, J.L.; Peña-Gil, C.; et al. A prospective study of the clinical outcomes and prognosis associated with comorbid COPD in the atrial fibrillation population. *Int. J. Chronic Obstr. Pulm. Dis.* **2019**, *14*, 371–380. [CrossRef] [PubMed] [PubMed Central]
11. Terzano, C.; Romani, S.; Conti, V.; Paone, G.; Oriolo, F.; Vitarelli, A. Atrial fibrillation in the acute, hypercapnic exacerbations of COPD. *Eur. Rev. Med. Pharmacol. Sci.* **2014**, *18*, 2908–2917. [PubMed]
12. Mehreen, T.; Ishtiaq, W.; Rasheed, G.; Kharadi, N.; Kiani, S.S.; Ilyas, A.; Kaleem, M.A.; Abbas, K. In-Hospital Mortality Among Critically Ill Patients with Atrial Fibrillation (AF) Versus Patients Without AF. *Cureus* **2021**, *13*, e18761. [CrossRef] [PubMed] [PubMed Central]
13. Xiao, X.; Han, H.; Wu, C.; He, Q.; Ruan, Y.; Zhai, Y.; Gao, Y.; Zhao, X.; He, J. Prevalence of Atrial Fibrillation in Hospital Encounters with End-Stage COPD on Home Oxygen: National Trends in the United States. *Chest* **2019**, *155*, 918–927. [CrossRef] [PubMed]

14. Schönhofer, B.; Wenzel, M.; Geibel, M.; Köhler, D. Blood transfusion and lung function in chronically anemic patients with severe chronic obstructive pulmonary disease. *Crit. Care Med.* **1998**, *26*, 1824–1828. [CrossRef] [PubMed]
15. Grymonprez, M.; Vakaet, V.; Kavousi, M.; Stricker, B.H.; Ikram, M.A.; Heeringa, J.; Franco, O.H.; Brusselle, G.G.; Lahousse, L. Chronic obstructive pulmonary disease and the development of atrial fibrillation. *Int. J. Cardiol.* **2019**, *276*, 118–124. [CrossRef] [PubMed]
16. Decker, C.; Garavalia, L.; Garavalia, B.; Simon, T.; Loeb, M.; Spertus, J.A.; Daniel, W.C. Exploring barriers to optimal anticoagulation for atrial fibrillation: Interviews with clinicians. *J. Multidiscip. Healthc.* **2012**, *5*, 129–135. [CrossRef] [PubMed]
17. Moudallel, S.; van den Bemt, B.J.F.; Zwikker, H.; de Veer, A.; Rydant, S.; Dijk, L.V.; Steurbaut, S. Association of conflicting information from healthcare providers and poor shared decision making with suboptimal adherence in direct oral anticoagulant treatment: A cross-sectional study in patients with atrial fibrillation. *Patient Educ. Couns.* **2021**, *104*, 155–162. [CrossRef] [PubMed]
18. Allen LaPointe, N.M.; Lokhnygina, Y.; Sanders, G.D.; Peterson, E.D.; Al-Khatib, S.M. Adherence to guideline recommendations for antiarrhythmic drugs in atrial fibrillation. *Am. Heart J.* **2013**, *166*, 871–878. [CrossRef] [PubMed]
19. O'Neal, W.T.; Sandesara, P.; Patel, N.; Venkatesh, S.; Samman-Tahhan, A.; Hammadah, M.; Kelli, H.M.; Soliman, E.Z. Echocardiographic predictors of atrial fibrillation in patients with heart failure with preserved ejection fraction. *Eur. Heart J. Cardiovasc. Imaging* **2017**, *18*, 725–729. [CrossRef] [PubMed]
20. Jover, E.; Marín, F.; Roldán, V.; Montoro-García, S.; Valdés, M.; Lip, G.Y. Atherosclerosis and thromboembolic risk in atrial fibrillation: Focus on peripheral vascular disease. *Ann. Med.* **2013**, *45*, 274–290. [CrossRef] [PubMed]

diagnostics MDPI

Article

Differences Between the 8th and 9th Editions of the TNM Staging System in Predicting Mortality in Non-Small Cell Lung Cancer Patients Staged with EBUS

Ezgi Demirdöğen [1,*], Orkun Eray Terzi [2], Özge Aydın Güçlü [1], Ahmet Ursavaş [1] and Mehmet Karadağ [1]

1 Department of Pulmonology, Faculty of Medicine, Bursa Uludağ University, Bursa 16059, Türkiye; ozgeguclu@uludag.edu.tr (Ö.A.G.); ahmetursavas@gmail.com (A.U.); karadag@uludag.edu.tr (M.K.)
2 Department of Pulmonology, University of Health Sciences Bursa Yüksek İhtisas Training and Research Hospital, Bursa 16310, Türkiye; terziorkun@gmail.com
* Correspondence: ezgicetin@uludag.edu.tr

Abstract: Background: The distinction between N2a and N2b in the lung cancer TNM 9th edition staging system has reduced the heterogeneity of prognosis using the previous staging system. Moreover, this distinction may enable new treatment approaches in non-small-cell lung cancer (NSCLC). We aimed to evaluate the differences in survival between 8th- and 9th-edition staging and the mortality prediction of the TNM 9th edition in NSCLC patients who did not undergo surgical staging and who were "N"-staged with solely endobronchial ultrasound–transbronchial needle aspiration (EBUS–TBNA) without endoscopic ultrasonography (EUS). **Methods:** Lung cancer patients who were newly diagnosed and staged with EBUS between May 2016 and January 2023 were retrospectively reviewed. Patients were divided into two groups, "All M0 = Model 1" and "T1–2 N1–2–3 M0 = Model 2", and compared according to their survival for both the 8th and 9th edition TNM staging systems. Cox regression analyses were performed for independent predictors of 2-year mortality. **Results:** In this retrospective study, a total of 90 patients were included. Most of the patients were male (84.4%), and the mean age of the study group was 64.0 ± 9.6; deceased patients were older (p = 0.024). There were no differences between groups in terms of smoking habit, comorbidities, tumor PET/CT localization, or 8th and 9th N-staging results with EBUS. The median follow-up period was 26 (0–100) months and longer for living patients than deceased patients in both groups (42 (23–100) vs. 18 (0–74), p = 0.03; 36 (24–100) vs. 20 (1–74), p < 0.001). According to the 8th edition of TNM staging, N2 stage (HR 2.26, 95% CI 1.01–5.05, p = 0.045) and N3 disease (HR 3.31, 95% CI 1.43–7.67, p = 0.005) are independent predictors of two-year mortality for Model 1 patients. When patients were staged according to the 9th edition TNM with EBUS, the relationship between N2a and mortality was not significant, while N2B disease increased the 2-year mortality risk by 2.78-fold (95% 1.07–7.22, p = 0.035), and N3 disease increased it by 3.31-fold (95% 1.43–7.67, p = 0.005). **Conclusions:** According to the TNM 9th edition staging system, we demonstrated that N2b disease significantly increases the risk of mortality in NSCLC cases using systematic mediastinal staging with EBUS–TBNA alone.

Keywords: non-small-cell lung cancer; EBUS; EBUS–TBNA; 9th edition TNM staging; mortality

1. Introduction

Lung cancer is the most common cancer worldwide, with almost 2.5 million new cases each year, and it is also the most common cause of cancer-related deaths, with 1.8 million

https://doi.org/10.3390/diagnostics15131570

deaths annually [1]. The International Association for the Study of Lung Cancer (IASLC) has established a standardized tumor node metastasis (TNM) staging classification that aims to bring together groups with similar prognoses and treatments based on the anatomic extent of malignancy. The "N" descriptors, categorized by anatomical location, had not been changed from the 4th edition until the recent 9th edition [2,3]. The TNM 8th edition lung cancer staging system has been widely used worldwide since 2017. The "N" descriptor of 8th-edition staging is based on the anatomic location of lymph node metastases, without taking into account the number of metastasized nodes. However, it was emphasized that factors other than the anatomical components of the lymph nodes were also effective in demonstrating prognostic heterogeneity, and it was stated that the lymph nodes should be reclassified into subgroups [4–7]. These changes in the 9th edition TNM "N" staging resulted from the significant survival difference between N2a and N2b [7]. According to the new staging system, which changed with the subdivision of N2 into N2a and N2b, T1N2a cases were found to have better survival rates than stage IIIA cases according to the 8th TNM edition, similar to stage IIB [8]. These findings led to the assignment of T1N2a tumors to stage IIB. In previous staging approaches, T1N1 tumors classified as stage IIB were downstaged to stage IIA because they had similar survival rates to stage IIA. In contrast, T2 tumors with multiple N2 in stage IIIA were upstaged to IIIB in the 9th edition for similar reasons. T3aN2a cases in the 8th TNM edition, which had been staged as IIIB, were downstaged to IIIA due to similar prognoses [8]. The new "N" descriptor in TNM staging, in addition to being a prognostic factor, plays a critical role in determining suitability for surgery and distinguishing between early-stage and locally advanced disease (N2/N3 status).

In recent guidelines with a high recommendation level, endosonography (EUS/EBUS or EBUS or EUS alone) is recommended as the initial procedure for mediastinal lymph node staging instead of surgical staging in patients with suspected or proven NSCLC with abnormal mediastinal and/or hilar nodes on CT and/or PET [9,10]. In fact, endoscopic staging is recommended even in certain patients with a negative PET [10]. Unlike EUS, EBUS provides unique access to the hilum and structures anterior to the large airways, but EBUS cannot access the lower mediastinal lymph nodes at stations 8 and 9 adjacent to the esophagus, which can be easily targeted with EUS or EUS-B [11]. The "SCORE" study showed that systematic EBUS combined with esophageal ultrasound using the same EBUS bronchoscope (EUS-B) increased the sensitivity of mediastinal lymph node staging in lung cancer patients by 9% compared to targeted EBUS [12]. The combined guidelines of ESGE, ERS, and ESTS recommend that the combination of EBUS/EUS or EBUS/EUS-B be preferred over either procedure alone [10]. In the same guidelines, EBUS alone is considered acceptable in cases in which combined endosonography is not available. In addition, compared to mediastinoscopy, EBUS is an isolated procedure that is associated with lower risks and costs [13]. There are no publications on the effect of "N" staging with EBUS–TBNA without EUS as a mortality predictor in the new TNM edition.

We planned to group the cases with non-metastatic NSCLC diagnoses as deceased and alive in two different models and to compare the demographic features, comorbidities, EBUS–TBNA-verified "N" stages, and PET/CT TNM stages according to both the 8th and 9th editions. We also aimed to evaluate the mortality prediction of the "N" descriptor in the 9th edition staging system in cases in which we performed complete systematic mediastinal staging with EBUS–TBNA without EUS or EUS-B.

2. Materials and Methods

2.1. Design, Patient Selection, and Data Collection

Patients who underwent EBUS for diagnosis and/or mediastinal staging of lung cancer between 1 May 2016 and 31 January 2023 were retrospectively reviewed. Patients with non-small-cell lung cancer (NSCLC) with no metastasis (M0) who underwent complete systematic mediastinal staging with EBUS were included in this study, as shown in the study flowchart (Figure 1). We collected data from medical records, including patient characteristics such as age, smoking history, presenting symptoms, and comorbidities. The duration of follow-up and timing of death were also noted. Tumor location and size, PET/CT findings, and the number of collected lymph nodes with EBUS and EBUS–TBNA pathological results were recorded. For pathological and clinical staging, the location of lymph nodes was categorized based on the IASLC lymph node map [14]. We classified the patients according to the 8th and 9th edition staging systems based on EBUS–TBNA for the "N" stage and PET/CT for the "T" and "M" descriptors. EBUS–TBNA was performed in each patient on all lymph nodes from N3 to N1 that could be reached and sampled, i.e., systematic mediastinal staging. All procedures were performed by one of two experienced bronchoscopists, and samples were processed by the same cytopathologist with no ROSE. EBUS–TBNA samples were air-sprayed into a mixture of 10% formalin and 96% alcohol and then sent to the cytology laboratory for cell block and cytoblock preparation. Air-dried smear preparations were prepared by staining with the May–Grunwald–Giemsa stain in the pathology laboratory.

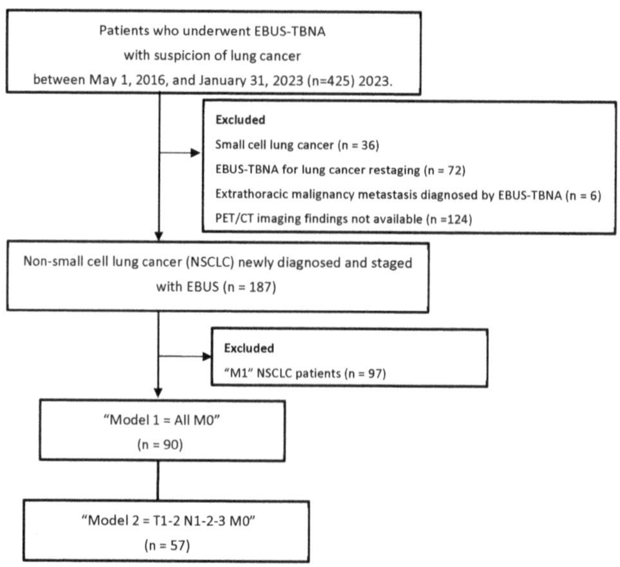

Figure 1. Flowchart of patient selection.

NSCLC cases with metastasis were excluded from the study population. Patients were examined using two models, "Model 1 = All M0" and "Model 2 = T1–2, N1–2–3, M0", and compared according to their survival in both the 8th and 9th edition TNM staging systems.

2.2. Statistical Analysis

The data were analyzed using IBM SPSS Statistics for Windows, Version 28.0., and $p < 0.05$ was considered statistically significant. The distribution of continuous data was validated using the Shapiro–Wilk test. The continuous data were presented as the mean ± standard deviation (SD), or median (minimum–maximum), while the categorical variables were expressed as n (%). For between-group comparisons, parametric independent sample t-tests or non-parametric Mann–Whitney U tests were used depending on the findings of the normality test. Pearson's chi-square test was performed to compare categorical variables. The multivariate Cox regression model was constructed with the variables that met the $p < 0.25$ threshold in the univariate analysis. The variables were selected using the backward stepwise LR approach, and the findings of the analysis were presented. A summary was provided for the hazard ratios (HRs) and 95% confidence intervals (95% CIs). Kaplan–Meier survival analysis, conducted using MedCalc® Statistical Software version 22.023 (MedCalc Software Ltd., Ostend, Belgium; https://www.medcalc.org; 2024), was employed to assess the impact of the 8th and 9th edition TNM staging systems on overall survival.

An a priori power analysis was conducted using G*Power version 3.1 to determine the required sample size for detecting a significant difference in age between two independent groups. The analysis was based on a one-tailed t-test, with an anticipated effect size (Cohen's d) of 0.577 (calculated from group means of 65.2 and 60.1 and standard deviations of 10.0 and 7.5, respectively). The alpha error probability was set at 0.05, and the desired statistical power was 0.80. The analysis indicated that a total sample size of 76 participants (38 per group) would be sufficient to detect the expected effect. The resulting noncentrality parameter was 2.52, and the critical t-value was 1.665. The actual power achieved with this configuration was 0.80, confirming the adequacy of the sample size to detect a medium effect.

3. Results

Out of the 90 included patients, 76 were male (84.4%), and the mean age of the study group was 64.0 ± 9.6. In the whole group, referred to as "Model 1", it was seen that the deceased patients were older than the living patients (65.2 ± 10.0 vs. 60.1 ± 7.6, $p = 0.024$, respectively), while in "Model 2", the age difference between the two groups was not statistically significant (64.8 ± 10.3 vs. 59.3 ± 7.9, $p = 0.06$, respectively). In addition, in "Model 1", it was seen that the deceased patients had more smoking pack years than the surviving patients [40 (8–144) vs. 33 (10–70), $p = 0.041$, respectively]. No differences were found between deceased and living patients in both models in terms of gender, tumor PET/CT localization, cancer stages by PET/CT, 8th and 9th EBUS "N" stages, and the presence of comorbidities such as hypertension, COPD, diabetes mellitus, coronary artery disease, cerebrovascular disease, asthma, and interstitial lung disease (Table 1). However, in both models, it was observed that subcarinal lymph nodes were sampled significantly more in cases with deceased patients than in cases with surviving patients, as shown in Table 1. The majority of the 90 patients included in this study were at the advanced stage, stage 3 (34 of them were stage 3A, 35 were stage 3B, and 10 were stage 3C). Histopathological subtypes detected in the study population consisting of NSCLC cases were adenocarcinoma ($n = 52$, 57.8%), squamous-cell carcinoma ($n = 35$, 38.9%), and undifferentiated non-small-cell carcinoma ($n = 3$, 3.3%). The distribution of pathological subtypes according to survival was similar in both models (Table 1). No difference was found when comparing patients who died and survived according to the "N" staging confirmed with EBUS using the 8th and 9th editions of the TNM (Table 1). The median follow-up period was 26 (0–100) months in all cases and longer for patients who were alive compared to patients who were deceased in both models, as expected [42 (23–100) vs. 18 (0–74), $p = 0.03$; 36 (24–100) vs. 20 (1–74), $p < 0.001$].

Table 1. Comparison of living and deceased patients' characteristics, cancer staging results, and factors contributing to overall mortality in two different non-metastatic NSCLC patient groups.

	All M0 NSCLC Patients (n = 90)	Model 1 = All M0 NSCLC Patients (n = 90)			Model 2 = T1–2 N1–2–3 M0 NSCLC Patients (n = 57)		
		Deceased (n = 65)	Alive (n = 25)	p-Value	Deceased (n = 40)	Alive (n = 17)	p-Value
Age, years	64.0 ± 9.6	65.2 ± 10.0	60.1 ± 7.6	0.024	64.8 ± 10.3	59.3 ± 7.9	0.06
Gender, male, n (%)	76 (84.4)	56 (86.2)	20 (80)	0.522	32 (80)	13 (76.5)	0.737
Smoking habits							
Current smoker	49 (54.4)	36 (55.4)	13 (52.0)	0.949	20 (50)	9 (52.9)	0.912
Ex-smoker	28 (31.1)	20 (30.8)	8 (32.0)		14 (35)	5 (29.4)	
Never smoked	13 (14.4)	9 (13.8)	4 (16.0)		6 (15)	3 (17.6)	
Smoking history, pack-year med (min–max)	40 (8–144)	40 (8–144)	33 (10–70)	0.041	40 (8–144)	40 (15–70)	0.689
Presence of comorbidities, n (%)	81 (90)	61 (93.8)	20 (80)	0.109	27 (67.5)	13 (76.5)	0.752
Comorbidity, n (%)							
Hypertension	30 (33.3)	21 (32.3)	9 (36.0)	0.805	13 (32.5)	7 (41.2)	0.557
COPD	22 (24.4)	18 (27.7)	4 (16.0)	0.288	10 (25)	3 (17.6)	0.734
Diabetes mellitus	17 (18.9)	13 (20.0)	4 (16.0)	0.771	7 (17.5)	4 (23.5)	0.598
Coronary artery disease	15 (16.7)	12 (18.5)	3 (12)	0.545	5 (12.5)	2 (11.8)	0.938
Extrathoracic malignancy	6 (6.7)	5 (7.7)	1 (4.0)	0.529	4 (10)	1 (5.9)	0.615
Cerebrovascular disease	3 (3.3)	3 (4.6)	0	0.557	2 (5)	0	0.348
Asthma	4 (4.4)	4 (6.2)	0	0.573	3 (7.5)	0	0.542
Interstitial lung disease	2 (2.2)	1 (1.5)	1 (4)	0.481	1 (2.5)	1 (5.9)	0.511
Congestive heart disease	1 (1.1)	1 (1.5)	0	0.533	1 (2.5)	0	0.511
Arrhythmia	1 (1.1)	1 (1.5)	0	0.533	1 (2.5)	0	0.511
Chronic renal failure	1 (1.1)	1 (1.5)	0	0.533	1 (2.5)	0	0.511
Autoimmune disease	1 (1.1)	1 (1.5)	0	0.533	1 (2.5)	0	0.511
Tuberculosis history	5 (5.6)	4 (6.2)	1 (4.0)	0.689	3 (7.5)	0	0.547
Lung cancer pathological classification n (%)							

Table 1. *Cont.*

	All M0 NSCLC Patients (n = 90)	Model 1 = All M0 NSCLC Patients (n = 90)			Model 2 = T1–2 N1–2–3 M0 NSCLC Patients (n = 57)		
		Deceased (n = 65)	Alive (n = 25)	p-Value	Deceased (n = 40)	Alive (n = 17)	p-Value
Adenocarcinoma	52 (57.8)	36 (55.4)	16 (64.0)	0.170	27 (67.5)	10 (58.8)	0.353
Squamous-cell carcinoma	35 (38.9)	28 (43.1)	7 (28.0)		12 (30)	5 (29.4)	
Undifferentiated non-small-cell carcinoma	3 (3.3)	1 (1.5)	2 (8.0)		1 (2.5)	2 (11.8)	
PET/CT tumor localization							
Right upper lobe	32 (35.6)	26 (40.0)	6 (24.0)	0.295	18 (45)	4 (23.5)	0.112
Right middle lobe	14 (15.6)	7 (10.8)	7 (28.0)		2 (5)	4 (23.5)	
Right lower lobe	10 (11.1)	7 (10.8)	3 (12.0)		5 (12.5)	3 (17.6)	
Left upper lobe	29 (32.2)	21 (32.3)	8 (32.0)		11 (27.5)	6 (35.3)	
Left lower lobe	5 (5.6)	4 (6.2)	1 (4.0)		4 (10)	0	
Mass long axis (mm) med (min–max)	30 (12–90)	30 (12–90)	30 (15–70)	0.220	25 (13–48)	23 (15–44)	0.426
Mass short axis (mm) med (min–max)	25 (10–73)	25 (10–73)	25 (12–60)	0.151	23 (10–42)	20 (12–40)	0.238
Mass SUV med (min–max)	10.0 (1.81–38.88)	10 (1.81–38.88)	10.9 (3.60–36.30)	0.405	9.80 (3.40–38.88)	7.68 (3.60–15.0)	0.005
PET staging, 8th TNM edition							
1A	1 (1.1)	1 (1.5)	0	0.172	1 (2.5)	0	0.684
1B	4 (4.4)	3 (4.6)	1 (4.0)		2 (5)	1 (5.9)	
2B	6 (6.7)	5 (7.7)	1 (4.0)		4 (10)	1 (5.9)	
3A	34 (37.8)	20 (30.8)	14 (56.0)		18 (45)	11 (64.7)	
3B	35 (38.9)	26 (40.0)	9 (36.0)		15 (37.5)	4 (23.5)	
3C	10 (11.1)	10 (15.4)	0		-	-	
PET staging, 9th TNM edition							
1A	1 (1.1)	1 (1.5)	0	0.110	1 (2.5)	0	0.344
1B	4 (4.4)	3 (4.6)	1 (4.0)		2 (5)	1 (5.9)	
2A	4 (4.4)	4 (6.2)	0		4 (10)	0	
2B	12 (13.3)	6 (9.2)	6 (24.0)		5 (12.5)	6 (35.3)	
3A	23 (25.6)	14 (21.5)	9 (36.0)		10 (25)	4 (23.5)	
3B	36 (40.0)	27 (41.5)	9 (36.0)		18 (45)	6 (35.3)	
3C	10 (11.1)	10 (15.4)	0		-	-	

Table 1. *Cont.*

	All M0 NSCLC Patients (n = 90)	Model 1 = All M0 NSCLC Patients (n = 90)			Model 2 = T1–2 N1–2–3 M0 NSCLC Patients (n = 57)		
		Deceased (n = 65)	Alive (n = 25)	p-Value	Deceased (n = 40)	Alive (n = 17)	p-Value
Number of malignant lymph nodes sampled with EBUS							
2R (n = 3)	2 (66.7)	1 (100)	1 (50)	0.386	1 (100)	1 (50)	0.386
4R (n = 66)	32 (48.5)	24 (50.0)	8 (44.4)	0.911	13 (44.8)	6 (50)	0.926
4L (n = 35)	17 (48.6)	15 (55.6)	2 (25.0)	0.316	10 (55.6)	2 (33.3)	0.315
7 (n = 82)	29 (35.4)	23 (39.7)	6 (25.0)	0.006	16 (44.4)	4 (25)	0.021
10R (n = 34)	10 (29.4)	9 (39.1)	1 (9.1)	0.146	7 (43.8)	1 (14.3)	0.373
11R (n = 49)	15 (30.6)	9 (27.3)	6 (37.5)	0.752	3 (13)	4 (36.4)	0.208
11L (n = 62)	14 (22.6)	12 (26.7)	2 (11.8)	0.228	7 (24.1)	2 (15.4)	0.570
Eighth TNM edition "N" stage, with EBUS							
N0	32 (35.6)	20 (30.8)	12 (48.0)	0.398	13 (32.5)	8 (47.1)	0.686
N1	4 (4.4)	3 (4.6)	1 (4.0)		2 (5)	1 (5.9)	
N2	35 (38.9)	26 (40.0)	9 (36.0)		16 (40)	6 (35.3)	
N3	19 (21.1)	16 (24.6)	3 (12.0)		9 (22.5)	2 (11.8)	
Ninth TNM edition "N" stage, with EBUS							
N0	32 (35.6)	20 (30.8)	12 (48.0)	0.549	13 (32.5)	8 (47.1)	0.781
N1	4 (4.4)	3 (4.6)	1 (4.0)		2 (5)	1 (5.9)	
N2a	21 (21.0)	16 (24.6)	5 (20.0)		10 (25.0)	3 (17.6)	
N2b	14 (15.6)	10 (15.4)	4 (16.0)		6 (15.0)	3 (17.6)	
N3	19 (21.1)	16 (24.6)	3 (12.0)		9 (22.5)	2 (11.8)	
Follow-up period, months	26 (0–100)	18 (0–74)	42 (23–100)	0.003	20 (1–74)	36 (24–100)	<0.001

Data are presented as means ± SD, median (min–max), and n (%).

In the multivariate Cox proportional hazards regression model including age, smoking habit, and N staging with EBUS, according to the eighth edition of TNM staging, N2 stage (HR 2.26, 95% CI 1.01–5.05, $p = 0.045$) and N3 disease (HR 3.31, 95% CI 1.43–7.67, $p = 0.005$) were independent predictors of 2-year mortality for all M0 patients (Model 1) (Table 2). When patients were staged according to the 9th edition TNM system with EBUS, in the same Cox analysis in non-metastatic patients (Model 1), no significant association was found between N2a and mortality, while N2b disease increased the 2-year mortality risk by 2.78 times (95% 1.07–7.22, $p = 0.035$) and N3 disease by 3.31 times (95% 1.43–7.67, $p = 0.005$) (Table 2). The Kaplan–Meier survival curves for 2-year mortality according to N stage in the previous (8th) and most recent TNM editions (9th) are shown in Figure 2.

Table 2. Independent variables affecting 2-year overall mortality in two models according to the 8th and 9th TNM editions.

	Model 1 NSCLC, All M0 Patients			Model 2 NSCLC, T1–2 N1–2–3 M0 Patients		
	HR	**95% CI**	***p*-Value**	**HR**	**95% CI**	***p*-Value**
N staging with EBUS according to the 8th edition of the TNM						
(Ref. N0)						
N1	1.94	0.42–9.01	0.395	4.65	0.85–25.50	0.076
N2	2.26	1.01–5.05	0.045	2.59	0.79–8.41	0.113
N3	3.31	1.43–7.67	0.005	6.37	1.94–20.83	0.002
N staging with EBUS according to the 9th edition of the TNM						
(Ref. N0)						
N1	1.94	0.42–9.01	0.394	4.65	0.85–25.50	0.076
N2a	1.97	0.80–4.86	0.139	2.51	0.67–9.38	0.169
N2b	2.78	1.07–7.22	0.035	2.67	0.67–10.74	0.163
N3	3.31	1.43–7.67	0.005	6.37	1.94–20.83	0.002

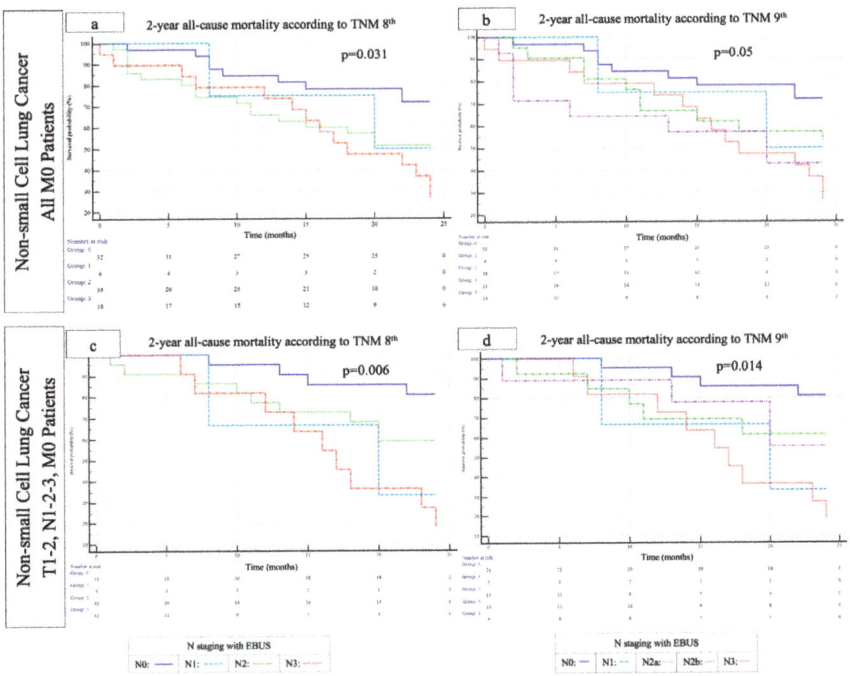

Figure 2. Kaplan–Meier survival analyses for 2-year all-cause mortality in patients with NSCLC according to N staging with EBUS–TBNA based on the 8th and 9th TNM editions. (**a**,**b**) For all M0 patients, mean survival times according to the 8th edition of the TNM classification were 20.6 ± 1.1 months for N0, 19.0 ± 3.2 months for N1, 16.6 ± 1.5 months for N2, and 16.7 ± 1.8 months for N3 (p = 0.031), while the corresponding values in the 9th edition TNM classification were 20.5 ± 1.1 months for N0, 19.0 ± 3.2 months for N1, 17.7 ± 1.8 months for N2a, 15.0 ± 12.5 months for N2b, and 16.7 ± 1.8 months for N3 (p = 0.05). (**c**,**d**) For patients with T1–2, N1–2–3, M0 disease, mean survival times according to the 8th edition of the TNM classification were 22.1 ± 0.9 months for N0, 17.3 ± 3.9 months for N1, 18.8 ± 1.6 months for N2, and 16.8 ± 2.0 months for N3 (p = 0.006), while the corresponding values in the 9th edition TNM classification were 22.1 ± 0.9 months for N0, 17.3 ± 3.9 months for N1, 18.4 ± 2.1 months for N2a, 19.3 ± 2.4 months for N2b, and 16.8 ± 2.0 months for N3 (p = 0.014).

4. Discussion

In our study, according to the 8th edition TNM staging system, cases of non-metastatic N2 and N3 disease confirmed with EBUS–TBNA were found to be independent predictors of mortality. In "N" staging confirmed with EBUS according to the 9th TNM edition, the N2a stage did not predict mortality, while N2b disease was shown to be an independent predictor of 2-year mortality, with a 2.78-fold increased risk in all M0 patients. However, the same prediction regarding N2b disease was not found in T1–2 N1–3 patients with no metastasis. EBUS-verified N3 disease, when evaluated according to both the 8th and 9th editions, increases the risk of mortality by 3.31 times and 6.37 times in Model 1 and Model 2, respectively.

In a systematic review, the overall mortality rate for lung cancer was found to be 10% (95% CI of 6–16%) [15]. Advanced age, male gender, stage, and the presence of comorbidities such as hypertension, cardiovascular disease, and diabetes mellitus have been shown to have a positive effect on mortality in lung cancer [15,16]. In our study group, no difference was observed in terms of comorbidities in the deceased and surviving groups in both models, and gender distribution and cancer stages according to PET/CT (for both the 8th and 9th editions) were similar. This provided a more homogeneous population for analysis in determining the risk factors affecting mortality. Long-term survival in NSCLC varies significantly according to the endoscopic "N" stage of the disease. In a study examining the prognostic impact of endoscopic N staging in over 1000 NSCLC cases, although patients with endoscopic N0–1/pathological N2–3 disease had worse survival rates than patients with pathological N0 disease, no significant difference was observed compared with patients with pathological N1 disease [17]. Additionally, EBUS has been shown to significantly increase the surgery rate in NSCLC patients with stage II–N1 disease, presumably by reducing the number of patients previously excluded as surgical candidates using PET scans [18]. In the first multicenter randomized controlled trial investigating the added value of mediastinoscopy after negative EBUS results (MEDIASTrial), the unforeseen N2 disease rate was 8.8% in those who underwent immediate resection, which was not lower than the rate of 7.7% in the mediastinoscopy group. As revealed in previous meta-analyses, the results of this study showed that EBUS is not inferior to mediastinoscopy even when determining early-stage cases that are candidates for surgery, and that it is preferable because of its time-saving nature and lower complication rates [19–21].

In patients with early-stage NSCLC, systematic mediastinal staging with EBUS–TBNA is known to increase the accuracy of staging. In a meta-analysis, the mean negative predictive value of EBUS–TBNA in detecting radiologically occult, unsuspected N2/N3 metastases was 91% (82–100%) [22]. However, although only targeted lymph node sampling is usually performed in patients with locally advanced NSCLC, even in these cases, systemic mediastinal staging has been shown to be superior to PET in determining the radiotherapy field. In the SEISMIC trial, PET-occult lymph node metastases were detected in 12% of patients following systematic endoscopic staging [23]. However, in patients with NSCLC who had an abnormal mediastinum detected with CT or PET/CT, even if systematic mediastinal staging is not performed, the complete evaluation of all lymph node stations and sampling from at least three different mediastinal lymph node stations, such as both lower paratracheal and subcarinal lymph nodes, are among the ESTS recommendations [10]. In our study, all accessible lymph nodes, not just those that were marked as suspicious using CT and/or PET imaging, were sampled with EBUS, as recommended in the lung cancer staging guidelines [8,10,24]. Complete systematic mediastinal lymph node staging, rather than targeted staging, was especially important for detecting multiple N2 disease, which is one of the significant changes in the new 9th edition of TNM staging. Changes to the TNM 9th edition classification of lung cancer were considered necessary because of the profound difference in survival between N2a and N2b. Analysis has shown that N2a and N2b clearly define prognostically different tumor groups, with data indicating a clear and consistent distinction between single and multiple N2 nodal metastases [25]. One study compared the ability of the 8th and 9th editions to discriminate overall survival and recurrence-free time in over 4000 patients with stage I–III NSCLC who underwent complete curative surgery and concurrent hilar/mediastinal lymphadenectomy [26]. In the aforementioned validation study, the survival difference between N1 and N2a was not as significant as the survival difference between N2a and N2b, suggesting that the number of lymph node metastases has a significant effect on prognosis, as shown in previous studies [27,28].

The guidelines prioritize the combination of EBUS/EUS for the mediastinal staging of NSCLC patients and emphasize the importance of combined endosonography to confirm multiple N2 disease in the new staging system. However, EUS–FNA could not be performed on the patients because it was not available in our center during the study period. In addition, EUS-B also was not performed due to the training requirement for EUS-B and the recent resolution of professional disagreements with gastroenterologists. Only EBUS–TBNA was performed for mediastinal staging in the included cases during the study period. We believe that it is important to demonstrate that EBUS–TBNA alone can maintain mortality prediction, even according to the new staging, when conditions are not suitable for combined endosonography. A meta-analysis has shown that adding EBUS to EUS increases sensitivity by 22%, but the increase in sensitivity is less when EUS is added to EBUS (12%) [29]. Therefore, if both endoscopic methods are not available, as was the case in our center in the recent past, EBUS–TBNA is recommended [10]. Furthermore, EBUS, which provides access to the hilar lymph nodes, should preferably be used if a single endoscopic method is available due to the existence of studies indicating the importance of the presence of N1 accompanying a single N2, which may also cause revisions to staging in the future [30]. In the aforementioned study, the prognosis for single N2 with N1 involvement was found to be worse than the prognosis for single N2 without N1. We found that N2b disease was an independent predictor of 2-year mortality, with a 2.78-fold increased risk in all M0 patients according to systematic 9th-edition TNM "N" staging performed with EBUS alone, without surgical staging or even EUS. In Model 1, which included all cases without metastasis and consisted of 90 patients, there were 33 patients with a more advanced T stage compared to Model 2. However, no 2-year mortality predictions for N2b were observed in Model 2, where T3–4 cases were not included. The reasons for this result may be that the number of cases was insufficient for analysis and that 10 patients who died were Stage 3C patients who were not included in Model 2. In addition, different tumor biology may have been a potential confounder for this result.

We would like to state that during our study, only molecular targeted therapy and immunotherapy are reimbursed for metastatic patients in our country. For this reason, no patient received targeted therapy or immunotherapy in our study, in which we included only non-metastatic cases. It was observed that there was no difference between the deceased and surviving groups in terms of "N" stages detected with EBUS according to 8th- and 9th-edition staging. For this reason, there may not be a significant difference in terms of treatments, especially since we compared relatively homogeneous groups in both models. Excluding metastatic cases and performing mortality analyses in this model makes our results noteworthy. However, difficulties in accessing medical records and not including treatment modalities due to missing data are among the most important limitations of our study. In addition, unlike the previous edition, the 9th edition study also included survival analyses of patients who underwent resection after induction therapy [8]. This situation led to an inconsistency in the survival analyses of patients who did not receive induction therapy. Therefore, the need for a revision with specific survival data for patients receiving induction therapy has been stated in another article by the corresponding author of the project study [31]. In addition to its prognostic significance, "N" staging is also important in deciding suitability for surgery and determining the extent of the disease. However, a change in classification does not automatically mean a therapeutic change, and it is important that treatment decisions are made based on evidence through well-designed clinical studies, as underlined in [31]. Other limitations of our study include its single-center and retrospective design.

5. Conclusions

This study showed that the importance of systematic mediastinal staging has increased following the new 9th edition TNM NSCLC staging system, and targeted staging should not be preferred for the detection of multiple N2 disease. Staging with EBUS–TBNA, without EUS, still maintains its place in NSCLC. However, it should not be forgotten that although staging with EBUS–TBNA alone maintains its importance, combined endosonography is the priority recommendation for staging, especially for the confirmation of multiple N2. Moreover, and most importantly, this is the first study to show that multiple N2 disease detected and confirmed with EBUS–TBNA alone, without the addition of EUS or EUS-B, is a significant predictor of mortality.

Author Contributions: Conceptualization, E.D., O.E.T. and Ö.A.G.; data curation, E.D., O.E.T. and A.U.; formal analysis, Ö.A.G.; investigation, E.D. and O.E.T.; methodology, E.D.; writing—original draft preparation, E.D.; writing—review and editing, Ö.A.G., A.U. and M.K.; supervision, M.K. All authors have read and agreed to the published version of the manuscript.

Funding: This research received no external funding.

Institutional Review Board Statement: This study was conducted in accordance with the Declaration of Helsinki and approved by the Bursa Uludag University Ethics Committee, approval date: 9 April 2025, approval number: 2025/7-36.

Informed Consent Statement: Patient consent was not obtained due to the retrospective design of this study and the use of anonymized data, as approved by the Ethics Committee.

Data Availability Statement: The original contributions presented in this study are included in the article. Further inquiries can be directed to the corresponding author.

Conflicts of Interest: The authors declare no conflicts of interest.

Abbreviations

The following alphabetically listed abbreviations are used in this article:

EBUS	Endobronchial ultrasound
EBUS–TBNA	Endobronchial ultrasound–transbronchial needle aspiration
ERS	European Respiratory Society
ESGE	European Society of Gastrointestinal Endoscopy
ESTS	European Society of Thoracic Surgery
EUS	Endoscopic (esophageal) ultrasound using a gastrointestinal scope
EUS-B	Endoscopic (esophageal) ultrasound using an EBUS scope
EUS-FNA	Endoscopic (esophageal) ultrasound with fine needle aspiration, with the use of a gastrointestinal scope
HR	Hazard ratio
NSCLC	Non-small-cell lung cancer
PET/CT	Positron emission tomography/computed tomography
ROSE	Rapid on-site evaluation
TNM	Tumor-node-metastasis

References

1. Bray, F.; Laversanne, M.; Sung, H.; Ferlay, J.; Siegel, R.L.; Soerjomataram, I.; Jemal, A. Global cancer statistics 2022: GLOBOCAN estimates of incidence and mortality worldwide for 36 cancers in 185 countries. *CA Cancer J. Clin.* **2024**, *74*, 229–263. [CrossRef] [PubMed]
2. Mountain, C.F. A new international staging system for lung cancer. *Chest* **1986**, *89* (Suppl. 4), 225S–233S. [CrossRef] [PubMed]
3. Asamura, H.; Nishimura, K.K.; Giroux, D.J.; Chansky, K.; Hoering, A.; Rusch, V.; Rami-Porta, R.; Members of the IASLC Staging and Prognostic Factors Committee and of the Advisory Boards, and Participating Institutions. IASLC Lung Cancer Staging Project: The New Database to Inform Revisions in the Ninth Edition of the TNM Classification of Lung Cancer. *J. Thorac. Oncol.* **2023**, *18*, 564–575. [CrossRef] [PubMed]
4. Asamura, H.; Chansky, K.; Crowley, J.; Goldstraw, P.; Rusch, V.W.; Vansteenkiste, J.F.; Watanabe, H.; Wu, Y.L.; Zielinski, M.; Ball, D.; et al. The International Association for the Study of Lung Cancer Lung Cancer Staging Project: Proposals for the Revision of the N Descriptors in the Forthcoming 8th Edition of the TNM Classification for Lung Cancer. *J. Thorac. Oncol.* **2015**, *10*, 1675–1684. [CrossRef]
5. Yun, J.K.; Lee, G.D.; Choi, S.; Kim, H.R.; Kim, Y.H.; Kim, D.K.; Park, S.I. Comparison between lymph node station- and zone-based classification for the future revision of node descriptors proposed by the International Association for the Study of Lung Cancer in surgically resected patients with non-small-cell lung cancer. *Eur. J. Cardiothorac. Surg.* **2019**, *56*, 849–857. [CrossRef]
6. Park, B.J.; Kim, T.H.; Shin, S.; Kim, H.K.; Choi, Y.S.; Kim, J.; Zo, J.I.; Shim, Y.M.; Cho, J.H. Recommended Change in the N Descriptor Proposed by the International Association for the Study of Lung Cancer: A Validation Study. *J. Thorac. Oncol.* **2019**, *14*, 1962–1969. [CrossRef]
7. Ahn, Y.; Lee, S.M.; Choe, J.; Choi, S.; Do, K.H.; Seo, J.B. Prognostic performance of the N category in the 9th edition of lung cancer staging. *Eur Radiol.* **2024**, *35*, 3788–3799. [CrossRef]
8. Rami-Porta, R.; Nishimura, K.K.; Giroux, D.J.; Detterbeck, F.; Cardillo, G.; Edwards, J.G.; Fong, K.M.; Giuliani, M.; Huang, J.; Kernstine KHSr Marom, E.M.; et al. The International Association for the Study of Lung Cancer Lung Cancer Staging Project: Proposals for Revision of the TNM Stage Groups in the Forthcoming (Ninth) Edition of the TNM Classification for Lung Cancer. *J. Thorac. Oncol.* **2024**, *19*, 1007–1027. [CrossRef]
9. Silvestri, G.A.; Gonzalez, A.V.; Jantz, M.A.; Margolis, M.L.; Gould, M.K.; Tanoue, L.T.; Harris, L.J.; Detterbeck, F.C. Methods for staging non-small cell lung cancer: Diagnosis and management of lung cancer, 3rd ed: American College of Chest Physicians evidence-based clinical practice guidelines. *Chest* **2013**, *143* (Suppl. 5), e211S–e250S. [CrossRef]
10. Vilmann, P.; Clementsen, P.F.; Colella, S.; Siemsen, M.; De Leyn, P.; Dumonceau, J.M.; Herth, F.J.; Larghi, A.; Vazquez-Sequeiros, E.; Hassan, C.; et al. Combined endobronchial and esophageal endosonography for the diagnosis and staging of lung cancer: European Society of Gastrointestinal Endoscopy (ESGE) Guideline, in cooperation with the European Respiratory Society (ERS) and the European Society of Thoracic Surgeons (ESTS). *Endoscopy* **2015**, *47*, 545–559. [CrossRef]
11. Vilmann, P.; Clementsen, P.F. Combined EUS and EBUS are complementary methods in lung cancer staging: Do not forget the esophagus. *Endosc. Int. Open* **2015**, *3*, E300–E301. [CrossRef] [PubMed]
12. Crombag, L.M.M.; Dooms, C.; Stigt, J.A.; Tournoy, K.G.; Schuurbiers, O.C.J.; Ninaber, M.K.; Buikhuisen, W.A.; Hashemi, S.M.S.; Bonta, P.I.; Korevaar, D.A.; et al. Systematic and combined endosonographic staging of lung cancer (SCORE study). *Eur. Respir. J.* **2019**, *53*, 1800800. [CrossRef] [PubMed]
13. Verdial, F.C.; Berfield, K.S.; Wood, D.E.; Mulligan, M.S.; Roth, J.A.; Francis, D.O.; Farjah, F. Safety and Costs of Endobronchial Ultrasound-Guided Nodal Aspiration and Mediastinoscopy. *Chest* **2020**, *157*, 686–693. [CrossRef] [PubMed]
14. Rusch, V.W.; Asamura, H.; Watanabe, H.; Giroux, D.J.; Rami-Porta, R.; Goldstraw, P.; Members of IASLC Staging Committee. The IASLC lung cancer staging project: A proposal for a new international lymph node map in the forthcoming seventh edition of the TNM classification for lung cancer. *J. Thorac. Oncol.* **2009**, *4*, 568–577. [CrossRef]
15. Tesfaw, L.M.; Dessie, Z.G.; Mekonnen Fenta, H. Lung cancer mortality and associated predictors: Systematic review using 32 scientific research findings. *Front. Oncol.* **2023**, *13*, 1308897. [CrossRef]
16. Malvezzi, M.; Santucci, C.; Boffetta, P.; Collatuzzo, G.; Levi, F.; La Vecchia, C.; Negri, E. European cancer mortality predictions for the year 2023 with focus on lung cancer. *Ann. Oncol.* **2023**, *34*, 410–419. [CrossRef]
17. Hwangbo, B.; Park, E.Y.; Yang, B.; Lee, G.K.; Kim, T.S.; Kim, H.Y.; Kim, M.S.; Lee, J.M. Long-term Survival According to N Stage Diagnosed by Endobronchial Ultrasound-Guided Transbronchial Needle Aspiration in Non-small Cell Lung Cancer. *Chest* **2022**, *161*, 1382–1392. [CrossRef]
18. Slavova-Azmanova, N.S.; Lizama, C.; Johnson, C.E.; Ludewick, H.P.; Lsester, L.; Karunarathne, S.; Phillips, M. Impact of the introduction of EBUS on time to management decision, complications, and invasive modalities used to diagnose and stage lung cancer: A pragmatic pre-post study. *BMC Cancer* **2016**, *16*, 44. [CrossRef]
19. Bousema, J.E.; Dijkgraaf, M.G.W.; van der Heijden, E.H.F.M.; Verhagen, A.F.T.M.; Annema, J.T.; van den Broek, F.J.C.; MEDIASTrial study group. Endosonography With or Without Confirmatory Mediastinoscopy for Resectable Lung Cancer: A Randomized Clinical Trial. *J. Clin. Oncol.* **2023**, *41*, 3805–3815. [CrossRef]

20. Figueiredo, V.R.; Cardoso, P.F.G.; Jacomelli, M.; Santos, L.M.; Minata, M.; Terra, R.M. EBUS-TBNA versus surgical mediastinoscopy for mediastinal lymph node staging in potentially operable non-small cell lung cancer: A systematic review and meta-analysis. *J. Bras. Pneumol.* **2020**, *46*, e20190221. (In English and Portuguese) [CrossRef]
21. Sehgal, I.S.; Dhooria, S.; Aggarwal, A.N.; Behera, D.; Agarwal, R. Endosonography Versus Mediastinoscopy in Mediastinal Staging of Lung Cancer: Systematic Review and Meta-Analysis. *Ann. Thorac. Surg.* **2016**, *102*, 1747–1755. [CrossRef] [PubMed]
22. Leong, T.L.; Loveland, P.M.; Gorelik, A.; Irving, L.; Steinfort, D.P. Preoperative Staging by EBUS in cN0/N1 Lung Cancer: Systematic Review and Meta-Analysis. *J. Bronchol. Interv. Pulmonol.* **2019**, *26*, 155–165. [CrossRef] [PubMed]
23. Steinfort, D.P.; Kothari, G.; Wallace, N.; Hardcastle, N.; Rangamuwa, K.; Dieleman, E.M.T.; Lee, P.; Li, P.; Simpson, J.A.; Yo, S.; et al. Systematic endoscopic staging of mediastinum to guide radiotherapy planning in patients with locally advanced non-small-cell lung cancer (SEISMIC): An international, multicentre, single-arm, clinical trial. *Lancet Respir. Med.* **2024**, *12*, 467–475. [CrossRef] [PubMed]
24. De Leyn, P.; Dooms, C.; Kuzdzal, J.; Lardinois, D.; Passlick, B.; Rami-Porta, R.; Turna, A.; Van Schil, P.; Venuta, F.; Waller, D.; et al. Revised ESTS guidelines for preoperative mediastinal lymph node staging for non-small-cell lung cancer. *Eur. J. Cardiothorac. Surg.* **2014**, *45*, 787–798. [CrossRef]
25. Huang, J.; Osarogiagbon, R.U.; Giroux, D.J.; Nishimura, K.K.; Bille, A.; Cardillo, G.; Detterbeck, F.; Kernstine, K.; Kim, H.K.; Lievens, Y.; et al. The International Association for the Study of Lung Cancer Staging Project for Lung Cancer: Proposals for the Revision of the N Descriptors in the Forthcoming Ninth Edition of the TNM Classification for Lung Cancer. *J. Thorac. Oncol.* **2024**, *19*, 766–785. [CrossRef]
26. Son, J.W.; Lee, J.; Jeon, J.H.; Cho, S.; Jung, W.; Shih, B.C.; Kim, K.; Jheon, S. Validation of IASLC 9th edition TNM classification for lung cancer: Focus on N descriptor. *BMC Cancer* **2024**, *24*, 1460. [CrossRef]
27. Yoo, C.; Yoon, S.; Lee, D.H.; Park, S.I.; Kim, D.K.; Kim, Y.H.; Kim, H.R.; Choi, S.H.; Kim, W.S.; Choi, C.M.; et al. Prognostic Significance of the Number of Metastatic pN2 Lymph Nodes in Stage IIIA-N2 Non-Small-Cell Lung Cancer After Curative Resection. *Clin. Lung Cancer* **2015**, *16*, e203–e212. [CrossRef]
28. Saji, H.; Tsuboi, M.; Shimada, Y.; Kato, Y.; Yoshida, K.; Nomura, M.; Matsubayashi, J.; Nagao, T.; Kakihana, M.; Usuda, J.; et al. A proposal for combination of total number and anatomical location of involved lymph nodes for nodal classification in non-small cell lung cancer. *Chest* **2013**, *143*, 1618–1625. [CrossRef]
29. Korevaar, D.A.; Crombag, L.M.; Cohen, J.F.; Spijker, R.; Bossuyt, P.M.; Annema, J.T. Added value of combined endobronchial and oesophageal endosonography for mediastinal nodal staging in lung cancer: A systematic review and meta-analysis. *Lancet Respir. Med.* **2016**, *4*, 960–968. [CrossRef]
30. Wu, T.; Cai, J.; Li, Y.; Xie, R.; Chen, K. Validation for revision of the stage IIIA(T1N2) in the forthcoming ninth edition of the TNM classification for lung cancer. *BMC Cancer* **2025**, *25*, 446. [CrossRef]
31. Rami-Porta, R. The TNM classification of lung cancer-a historic perspective. *J. Thorac. Dis.* **2024**, *16*, 8053–8067. [CrossRef]

diagnostics

MDPI

Article

Diagnostic Utility of Podoplanin Immunohistochemistry Combined with the NanoSuit-Correlative Light and Electron Microscopy Method for Thoracic Malignant Tumors

Shin-ya Katsuragi [1], Yuri Sakano [2], Isao Ohta [3], Hisami Kato [1], Rei Ishikawa [1], Hirofumi Watanabe [1,4], Ryosuke Miyazaki [1], Katsuhiro Yoshimura [1], Hidetaka Yamada [1], Yasuhiro Sakai [1], Yusuke Inoue [4], Yusuke Takanashi [5], Keigo Sekihara [5], Kazuhito Funai [5], Yoshiro Otsuki [6], Hideya Kawasaki [2,*] and Kazuya Shinmura [1,*]

1 Department of Tumor Pathology, Hamamatsu University School of Medicine, Hamamatsu 431-3192, Japan; skpath@hama-med.ac.jp (S.-y.K.); hisami@hama-med.ac.jp (H.K.); reipatho@hama-med.ac.jp (R.I.); d19033@hama-med.ac.jp (H.W.); 07486215@hama-med.ac.jp (R.M.); ky@hama-med.ac.jp (K.Y.); h-yamada@hama-med.ac.jp (H.Y.); ya-sakai@hama-med.ac.jp (Y.S.)
2 NanoSuit Research Laboratory, Division of Preeminent Bioimaging Research, Institute of Photonics Medicine, Hamamatsu University School of Medicine, Hamamatsu 431-3192, Japan; a20044@hama-med.ac.jp
3 Advanced Research Facilities and Services, Division of Preeminent Research Supports, Institute of Photonics Medicine, Hamamatsu University School of Medicine, Hamamatsu 431-3192, Japan; ohtaisao@hama-med.ac.jp
4 Second Division, Department of Internal Medicine, Hamamatsu University School of Medicine, Hamamatsu 431-3192, Japan; yinoue@hama-med.ac.jp
5 Department of Surgery 1, Hamamatsu University School of Medicine, Hamamatsu 431-3192, Japan; nashimed1@gmail.com (Y.T.); sekihara@hama-med.ac.jp (K.S.); kfunai@hama-med.ac.jp (K.F.)
6 Department of Pathology, Seirei Hamamatsu General Hospital, Hamamatsu 430-8558, Japan; otsuki@sis.seirei.or.jp
* Correspondence: gloria@hama-med.ac.jp (H.K.); kzshinmu@hama-med.ac.jp (K.S.); Tel.: +81-53-435-2504 (H.K.); +81-53-435-2220 (K.S.)

Abstract: Background/Objectives: Differentiating thoracic malignant tumors, such as epithelioid malignant pleural mesothelioma (EMPM) and non-small-cell lung carcinoma (NSCLC), primarily comprising lung adenocarcinoma (LAC) and lung squamous cell carcinoma (LSCC), remains a challenge in routine pathological diagnosis. This study aimed to evaluate whether podoplanin (PDPN) immunohistochemistry combined with scanning electron microscopy (SEM) using the NanoSuit-correlative light and electron microscopy (CLEM) methods could serve as a reliable tool for distinguishing these thoracic malignancies. **Methods/Results:** Initially, PDPN expression was assessed by immunohistochemical analysis in 11 EMPM, 100 LAC, and 23 LSCC cases. PDPN positivity was predominantly observed in the cell membrane and was significantly more frequent in EMPM (100%) than in LAC (2%; $p < 0.0001$) or LSCC (43.5%; $p = 0.0018$). Subsequently, field emission–SEM (FE-SEM) observations of PDPN-positive sites on immunohistochemical slides, conducted using the NanoSuit-CLEM method, revealed distinctive ultrastructural features. EMPM exhibited densely packed, elongated microvilli, whereas such structures were absent in LAC and LSCC. Furthermore, analysis of thick-cut sections (20 μm) demonstrated extensive microvilli coverage characteristic of EMPM. **Conclusions:** These findings suggest that the combined approach of PDPN immunohistochemistry and FE-SEM observation of PDPN-positive sites, using the NanoSuit-CLEM method, constitutes an effective diagnostic strategy for enhancing the accuracy of distinguishing EMPM from NSCLCs.

Keywords: epithelioid malignant pleural mesothelioma (EMPM); lung adenocarcinoma (LAC); lung squamous cell carcinoma (LSCC); microvilli; NanoSuit-correlative light and electron microscopy (CLEM) method; non-small-cell lung carcinoma (NSCLC); podoplanin (PDPN); field emission–scanning electron microscope (FE-SEM)

1. Introduction

Thoracic malignant tumors, including epithelioid malignant pleural mesothelioma (EMPM) and non-small cell lung carcinoma (NSCLC), primarily comprising lung adenocarcinoma (LAC) and lung squamous cell carcinoma (LSCC), present significant diagnostic challenges owing to their overlapping clinical and histopathological features in certain clinical contexts [1–3]. Lung cancer remains the leading cause of cancer-related mortality worldwide, and NSCLC accounts for approximately 85% of all cases [4–7]. Although relatively uncommon, EMPM is a highly aggressive malignancy that is primarily associated with asbestos exposure [1,4]. However, non-asbestos etiological factors, including exposure to naturally occurring mineral fibers such as erionite and previous therapeutic radiation, have been implicated in its pathogenesis [8]. EMPM is characterized by rapid disease progression, limited treatment options, and poor prognosis, underscoring the need for early detection and effective therapeutic strategies [1–3]. Accurate differentiation among these tumors is essential because their respective treatment strategies and prognoses differ considerably. Podoplanin (PDPN), a transmembrane glycoprotein, has emerged as a valuable biomarker for distinguishing EMPM from NSCLCs [9–11]. According to recent guidelines for the pathologic diagnosis of mesothelioma [9], PDPN is expressed in approximately 93% of epithelioid mesotheliomas, but in only approximately 3% of LAC cases. This differential expression permits employing PDPN immunohistochemistry for diagnostic pathology. However, previous studies have reported variations in PDPN positivity. Amatya et al. [11] observed a high positivity rate of 7% in LAC, whereas PDPN expression in LSCC ranged from 15 to 66% [12–14]. These findings indicated that PDPN immunohistochemistry alone may not reliably differentiate between EMPM and NSCLCs. Consequently, the current diagnostic consensus for EMPM relies on the use of multiple antibodies in immunohistochemistry, which increases diagnostic costs, demands more resources for slide preparation, and lengthens the time required for diagnosis.

Advancements in electron microscopy have significantly enhanced the ability to visualize biological specimens at nanometer-scale resolution [15]. However, conventional sample preparation techniques for scanning electron microscopy (SEM) involve chemical fixation, dehydration, and metal coating, which can introduce structural artifacts and compromise the hydration state of biological samples. These limitations hinder the accurate representation of native biological structures. To address these challenges, the NanoSuit method was developed as a biomimetic technology that enables high-resolution imaging of biological specimens, such as *Drosophila* larvae, using field emission–scanning electron microscopy (FE–SEM) without the need for conventional dehydration or fixation [16].

This technique involves the formation of an ultrathin polymer layer on the surface of a specimen to preserve its hydration state and structural integrity, thereby mitigating the key limitations associated with traditional SEM sample preparation methods. The NanoSuit approach has proven to be highly advantageous in biomedical and biological research because it facilitates the observation of hydrated biological specimens with minimal artifacts [17]. A key advantage of this method is its ability to suppress the charging effects caused by electron beam irradiation, while maintaining the natural morphology of biological samples [17]. This enables the high-resolution imaging of cells, tissues, and microorganisms with improved structural fidelity. Moreover, integrating the NanoSuit technique with correlative light and electron microscopy (CLEM) provides a powerful approach for correlating optical light microscopy with electron microscopy images, allowing comprehensive structural and compositional analyses of biomedical specimens [18].

The integration of NanoSuit technology with CLEM, referred to as NanoSuit-CLEM, enhances the precision of SEM imaging by incorporating immunohistochemical labeling techniques. The combination of osmium or gold chloride with 3,3′-diaminobenzidine (DAB) staining increases contrast in backscattered electron (BSE) mode, enabling selective visualization of specific cellular components [17,18]. This method is particularly beneficial for analyzing formalin-fixed paraffin-embedded (FFPE) tissue specimens and facilitating the precise correlation of pathological lesions between different imaging modalities. By combining light microscopy and SEM images, NanoSuit-CLEM can be a powerful tool for detailed histopathological analysis. One of the critical challenges in SEM-based pathology is the difficulty in correlating grayscale SEM images with light microscopy-based identification of pathological lesions. To address this issue, NanoSuit-CLEM incorporates fiducial marking techniques [18]. In this process, the target regions are first identified using light microscopy, and digital images are acquired. After removing the cover glass, a thin film of NanoSuit solution is applied to the specimen surface using a spin coater. An additional NanoSuit solution is placed around the target area to serve as a fiducial marker, improving positional accuracy during SEM observation. Unlike traditional CLEM approaches that require expensive custom sample holders and specialized imaging systems, this technique is cost-effective and minimizes artifacts. Furthermore, the NanoSuit membrane was removed after SEM imaging, allowing subsequent hematoxylin and eosin (H&E) staining for re-evaluation of the same tissue section. This feature offers the unique advantage of enabling sequential multimodal imaging, thereby enhancing the reliability of histopathological diagnoses. The NanoSuit-CLEM method has been used to observe various ultrastructures, including primary cilia in cancer cells, human papillomavirus in the mesopharynx, and lanthanum phosphate deposition in the gastrointestinal mucosa [19–22]. Its greatest advantage lies in its ability to perform SEM observations on FFPE specimens while correlating the findings with corresponding H&E-stained images or immunohistochemical images [18]. This versatility enables a wide range of applications and offers the potential for new discoveries.

In this study, we hypothesized that the ultrastructural differences in the immunohistochemically positive region for PDPN between EMPM and NSCLCs facilitate their differential diagnosis. To test this hypothesis, we performed SEM observations of PDPN-positive sites in these thoracic malignant tumors using the NanoSuit-CLEM method to develop a simple and practical approach for routine pathological diagnosis.

2. Materials and Methods

2.1. Case Collection

FFPE tissue blocks were obtained from 11 primary EMPM (mean age ± standard deviation: 59.2 ± 10.6 years; gender distribution: male: female = 9:2), 100 primary LAC (mean age ± standard deviation: 70.3 ± 7.65 years; gender distribution: male: female = 54:46; pathological tumor stage [pT]: pT1:pT2–4 = 60:40; pathological node stage [pN]: pN0:pN1–2 = 76:15), and 23 primary LSCC cases (mean age ± standard deviation: 71.9 ± 5.37 years; gender distribution: male: female = 20:3; pT: pT1:pT2–4 = 6:17; pN: pN0:pN1–2 = 19:3) at Hamamatsu University Hospital. Detailed case information is provided in Supplementary Tables S1–S3. For the pathological diagnosis of EMPM, immunohistochemical slides with several antibodies were used in addition to H&E-stained specimens (representative images are provided in Supplementary Figure S1). Thoracic malignant tumor samples were used for immunohistochemical analyses. The study protocol was reviewed and approved by the Institutional Review Board of the Hamamatsu University School of Medicine [approval numbers: 15-067 (17 July 2015) and 23-348 (27 February 2024)].

2.2. Immunohistochemical Analysis

Immunohistochemical staining was performed on the FFPE tissue sections using an automated system (Autostainer; DAKO, Carpinteria, CA, USA) to ensure standardized and reproducible staining conditions. Initially, FFPE sections were deparaffinized in xylene, rehydrated through a graded ethanol series, and rinsed in distilled water. As antigen retrieval was not required for the mouse anti-PDPN monoclonal antibody (clone D2-40, DAKO), the sections were directly treated with a hydrogen peroxide solution to block endogenous peroxidase activity, thereby minimizing nonspecific background staining. Subsequently, the sections were incubated with the primary antibody at a dilution of 1:200 under optimized conditions. Following primary antibody incubation, sections were rinsed with buffer and treated with a horseradish peroxidase-conjugated polymer detection system (Histofine Simple Stain MAX PO; Nichirei Biosciences, Tokyo, Japan) to facilitate signal amplification. For visualization, the sections were incubated with DAB as a chromogenic substrate, enabling the detection of immunoreactivity. After DAB development, the slides were counterstained with hematoxylin, dehydrated using graded ethanol and xylene, and coverslipped for microscopic examination. Positive and negative controls were included in each staining batch to ensure the specificity and reliability of the staining procedure. Immunostaining and H&E staining were performed using a Leica DMD108 digital microimaging system (Leica Microsystems, Wetzlar, Germany).

In this study, tumors were classified as PDPN-positive if more than 3% of the tumor cells were positive under conventional light microscopy. Tumors that did not meet this threshold were classified as PDPN-negative. Although 3% is a relatively low cut-off value, this threshold was adopted considering the limited nature of biopsy samples, which often contain only a small portion of the tumor. Even when the overall expression of the marker is low, there remains a possibility that a few positive cells may be present within the sampled area by chance. Therefore, this cut-off was established to account for such sampling variability. Although this specific cut-off is not established for PDPN, similarly low thresholds have occasionally been applied in the immunohistochemical evaluation of other biomarkers in the literature to accommodate sampling limitations and heterogeneous expression patterns [23–25].

2.3. FE-SEM Analysis of PDPN-Positive Site Using the NanoSuit-CLEM Method

Sections prepared for PDPN immunohistochemistry were also subjected to FE-SEM observations using the NanoSuit method [16–19]. A schematic overview of the procedure is presented in Figure 1. After acquiring the digital image data, a tissue section containing the region of interest was marked on both the front of the coverslip and the back of the microscope slide using a water- and organic solvent-resistant pen. The mark on the back of the slide served as a reference point for subsequent markings with the surface shield enhancer (SSE) solution. The diaphane and coverslip were subsequently removed from the slide. Following the enhancement of DAB signal using osmium, a diluted SSE solution was applied to the surface of the tissue section, covering the entire slide. The sample was left to stand for 1 min, after which excess SSE solution was removed via spin coating (2000 rpm, 15 s). SSE stock solution was prepared by dissolving sucrose, fructose, and sodium chloride in distilled water, followed by the addition of citric acid and sodium glutamate (pH 7.4). The resulting aqueous solution was then mixed with glycerin in a 1:2 ratio and subsequently diluted 20-fold for use. An on-demand droplet spotter (Hamamatsu Nanotechnology Co., Hamamatsu, Japan) was used to mark the designated pathological site precisely. This system can dispense minute volumes (pL–fL) of solution and is compatible with high-viscosity liquids such as SSE [18]. Upon the application of a pulse voltage between the liquid and substrate, a jet was ejected from the apex of the glass capillary to form a droplet.

The SSE stock solution was applied to minute droplets at multiple precise locations using a computer-controlled droplet spotter system, targeting the black mark on the slide.

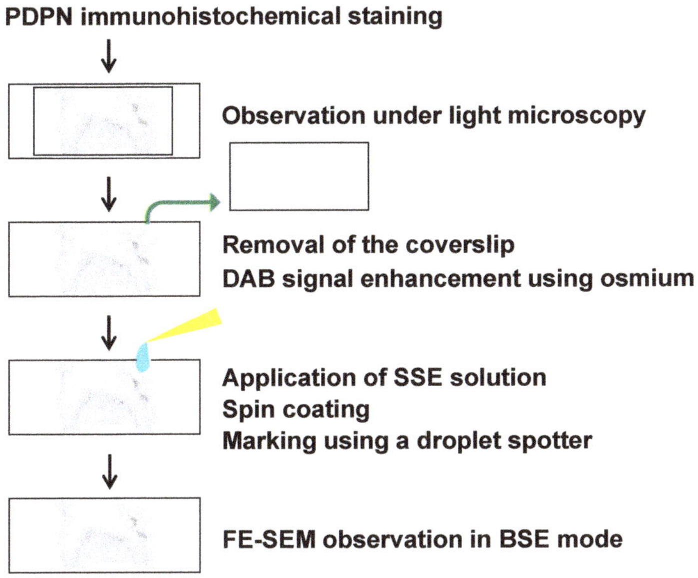

Figure 1. Schematic overview of the FE-SEM analysis of a PDPN-positive site using the NanoSuit-CLEM method. Detailed procedures are described in the main text.

Following rehydration, the sections were directly introduced into the FE-SEM system (S-4800; HITACHI, Tokyo, Japan), where NanoSuits were formed in situ under electron beam irradiation. FE-SEM observations using the NanoSuit-CLEM method were conducted using a BSE detector equipped with an yttrium aluminum garnet (YAG) crystal (YAG-BSE) detector. This approach provides a stable, conductive, and artifact-free imaging environment, ensuring that high-resolution electron microscopy is suitable for detailed ultrastructural analyses.

2.4. Statistical Analysis

Statistical analyses were performed using Fisher's exact test. GraphPad QuickCalcs (GraphPad Software Inc., San Diego, CA, USA) was used for the statistical analyses. p-values of less than 0.05 were considered statistically significant.

3. Results

3.1. Differences in PDPN Immunostaining Positivity Among Thoracic Malignant Tumors

Immunostaining for PDPN is a well-established marker for differentiating EMPM from NSCLCs. In this study, we analyzed the PDPN immunostaining profiles of malignant thoracic tumors diagnosed at our institution, including EMPM (n = 11), LAC (n = 100), and LSCC (n = 23). Tumor cell positivity for PDPN immunoreactivity was significantly more frequent in EMPM (11/11, 100%) than in LAC (2/100, 2%; $p < 0.0001$) or LSCC (10/23, 43.5%; $p = 0.0018$) (Figure 2a,b: representative immunohistochemical results; Figure 2c: bar graph). No significant associations were identified between PDPN immunoreactivity and clinicopathological factors in LAC and LSCC, as detailed in Supplementary Tables S4 and S5, respectively. In the analyzed specimens, membranous PDPN staining was observed across all three tumor types; however, the proportion of stained areas differed between EMPM and NSCLCs. Specifically, EMPM showed a relatively diffuse staining pattern, whereas LAC and LSCC showed focal staining. These findings are consistent with those of a previous study [10]. Furthermore, as is widely recognized [26], positive PDPN signals were identified in the stromal lymphatic vessels of all three tumor types (Supplementary Figure S2).

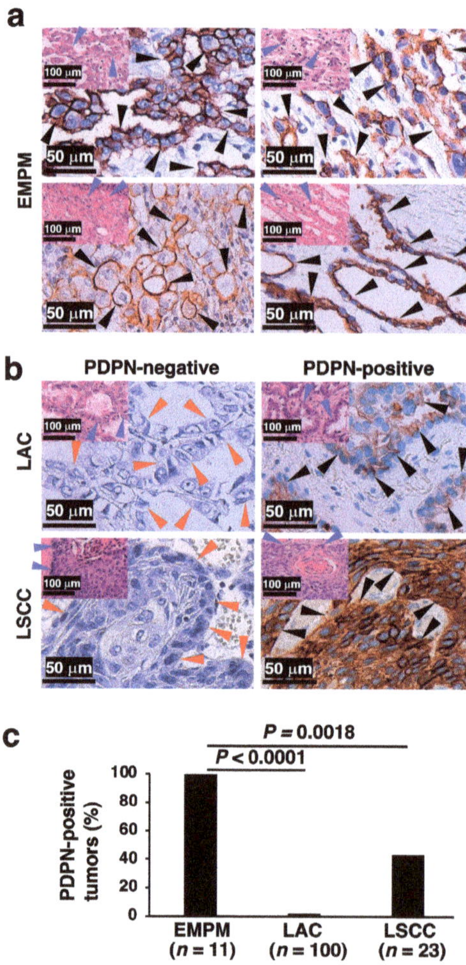

Figure 2. PDPN immunohistochemical status in thoracic malignant tumors. (**a**) Representative PDPN immunohistochemical staining images of four EMPM cases, demonstrating tubulopapillary (top left), trabecular (top right), deciduoid (bottom left), and acinar (bottom right) proliferation patterns. Insets in the upper left corner of each panel show corresponding H&E-stained images. (**b**) PDPN immunohistochemical staining images of two LAC and two LSCC cases. The left panels represent PDPN-negative cases, while the right panels represent PDPN-positive cases. Insets in the upper left corner of each panel show corresponding H&E-stained images. (**c**) Proportion of PDPN-positive tumors among EMPM, LAC, and LSCC cases. Statistical analysis was conducted using Fisher's exact test. Black arrowheads indicate PDPN-positive tumor cells in immunohistochemically stained images, and red and blue arrowheads denote the architectural patterns of tumor cell proliferation in the immunohistochemical and H&E-stained images, respectively.

Our observation of cell membrane-associated PDPN distribution, which is common across the three tumor types, along with the difference in PDPN positivity rates and areas between EMPM and NSCLCs, led us to hypothesize that structural differences at PDPN-positive sites may exist between these tumor types. Such structural variations could further enhance the diagnostic utility of PDPN immunostaining in distinguishing EMPM from NSCLCs.

3.2. FE-SEM Observation of PDPN-Positive Sites in Thoracic Malignant Tumors Using the NanoSuit-CLEM Method

We subsequently investigated the structural differences between EMPM and NSCLCs. For this, we used the NanoSuit-CLEM method for electron microscopic observation of membranous PDPN staining sites. DAB, commonly used in conventional immunohistochemical staining and optical microscopy, has an affinity for osmium, a component of electron microscopy. Consequently, the DAB deposition sites identified by optical microscopy correspond to regions of high electron reflection under an electron microscope. We performed PDPN immunostaining in EMPM and NSCLC samples, followed by osmium staining and FE-SEM observations using the NanoSuit-CLEM method. Paired images of PDPN immunostaining and FE-SEM with a YAG-BSE detector were successfully obtained (representative images are shown in Figure 3a; Case 1 of EMPM). Using this technique, high electron reflection sites in the YAG-BSE images, corresponding to the DAB deposition sites, were observed in the three EMPM cases (Cases 1–3) (Figure 3b). Under magnification, these regions revealed densely packed, thick, and elongated villous structures, indicating well-developed microvilli (Figure 3b, bottom row).

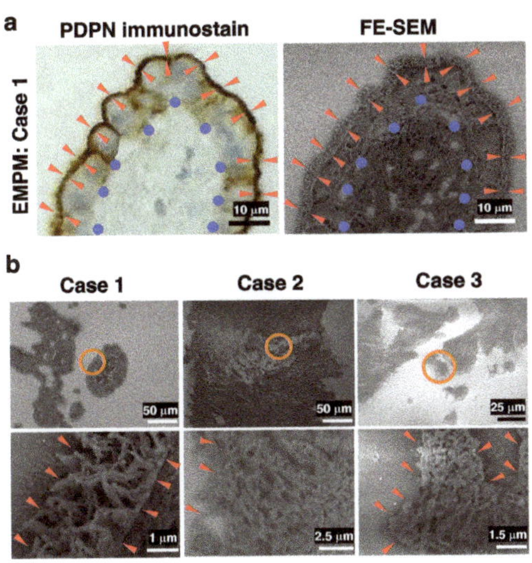

Figure 3. FE-SEM observation of immunohistochemical PDPN-positive sites in three cases of EMPM using the NanoSuit-CLEM method. (**a**) Representative paired images of PDPN immunohistochemical staining (left) and FE-SEM imaging (right) obtained using a YAG-BSE detector via the CLEM technique in Case 1 of EMPM. Briefly, PDPN-stained immunohistochemical slides—with coverslips removed—are treated with SSE solution following DAB enhancement with osmium. After region marking using a droplet spotter, the slides are directly introduced into the FE-SEM system, where NanoSuit formation is induced by electron beam irradiation. FE-SEM imaging is performed in the BSE mode. Red arrowheads indicate the PDPN-positive, exposed surface of mesothelial cells surrounding fibrovascular connective tissue, which is denoted by blue circles. (**b**) FE-SEM images showing PDPN-immunopositive sites in Cases 1–3 of EMPM. The regions enclosed by orange circles in the upper panels are shown at higher magnification in the lower panels. Red arrowheads highlight regions where the surface of tumor cells is exposed to the extracellular space.

In contrast, a similar analysis of LAC and LSCC revealed poorly nourished microvilli in DAB deposition areas in both LAC and LSCC in the YAG-BSE mode (Figure 4). These findings indicate that PDPN immunostaining effectively highlights well-developed microvillus structures in EMPM, whereas such structures were not observed in LAC or LSCC. These findings suggest that PDPN immunohistochemistry is effective for distinguishing between EMPM and NSCLC. Moreover, when tumor cells are PDPN-positive, observing the corresponding site on the immunohistochemical slide using FE-SEM with the NanoSuit-CLEM method further enhances the accuracy of the differential diagnosis.

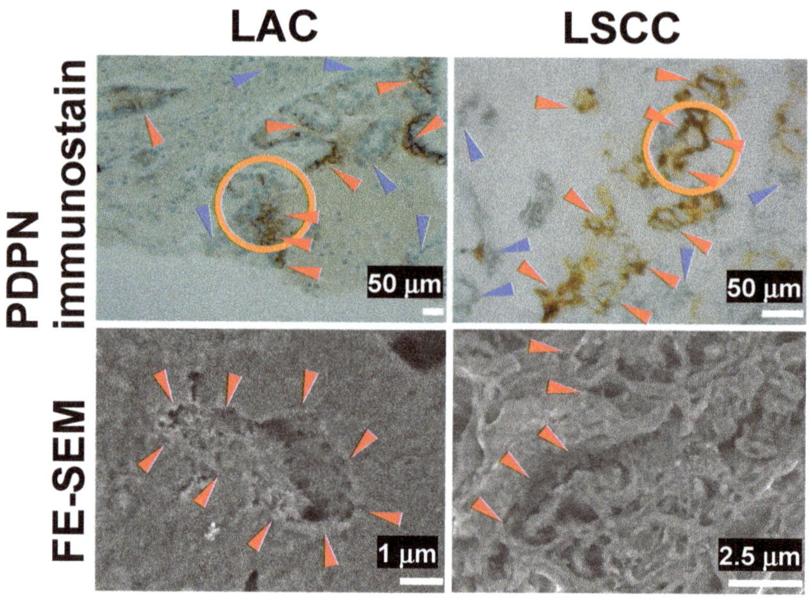

Figure 4. FE-SEM observation of immunohistochemical PDPN-positive sites in LAC and LSCC using the NanoSuit-CLEM method. The upper panels show PDPN immunohistochemical staining, while the lower panels present FE-SEM images captured using a YAG-BSE detector for a case of LAC and a case of LSCC. The areas enclosed by orange circles in the upper panels are presented as higher-magnification FE-SEM images in the corresponding lower panels. Red and blue arrowheads in the immunohistochemical images indicate PDPN-positive and PDPN-negative tumor cells, respectively. Red arrowheads in the FE-SEM images highlight regions where the surface of tumor cells is exposed to the extracellular space.

3.3. FE-SEM Using the NanoSuit-CLEM Method for the Detailed Characterization of Surface Structures in Thick-Cut Sections of EMPM

To emphasize cell surface morphology rather than thin cross-sectional views, we prepared thick-cut sections (20 µm) of EMPM. Subsequent FE-SEM observations of PDPN-positive sites in the YAG-BSE mode, employing the NanoSuit-CLEM method, revealed that the entire surface of individual EMPM cells was covered with dense, elongated microvillar structures (Figure 5). In contrast, such microvillar structures were absent on the surface of tumor cells in thick-cut sections of LSCC, where the surface appeared rugged and lacked organized and elongated projections (Figure 5). These findings indicate that the NanoSuit-CLEM method, in conjunction with FE-SEM analysis of thick-cut sections, facilitates detailed characterization of the fine surface morphology of EMPM cells.

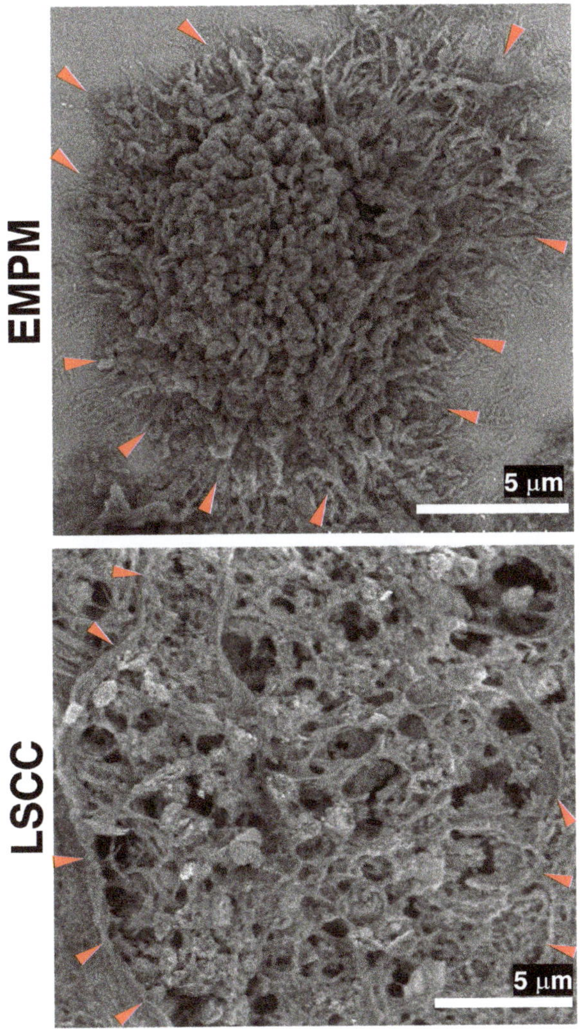

Figure 5. FE-SEM analysis of thick-cut section samples from EMPM and LSCC using the NanoSuit-CLEM method. FE-SEM images of a tumor cell from EMPM and a tumor cell from LSCC are obtained using a YAG-BSE detector. Red arrowheads indicate regions where the surface of tumor cells is exposed to the extracellular space.

3.4. Proposed Workflow for Differential Diagnosis of Thoracic Malignant Tumors

Based on our results, we prepared a workflow for the differential diagnosis of malignant thoracic tumors using PDPN immunohistochemistry and subsequent FE-SEM analysis of PDPN-positive sites using the NanoSuit-CLEM method (Figure 6). In this workflow, PDPN immunohistochemistry alone, rather than a multi-antibody panel, served as the initial diagnostic step for epithelioid-type thoracic malignant tumors located in the pleural region. Upon detection of PDPN positivity on the tumor cell membrane, the slide was directly subjected to FE-SEM analysis using the NanoSuit-CLEM method. This process involved the removal of the coverslip, DAB enhancement with osmium, and coating of the immunostained tissue with SSE solution. The presence of well-developed microvilli at immunohistochemical PDPN-rich regions confirmed the mesothelial lineage, thereby leading to a diagnosis of EMPM. Conversely, the absence of these structures suggested the presence of NSCLCs. In patients with negative PDPN immunohistochemistry results, conventional immunohistochemical panels, as recommended by established guidelines [9], should be used. The NanoSuit-CLEM method facilitates rapid FE-SEM analysis of the same immunohistochemical slide and allows reapplication of the cover glass post-analysis for archival purposes. Therefore, this approach offers significant advantages over the current multi-antibody immunohistochemistry systems.

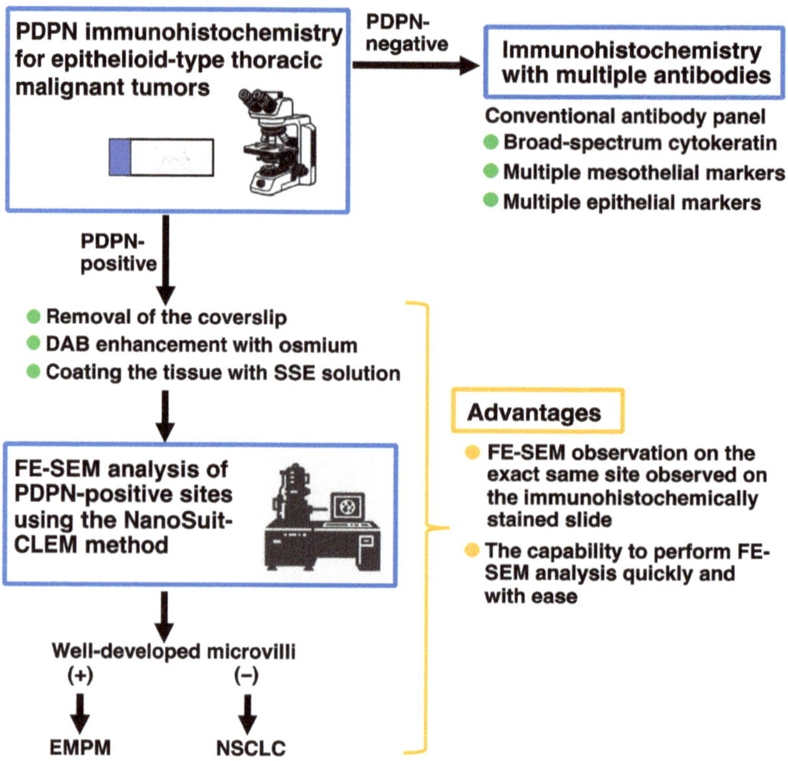

Figure 6. Proposed workflow for the differential diagnosis of thoracic malignant tumors using PDPN immunohistochemistry and subsequent FE-SEM analysis of PDPN-positive sites via the NanoSuit-CLEM method.

4. Discussion

In this study, the immunohistochemical analysis of PDPN expression in 11 cases of EMPM, 100 cases of LAC, and 23 cases of LSCC revealed significant differences in PDPN positivity among these tumor types. PDPN positivity in the tumor cell membranes was observed in 100%, 2%, and 43.5% of EMPM, LAC, and LSCC cases, respectively. Subsequent FE-SEM observations of PDPN-positive sites using the NanoSuit-CLEM method demonstrated densely packed, elongated microvilli in EMPM, whereas such structures were absent in NSCLCs. In addition, analysis of thicker sections revealed extensive microvilli coverage in the EMPM. These findings suggest that PDPN immunohistochemistry coupled with FE-SEM using the NanoSuit-CLEM method provides a practical approach for distinguishing EMPM from NSCLCs. This study is the first to demonstrate the presence of well-developed microvilli at membranous PDPN-stained sites in EMPM of FFPE sections and establish their utility in the pathological differential diagnosis of thoracic malignant tumors. Furthermore, it serves as a compelling example of how previously recognized ultrastructural findings, which were historically difficult to integrate into routine diagnostics owing to the labor-intensive nature of traditional techniques, can now be readily applied in contemporary pathological practice owing to recent methodological advancements.

Mesotheliomas, including EMPM, are rare tumors that present significant diagnostic challenges. The primary differential diagnosis for EMPM often includes NSCLCs, and diagnostic workflows typically rely on panels of mesothelial and epithelial immunohistochemical markers to establish the mesothelial lineage [9]. The most widely used mesothelial markers include calretinin, PDPN, WT1, and CK5/6, all of which demonstrate more than 80% sensitivity for EMPM [3,9,27]. However, none of these markers are entirely specific to the mesothelial origin, as they can also show positivity in subsets of carcinomas. This lack of perfect sensitivity and specificity extends to newly identified immunohistochemical markers [28,29]. Consequently, current guidelines recommend a first-line immunohistochemical panel comprising broad-spectrum cytokeratin, along with at least two mesothelial and two epithelial markers [9]. In this study, we demonstrated the utility of PDPN immunohistochemistry and subsequent FE-SEM analysis of PDPN-positive sites using the NanoSuit-CLEM method. Based on our findings, we propose a simplified diagnostic workflow in which PDPN immunohistochemistry serves as the initial diagnostic test for epithelioid-type malignant thoracic tumors located in the pleural region (Figure 6). If PDPN positivity is detected on the tumor cell membrane, the slide can be directly processed for FE-SEM analysis using the NanoSuit-CLEM method. The presence of well-developed microvilli confirms a mesothelial lineage, supporting the diagnosis of EMPM, whereas their absence suggests NSCLCs. For cases in which PDPN is negative on the initial immunohistochemical analysis, conventional immunohistochemical panels, as recommended by existing guidelines [9], should be utilized. The NanoSuit-CLEM method facilitates rapid SEM analysis of the same immunohistochemical slide, offering significant practical advantages, including streamlined diagnostic workflows and reduced reliance on extensive antibody panels.

The ultrastructural characteristics of mesothelioma are well documented, with the presence of well-developed microvilli on the cell membranes of EMPM, which serve as a distinguishing feature that is absent in LAC [30–32]. PDPN, a type I transmembrane glycoprotein with an extracellular domain, a transmembrane region, and a short cytoplasmic tail, is concentrated in plasma membrane extensions, such as microvilli [33]. Consequently, SEM observation of membranous PDPN-rich sites in the EMPM highlights the presence of microvilli. In contrast, PDPN expression in NSCLCs is likely independent of microvilli, as it is localized to irregular plasma membrane structures other than microvilli. Our study successfully demonstrates that the presence or absence of microvilli in PDPN-rich regions facilitates the differential diagnosis of EMPM and NSCLCs. Furthermore, although microvilli in mesotheliomas have traditionally been ultrastructurally observed in samples fixed using conventional methods (e.g., glutaraldehyde fixation) [30–32], their preservation in FFPE sections has not been previously established. Using the NanoSuit-CLEM method, we demonstrated that microvilli were preserved in mesothelioma cells within FFPE sections, making this approach highly valuable for the differential diagnosis of thoracic malignancies.

An approach combining the NanoSuit-CLEM method with energy-dispersive X-ray spectroscopy (EDS) was recently developed [17]. This technique enables the elemental analysis of FFPE tissue specimens, facilitates the identification of metal deposits, and contributes to advancements in medical diagnostics. For instance, lanthanum phosphate deposits in the gastrointestinal tract can be rapidly detected in patients undergoing lanthanum carbonate treatment, facilitating pathological diagnosis [19]. Additionally, this approach provided insights into gastric black spots observed in patients with a history of *Helicobacter pylori* (*H. pylori*) eradication therapy [34]. In contrast to individuals with current *H. pylori* infection who rarely exhibit black spots, those with a history of eradication therapy frequently present with these lesions. Elemental analysis of the gastric black spots using this method revealed that iron was the primary component. The applicability of the NanoSuit-CLEM combined with SEM-EDS extends beyond the gastrointestinal tract and is promising for use in other organs. For instance, exogenous substances are constantly inhaled into the lungs, and elemental analysis of FFPE lung specimens may help elucidate previously unidentified aspects of pulmonary pathology [35–38]. Moreover, this method could potentially contribute to differentiating EMPM from NSCLC using an alternative approach to our proposed workflow (Figure 6), further enhancing its diagnostic utility in pulmonary pathology.

A limitation of the present study is the relatively small number of EMPM cases ($n = 11$). Although clear differences in PDPN positivity and the ultrastructural appearance of microvilli were demonstrated in comparison with LAC and LSCC, further validation is required. A future multi-institutional study involving a larger cohort of EMPM cases and the application of FE-SEM using the NanoSuit-CLEM method would enable a concrete conclusion regarding the utility of this approach in distinguishing EMPM from NSCLCs.

Advances in electron microscopy technology have led to the development of desktop and tabletop electron microscopes, as well as high-performance instruments [39,40]. As demonstrated in this study, SEM analysis using the NanoSuit-CLEM method on conventional pathological slides, including those stained with H&E, special stains, and immunohistochemistry, offers significant potential in clinical pathology. The combination of such accessible electron microscopy technologies with the NanoSuit-CLEM method could greatly expand the use of electron microscopy in routine pathological diagnostics.

5. Conclusions

A combined application of PDPN immunohistochemistry and FE-SEM using the NanoSuit-CLEM method provides an effective approach for distinguishing EMPM from NSCLCs. Consistent PDPN positivity in the EMPM and the presence of densely packed elongated microvilli at PDPN-positive sites are valuable diagnostic markers. This method bridges the gap between conventional immunohistochemistry and ultrastructural pathology, enabling a more precise tumor classification.

Supplementary Materials: The following is available online at https://www.mdpi.com/article/10.3390/diagnostics15101298/s1, Figure S1: Representative immunohistochemical images of an EMPM case, Figure S2: Representative immunohistochemical image showing lymphatic vessel detection in a thoracic malignant tumor using PDPN immunohistochemistry, Figure S3: Original data, Table S1: Age and sex distribution of patients with EMPM, Table S2: Clinicopathological characteristics of patients with LAC, Table S3: Clinicopathological characteristics of patients with LSCC, Table S4: Clinicopathological factors in 100 LAC cases stratified by PDPN immunostaining status, Table S5: Clinicopathological factors in 23 LSCC cases stratified by PDPN immunostaining status.

Author Contributions: Conceptualization, K.S. (Kazuya Shinmura); methodology, I.O., H.K. (Hideya Kawasaki) and K.S. (Kazuya Shinmura); formal analysis, K.S. (Kazuya Shinmura); investigation, S.-y.K., Y.S. (Yuri Sakano), I.O., H.K. (Hisami Kato), R.I., H.W., R.M., K.Y., H.Y., Y.S. (Yasuhiro Sakai), H.K. (Hideya Kawasaki) and K.S. (Kazuya Shinmura); resources, Y.I., Y.T., K.S. (Keigo Sekihara) and K.F.; writing—original draft preparation, S.-y.K. and K.S. (Kazuya Shinmura); writing—review and editing, S.-y.K., Y.S. (Yuri Sakano), I.O., H.K. (Hisami Kato), R.I., H.W., R.M., K.Y., H.Y., Y.S. (Yasuhiro Sakai), Y.I., Y.T., K.S. (Keigo Sekihara), K.F., Y.O., H.K. (Hideya Kawasaki) and K.S. (Kazuya Shinmura); supervision, Y.O., H.K. (Hideya Kawasaki) and K.S. (Kazuya Shinmura); funding acquisition, H.K. (Hideya Kawasaki) and K.S. (Kazuya Shinmura). All authors have read and agreed to the published version of the manuscript.

Funding: This work was supported in part by a Grant-in-Aid from the Japan Society for the Promotion of Science (20K07445 to K.S. and 24K02249 to H.K. and K.S.).

Institutional Review Board Statement: The study was conducted according to the guidelines of the Declaration of Helsinki, and approved by the Institutional Review Board of Hamamatsu University School of Medicine [approval numbers: 15-067 (17 July 2015) and 23-348 (27 February 2024)].

Informed Consent Statement: Patient consent was waived due to the retrospective nature of the study, use of anonymized samples from pathologic archives, and implementation of an opt-out procedure through a public information notice.

Data Availability Statement: The data presented in this study are available on request from the corresponding author.

Acknowledgments: We thank Takaharu Kamo (Hamamatsu University School of Medicine) for his technical assistance. We also thank the Advanced Research Facilities and Services, Hamamatsu University School of Medicine for allowing the use of their equipment.

Conflicts of Interest: The authors declare no conflicts of interest.

Abbreviations

The following abbreviations are used in this manuscript:

BSE	Backscattered electron
CLEM	Correlative light and electron microscopy
DAB	3,3′-diaminobenzidine tetrahydrochloride
EDS	Energy-dispersive X-ray spectroscopy
EMPM	Epithelioid malignant pleural mesothelioma
FE-SEM	Field emission–scanning electron microscopy
FFPE	Formalin-fixed paraffin-embedded
H&E	Hematoxylin and eosin
H. pylori	*Helicobacter pylori*
LAC	Lung adenocarcinoma
LSCC	Lung squamous cell carcinoma
PDPN	Podoplanin
SSE	Surface shield enhancer
YAG	Yttrium aluminum garnet

References

1. Beasley, M.B.; Galateau-Salle, F.; Dacic, S. Pleural Mesothelioma Classification Update. *Virchows Arch.* **2021**, *478*, 59–72. [CrossRef] [PubMed]
2. Dacic, S. Pleural Mesothelioma Classification-Update and Challenges. *Mod. Pathol.* **2022**, *35*, 51–56. [CrossRef]
3. Chapel, D.B.; Schulte, J.J.; Husain, A.N.; Krausz, T. Application of Immunohistochemistry in Diagnosis and Management of Malignant Mesothelioma. *Transl. Lung Cancer Res.* **2020**, *9*, S3–S27. [CrossRef] [PubMed]
4. Bray, F.; Laversanne, M.; Sung, H.; Ferlay, J.; Siegel, R.L.; Soerjomataram, I.; Jemal, A. Global cancer statistics 2022: GLOBOCAN estimates of incidence and mortality worldwide for 36 cancers in 185 countries. *CA Cancer J. Clin.* **2024**, *74*, 229–263. [CrossRef]
5. Thai, A.A.; Solomon, B.J.; Sequist, L.V.; Gainor, J.F.; Heist, R.S. Lung Cancer. *Lancet* **2021**, *398*, 535–554. [CrossRef]
6. Alkarn, A.; Stapleton, L.J.; Eleftheriou, D.; Stewart, L.; Chalmers, G.W.; Hamed, A.; Hussein, K.; Blyth, K.G.; van der Horst, J.C.; Maclay, J.D. Real-Life Pre-Operative Nodal Staging Accuracy in Non-Small Cell Lung Cancer and Its Relationship with Survival. *Diagnostics* **2025**, *15*, 430. [CrossRef]
7. Steinestel, K.; Arndt, A. Current Biomarkers in Non-Small Cell Lung Cancer—The Molecular Pathologist's Perspective. *Diagnostics* **2025**, *15*, 631. [CrossRef]
8. Attanoos, R.L.; Churg, A.; Galateau-Salle, F.; Gibbs, A.R.; Roggli, V.L. Malignant Mesothelioma and Its Non-Asbestos Causes. *Arch. Pathol. Lab. Med.* **2018**, *142*, 753–760. [CrossRef]
9. Husain, A.N.; Chapel, D.B.; Attanoos, R.; Beasley, M.B.; Brcic, L.; Butnor, K.; Chirieac, L.R.; Churg, A.; Dacic, S.; Galateau-Salle, F.; et al. Guidelines for Pathologic Diagnosis of Mesothelioma: 2023 Update of the Consensus Statement from the International Mesothelioma Interest Group. *Arch. Pathol. Lab. Med.* **2024**, *148*, 1251–1271. [CrossRef]

10. Suzuki, H.; Kaneko, M.K.; Kato, Y. Roles of Podoplanin in Malignant Progression of Tumor. *Cells* **2022**, *11*, 575. [CrossRef]
11. Amatya, V.J.; Kushitani, K.; Kai, Y.; Suzuki, R.; Miyata, Y.; Okada, M.; Takeshima, Y. Glypican-1 immunohistochemistry is a novel marker to differentiate epithelioid mesothelioma from lung adenocarcinoma. *Mod. Pathol.* **2018**, *31*, 809–815. [CrossRef] [PubMed]
12. Ordóñez, N.G. The diagnostic utility of immunohistochemistry in distinguishing between epithelioid mesotheliomas and squamous carcinomas of the lung: A comparative study. *Mod. Pathol.* **2006**, *19*, 417–428. [CrossRef] [PubMed]
13. Shimada, Y.; Ishii, G.; Nagai, K.; Atsumi, N.; Fujii, S.; Yamada, A.; Yamane, Y.; Hishida, T.; Nishimura, M.; Yoshida, J.; et al. Expression of podoplanin, CD44, and p63 in squamous cell carcinoma of the lung. *Cancer Sci.* **2009**, *100*, 2054–2059. [CrossRef] [PubMed]
14. Comin, C.E.; Novelli, L.; Cavazza, A.; Rotellini, M.; Cianchi, F.; Messerini, L. Expression of thrombomodulin, calretinin, cytokeratin 5/6, D2-40 and WT-1 in a series of primary carcinomas of the lung: An immunohistochemical study in comparison with epithelioid pleural mesothelioma. *Tumori J.* **2014**, *100*, 559–567.
15. Shah, F.A.; Ruscsák, K.; Palmquist, A. 50 Years of Scanning Electron Microscopy of Bone—A Comprehensive Overview of the Important Discoveries Made and Insights Gained into Bone Material Properties in Health, Disease, and Taphonomy. *Bone Res.* **2019**, *7*, 7–15.
16. Takaku, Y.; Suzuki, H.; Ohta, I.; Ishii, D.; Muranaka, Y.; Shimomura, M.; Hariyama, T. A thin polymer membrane, nano-suit, enhancing survival across the continuum between air and high vacuum. *Proc. Natl. Acad. Sci. USA* **2013**, *110*, 7631–7635. [CrossRef]
17. Hariyama, T.; Takaku, Y.; Kawasaki, H.; Shimomura, M.; Senoh, C.; Yamahama, Y.; Hozumi, A.; Ito, S.; Matsuda, N.; Yamada, S.; et al. Microscopy and Biomimetics: The Nano-Suit® Method and Image Retrieval Platform. *Microscopy* **2022**, *71*, 1–12. [CrossRef]
18. Kawasaki, H.; Itoh, T.; Takaku, Y.; Suzuki, H.; Kosugi, I.; Meguro, S.; Iwashita, T.; Hariyama, T. The NanoSuit method: A novel histological approach for examining paraffin sections in a nondestructive manner by correlative light and electron microscopy. *Lab. Investig.* **2020**, *100*, 161–173. [CrossRef]
19. Shinmura, K.; Kawasaki, H.; Baba, S.; Ohta, I.; Kato, H.; Yasuda, H.; Yamada, S.; Misawa, K.; Sugimoto, K.; Osawa, S.; et al. Utility of Scanning Electron Microscopy Elemental Analysis Using the 'NanoSuit' Correlative Light and Electron Microscopy Method in the Diagnosis of Lanthanum Phosphate Deposition in the Esophagogastroduodenal Mucosa. *Diagnostics* **2019**, *10*, 1. [CrossRef]
20. Shinmura, K.; Kusafuka, K.; Kawasaki, H.; Kato, H.; Hariyama, T.; Tsuchiya, K.; Kawanishi, Y.; Funai, K.; Misawa, K.; Mineta, H.; et al. Identification and characterization of primary cilia-positive salivary gland tumours exhibiting basaloid/myoepithelial differentiation. *J. Pathol.* **2021**, *254*, 519–530. [CrossRef]
21. Yamada, S.; Itoh, T.; Ikegami, T.; Imai, A.; Mochizuki, D.; Nakanishi, H.; Ishikawa, R.; Kita, J.; Nakamura, Y.; Takizawa, Y.; et al. Association between human papillomavirus particle production and the severity of recurrent respiratory papillomatosis. *Sci. Rep.* **2023**, *13*, 5514. [CrossRef]
22. Shinmura, K.; Kato, H.; Kawasaki, H.; Hariyama, T.; Yoshimura, K.; Tsuchiya, K.; Watanabe, H.; Ohta, I.; Asahina, E.; Sumiyoshi, F.; et al. Primary Cilia Are Frequently Present in Small Cell Lung Carcinomas but Not in Non-Small Cell Lung Carcinomas or Lung Carcinoids. *Lab. Investig.* **2023**, *103*, 100007. [CrossRef] [PubMed]
23. Suresh, P.K.; Sahu, K.K.; Pai, R.R.; Sridevi, H.B.; Ballal, K.; Khandelia, B.; Minal, J.; Annappa, R. The prognostic significance of neuroendocrine differentiation in colorectal carcinomas: Our experience. *J. Clin. Diagn. Res.* **2015**, *9*, EC01–EC04. [CrossRef]
24. Haraldsson, S.; Klarskov, L.; Nilbert, M.; Bernstein, I.; Bonde, J.; Holck, S. Differential expression of CK20, β-catenin, and MUC2/5AC/6 in Lynch syndrome and familial colorectal cancer type X. *BMC Clin. Pathol.* **2017**, *17*, 11. [CrossRef]
25. Jun, S.Y.; Kim, J.; Yoon, N.; Maeng, L.S.; Byun, J.H. Prognostic potential of Cyclin D1 expression in colorectal cancer. *J. Clin. Med.* **2023**, *12*, 572. [CrossRef]
26. Kalof, A.N.; Cooper, K. D2-40 immunohistochemistry—So far! *Adv. Anat. Pathol.* **2009**, *16*, 62–64. [CrossRef]
27. Sauter, J.L.; Dacic, S.; Galateau-Salle, F.; Attanoos, R.L.; Butnor, K.J.; Churg, A.; Husain, A.N.; Kadota, K.; Khoor, A.; Nicholson, A.G.; et al. The 2021 WHO Classification of Tumors of the Pleura: Advances Since the 2015 Classification. *J. Thorac. Oncol.* **2022**, *17*, 608–622. [CrossRef]
28. Sahin, N.; Akatli, A.N.; Celik, M.R.; Ulutas, H.; Samdanci, E.T.; Colak, C. The Role of CD90 in the Differential Diagnosis of Pleural Malignant Mesothelioma, Pulmonary Carcinoma and Comparison with Calretınin. *Pathol. Oncol. Res.* **2017**, *23*, 487–491. [CrossRef]
29. Mawas, A.S.; Amatya, V.J.; Kushitani, K.; Kai, Y.; Miyata, Y.; Okada, M.; Takeshima, Y. MUC4 immunohistochemistry is useful in distinguishing epithelioid mesothelioma from adenocarcinoma and squamous cell carcinoma of the lung. *Sci. Rep.* **2018**, *8*, 134. [CrossRef]
30. Dewar, A.; Valente, M.; Ring, N.P.; Corrin, B. Pleural mesothelioma of epithelial type and pulmonary adenocarcinoma: An ultrastructural and cytochemical comparison. *J. Pathol.* **1987**, *152*, 309–316. [CrossRef]
31. Wick, M.R.; Loy, T.; Mills, S.E.; Legier, J.F.; Manivel, J.C. Malignant epithelioid pleural mesothelioma versus peripheral pulmonary adenocarcinoma: A histochemical, ultrastructural, and immunohistologic study of 103 cases. *Hum. Pathol.* **1990**, *21*, 759–766. [CrossRef] [PubMed]

32. Hammar, S.P. Macroscopic, histologic, histochemical, immunohistochemical, and ultrastructural features of mesothelioma. *Ultrastruct. Pathol.* **2006**, *30*, 3–17. [CrossRef]
33. Carrasco-Ramírez, P.; Greening, D.W.; Andrés, G.; Gopal, S.K.; Martín-Villar, E.; Renart, J.; Simpson, R.J.; Quintanilla, M. Podoplanin is a component of extracellular vesicles that reprograms cell-derived exosomal proteins and modulates lymphatic vessel formation. *Oncotarget* **2016**, *7*, 16070–16089. [CrossRef] [PubMed]
34. Sugiura, K.; Kawasaki, H.; Egami, T.; Kaneko, M.; Ishida, N.; Tamura, S.; Tani, S.; Yamade, M.; Hamaya, Y.; Osawa, S.; et al. Iron Deposition in Gastric Black Spots: Clinicopathological Insights and NanoSuit-Correlative Light and Electron Microscopy Analysis. *DEN Open* **2024**, *5*, e398. [CrossRef]
35. Crawford, J.A.; Sanyal, S.; Burnett, B.R.; Wiesenfeld, S.L.; Abraham, J.L. Accelerated silicosis in sandblasters: Pathology, mineralogy, and clinical correlates. *Am. J. Ind. Med.* **2024**, *67*, 179–199. [CrossRef]
36. Visonà, S.D.; Untalan, M.; Bertoglio, B.; Capella, S.; Belluso, E.; Billò, M.; Ivic-Pavlicic, T.; Taioli, E. Asbestos burden in lungs of subjects deceased from mesothelioma who lived in proximity to an asbestos factory: A topographic post-mortem SEM-EDS study. *Am. J. Ind. Med.* **2025**, *68*, 112–121. [CrossRef]
37. Newbury, D.E.; Ritchie, N.W. Performing Elemental Microanalysis with High Accuracy and High Precision by Scanning Electron Microscopy/Silicon Drift Detector Energy-Dispersive X-ray Spectrometry (SEM/SDD-EDS). *J. Mater. Sci.* **2015**, *50*, 493–518. [CrossRef]
38. Ruaro, B.; Salton, F.; Braga, L.; Wade, B.; Confalonieri, P.; Volpe, M.C.; Baratella, E.; Maiocchi, S.; Confalonieri, M. The History and Mystery of Alveolar Epithelial Type II Cells: Focus on Their Physiologic and Pathologic Role in Lung. *Int. J. Mol. Sci.* **2021**, *22*, 2566. [CrossRef]
39. Pretorius, E. The use of a desktop scanning electron microscope as a diagnostic tool in studying fibrin networks of thrombo-embolic ischemic stroke. *Ultrastruct. Pathol.* **2011**, *35*, 245–250. [CrossRef]
40. Marchand, E.; Bertrand, B.; Hedouin, V.; Demondion, X.; Becart, A. Study of Root Transparency in Different Postmortem Intervals Using Scanning Electron Microscopy. *Diagnostics* **2023**, *13*, 2808. [CrossRef] [PubMed]

diagnostics

MDPI

Article

Transthoracic Lung Ultrasound in Systemic Sclerosis-Associated Interstitial Lung Disease: Capacity to Differentiate Chest Computed-Tomographic Characteristic Patterns

Cinzia Rotondo [1,*], Giuseppe Busto [1], Valeria Rella [1], Raffaele Barile [1], Fabio Cacciapaglia [2,3], Marco Fornaro [4], Florenzo Iannone [4], Donato Lacedonia [5], Carla Maria Irene Quarato [5], Antonello Trotta [1], Francesco Paolo Cantatore [1] and Addolorata Corrado [1]

1 Rheumatology Unit, Department of Medical and Surgical Sciences, University of Foggia, 71122 Foggia, Italy
2 Rheumatology Service, Internal Medicine Unit "F. Miulli" General Hospital, Acquaviva delle Fonti, 70021 Bari, Italy
3 Department of Medicine and Surgery, LUM "G De Gennaro", Casamassima, 70010 Bari, Italy
4 Reumatology Unit, Dipartimento di Medicina di Precisione e Rigenerativa e Area Jonica, Università degli Studi di Bari Aldo Moro, 70121 Bari, Italy
5 Institute of Respiratory Diseases, Department of Medical and Surgical Sciences, University of Foggia, 71122 Foggia, Italy
* Correspondence: cinzia.rotondo@gmail.com

Abstract: Background/Objectives: Even today, interstitial lung disease (ILD) is diagnosed by chest high-resolution computed tomography (lung HR-CT). Large amounts of data are available about the usefulness of transthoracic lung ultrasound (LUS) in ILD. This study aimed to evaluate the transthoracic LUS capacity to discriminate different ILD patterns in systemic sclerosis (SSc) patients, such as usual interstitial pneumonia (UIP), non-specific interstitial pneumonia (NSIP) with ground glass opacification/opacity (GGO), and NSIP with GGO and reticulations, as well as the possibility of identifying progressive fibrosing ILD. **Methods**: We enrolled SSc-patients attending the outpatient Clinic of the Rheumatology Unit of Policlinico of Foggia and the Rheumatology Unit of Policlinico of Bari who satisfied these inclusion criteria: age older than 18 years; the satisfaction of ACR/EULAR 2013 classification criteria for SSc; chest HR-CT scan within three months before or three months after transthoracic LUS evaluation; and availability of recent and complete pulmonary function test. The exclusion criteria were as follows: history or recent reactivation of chronic obstructive pulmonary disease, lung cancer, lung infection, heart failure, pulmonary oedema, pulmonary arterial hypertension, acute respiratory distress syndrome and diffuse alveolar haemorrhage and thoracic surgery. All enrolled SSc-patients underwent transthoracic LUS, performed by an experienced sonographer. The ILD diagnosis and the respective patterns were assessed by chest HR-CT, which still represents the best diagnostic tool. **Results**: ILD was observed in 99 (63.5%) patients. Of these, 25% had the UIP pattern and 75% the NSIP pattern (46 with GGO, 28 with GGO and reticulations). By receiver operating characteristic (ROC) curve analysis, higher values of accuracy, sensitivity, specificity, and negative clinical utility index (CUI) were found for pleural line irregularity (0.84 (95% CI: 0.75–0.91), 96%, and 73.6%, p = 0.0001; 0.72), and pleural line thickness (0.84 (95% CI: 0.74–0.91), 72%, and 96.4%, p = 0.0001; 0.85) for detecting the UIP pattern. The best performance among transthoracic LUS signs for NSIP with the GGO pattern was observed for B-lines (accuracy: 0.88 (95% CI: 0.80–0.93), sensitivity: 93.4% and specificity: 82.4, p = 0.0001; CUI+: 0.75, CUI−: 0.77). LUS signs with higher accuracy, sensitivity, and specificity for NSIP with GGO and reticulations were pleural line irregularity (0.89 (95% CI: 0.80–0.95), 96.4%, and 82.4%, p = 0.0001) with CUI−: 0.72, and B-lines (0.89 (95% CI: 0.80–0.95), 96.4%, 82.4%, p = 0.0001), with CUI+: 0.80 and

https://doi.org/10.3390/diagnostics15040488

CUI−: 0.70. Furthermore, a total number of B-lines > 10 maximises LUS performance with 92.3% sensitivity, and an accuracy of 0.83 ($p = 0.0001$) for detecting the NSIP pattern, particularly GGO. A sample-restricted analysis (66 SSc patients) evidenced the presence of progressive fibrosing ILD in 77% of these patients. By binary regression analysis, the unique LUS sign associated with progressive fibrosing ILD was the presence of pleural line irregularity (OR: 3.6; 95% CI 1.08–11.9; $p = 0.036$). **Conclusions**: Our study demonstrated that transthoracic LUS presented a high capacity to discriminate the different patterns of SSc-ILD. Therefore, the hypothesis that transthoracic LUS is an effective screening method for the evaluation of the presence of SSc-ILD and establishing the correct timing of chest HR-CT, in order to avoid patients receiving excessive exposure to ionising radiation, is supported.

Keywords: interstitial lung disease; usual interstitial pneumonia; non-specific interstitial pneumonia; lung ultrasound (LUS); B-lines

1. Introduction

Interstitial lung disease (ILD) is one of the most frequent clinical features in systemic sclerosis (SSc), estimated in 53% of patients with SSc with a diffuse cutaneous subset and in 35% of patients with a limited cutaneous subset [1]. Together with pulmonary arterial hypertension, despite advances in therapeutic management, it currently represents the most frequent cause of death in patients with SSc (ranging between 17% and 35%) [2–6].

The American College of Rheumatology (ACR) and the American College of Chest Physicians (CHEST) Guideline have recently confirmed that the best instrumental method for diagnosing and following ILD in rheumatic disorders, including SSc, is chest high-resolution computed tomography (chest HR-CT) [7], due to its capacity to recognise SSc pulmonary disease [8,9] and to assess the ILD extension [10–12]. Given the subclinical onset and the need to diagnose SSc-ILD as early as possible [13], in order to try to slow down its progression through pharmacological therapy [14], a pulmonary function test (PFT) every 3–6 months in the first year of diagnosis, and then less frequently once stable, and lung HR-CT, when necessary on the base of reported symptoms, clinical examination, and PFT, are recommended to be performed, especially for the population considered to be at the highest risk (anti–Scl-70 positivity, antinuclear antibody with nucleolar pattern, diffuse cutaneous subtype, male sex, African American race, early disease (first 5–7 y after onset), and elevated acute phase reactants) [7].

The most frequent (70–78%) lung CT pattern of ILD in SSc, above all in early phases of the disease, is non-specific interstitial pneumonia (NSIP) characterised by the presence of ground glass opacities (GGO) and reticulations, and less commonly (10–13.8%) by usual interstitial pneumonia (UIP), identified by honeycombing and traction bronchiectasis [15,16].

In the last decade, however, the issues of high cost and the risk related to excessive exposure to ionising radiation, which every CT scan induces, as well as the increased risk of developing neoplasms (especially breast carcinomas) in patients with SSc [17], have induced the scientific community to search for new instrumental methods that are less harmful to the patient. Among these, transthoracic lung ultrasound (LUS) has assumed considerable importance, due to preliminary data supporting a positive association between LUS signs and CT findings [18–26], but no study has ever compared specific tomographic patterns of ILD with particular signs of transthoracic LUS in SSc patients.

This study aimed to evaluate the LUS capacity to discriminate different ILD patterns, such as UIP, NSIP with GGO, and NSIP with GGO and reticulations, in SSc patients.

Furthermore, we wanted to evaluate the presence of any characteristic LUS signs to identify progressive fibrosing ILD.

2. Materials and Methods

We performed a cross-sectional study, conducted from January 2023 to November 2024. The enrolled patients were followed at the outpatient clinic of the Rheumatology Unit of Policlinico of Foggia and the Rheumatology Unit of Policlinico of Bari. All the patients satisfied the ACR/EULAR 2013 classification criteria for SSc [27]. The clinical examination at enrolled visit included age, disease history, medications, visceral involvement, and laboratory test for reactants of the acute phase of inflammation (erythrocyte sedimentation rate, C-reactive protein), antinuclear antibodies (assessed by immunofluorescence) and extractable nuclear antigens antibodies (evaluated by immunoblotting), the presence of ILD and its pattern at chest HR-CT, the respiratory capacity by pulmonary function test, evaluation of the presence of main LUS signs of ILD, and the assessment of nailfold capillaroscopy patterns. The extent and severity of skin thickening were evaluated by the modified Rodnan skin score.

The inclusion criteria were as follows: age older than 18 years, the satisfaction of ACR/EULAR 2013 classification criteria for SSC [27]; chest HR-CT scan within three months before or three months after transthoracic LUS evaluation; and availability of recent and complete pulmonary function test.

The exclusion criteria were as follows: history or recent reactivation of chronic obstructive pulmonary disease, lung cancer, lung infection, heart failure, pulmonary oedema, pulmonary arterial hypertension, acute respiratory distress syndrome, and diffuse alveolar haemorrhage and thoracic surgery.

2.1. Chest High-Resolution Computed Tomography

Chest HR-TC evaluation was performed with a high-resolution spiral technique at the patients' reference centres, using a multi-detector CT scanner with 64 channels (Toshiba, Tokyo, Japan). The standard parameters for CT acquisition were as follows: tube voltage, 120 kVp; tube current, standard (reference mAs, 60–120); slice thickness, 0.5 mm; and reconstruction interval, 0.5–1.0 mm. All chest HR-CT images were captured at full inspiration, with the patient in the supine position and without contrast medium. Each exam, recorded in digital format, was acquired and subsequently examined by the same operator. The lung field was divided into three segments (basal, middle and apical), which did not correspond to the lung lobes. The inferior angle of the scapula was considered a landmark for the subdivision between basal and middle segments, while the middle and apical segments were divided to be constituted by the same number of intercostal spaces. The segmentation was used to standardize the subdivision of transthoracic LUS areas and tomographic lung segments. For each lung field, the examiner assessed the presence of characteristic lesions (covering at least 20% of the segment area) for NSIP with GGO only, NSIP with GGO and reticulations, and UIP [28]. In patients with ILD, the lung segments characterised by the presence of lesions were considered and compared with the transthoracic LUS findings.

The presence of progressive fibrosing ILD was defined, as suggested by Raghu G et al. [29], by worsening of respiratory symptoms, the evidence of radiographic progression, and/or absolute changes in forced vital capacity (FVC) or diffusing capacity of the lung for carbon monoxide (DLCO) at the enrolled visit compared to those performed by the patient in the previous year (only in patients who had such data available).

2.2. Transthoracic Lung Ultrasound

Transthoracic lung ultrasound (LUS) was assessed according to the principal recommendation [30], evaluating all scanning lung fields on the anterolateral (28 sites) and the posterior thorax (30 sites). All transthoracic LUS were performed by a blinded certified operator at the University Hospital of Foggia. The examination was conducted with the patient sitting and their hands resting on their knees for posterior scanning, while the patient adopted a supine position for anterolateral acquisitions. The transthoracic LUS was performed, by an expert sonographer, with a Philips Epiq 7 ultrasound system (Philips Ultrasound; Bothell, WA, USA), using a convex probe (3.5–5 MHz) or a linear transducer (8.0–12.0 MHz), where appropriate, to improve the identification of pleural alterations.

In each intercostal space examined, the following ultrasonographic characteristics of the pleural line were evaluated:

- Thickness of the pleural line, abnormal if >2.8 mm (Figure 1a,b) [31];
- Irregularities of the pleural line, abnormal if extended more than 3 mm (Figure 1a,b) [31,32];
- Mobility of the pleural line, in particular, the presence of sliding signs;
- The presence and total number of B-lines (abnormal in the presence of 3 or more B-lines in at least two consecutive scans or the presence of at least 5 B-lines in the entire intercostal space examined) (Figure 1c) [31,33]. The total number of B-lines was given by the sum of all the B-lines found in the different fields examined;
- The presence of at least one subpleural cyst (Figure 1d), defined as hypo-echoic lesions that interrupt the pleural line;
- The presence of the pleural effusion.

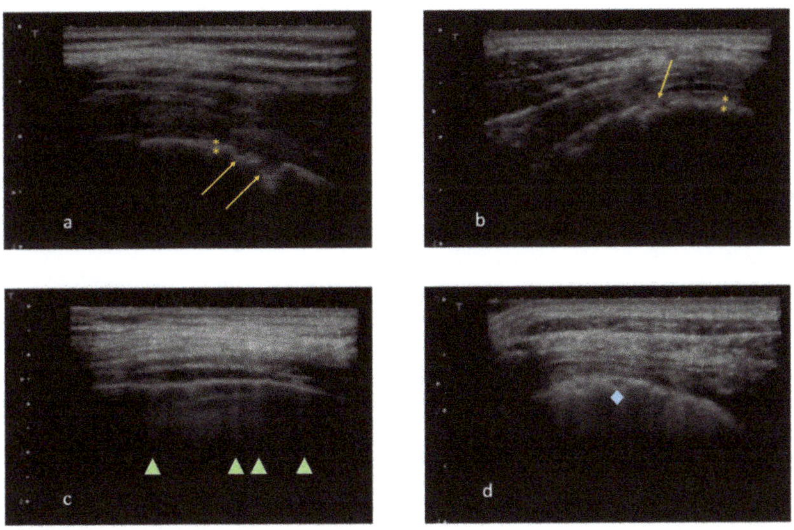

Figure 1. Echograms: signs detected by transthoracic lung ultrasound (linear probe between 7 and 10 MHz); (**a**) thickening (yellow asterisk) and irregularity (yellow arrow) of the pleural line; (**b**) thickening (yellow asterisk) and irregularity (yellow arrow) of the pleural line; (**c**) B-lines (green arrowheads); (**d**) subpleural cysts (blue rhombus).

2.3. Pulmonary Function Test

Forced expiratory volume in 1 s (FEV1) and FVC were measured with standard spirometry Sensormedics (Milano, Italia). The PFT was conducted with three inspiratory and expiratory manoeuvres and was considered reliable when the values obtained were repeatable. The values were expressed as the percentage of predicted, using the Quanjer and Stocks equations.

DLCO was assessed with the "single breath" technique and adjusted for haemoglobin and carbon monoxide (CO) levels. The results were recorded as percentages of predicted values.

2.4. Statistical Analysis

The results were expressed as means (m) ± standard deviation (sd); categorical variables were expressed as numbers (percentage). The normality of the distribution of the study variables was verified with the Kolmogorov–Smirnov test.

Statistical differences between the two groups were analyzed with the t-student test for unpaired data. Multiple comparisons of continuous variables were analyzed with analysis of variance (ANOVA) followed by the Bonferroni test.

Pearson $\chi 2$ and Fisher's exact test, followed by Z-test, were used to compare categorical variables and percentages. Pearson Correlation was used to evaluate the correlation between the number of B-lines and the Warrick score.

The ROC curve was used to compare the diagnostic performance of each transthoracic LUS sign with chest HR-CT patterns of ILD as NSIP with GGO, NSIP with GGO and reticulations, and UIP, assessing the area under the curve (AUC). The AUC values of 0.50–0.59, 0.60–0.69, 0.70–0.79, and ≥0.80 were defined as none, poor, acceptable, and excellent accuracy, respectively [34].

The optimal cutoff for the number of B-lines in the patterns NSIP with GGO, and NSIP with GGO and reticulations, corresponding to the maximum sum of sensitivity and specificity, were computed by ROC curve.

Positive and negative clinical utility index (CUI) values were calculated as the product between sensitivity and positive predictive value (PPV), and the product between specificity and negative predictive value (NPV), respectively. Grading was performed using the categories excellent utility ≥ 0.81, good utility ≥ 0.64 and fair utility ≥ 0.49 and poor utility < 0.49, as suggested by Mitchell [35].

Binary logistic regression was used to evaluate transthoracic LUS signs (independent variables) that were associated with progressive fibrosing ILD (dependent variable).

Statistical significance was defined as a value of $p \leq 0.05$.

Statistical analysis was performed with IBM SPSS Statistics 26.

2.5. Ethics Approval

Our study was approved by the local Ethics Committee (115/C.E.—21 September 2021). All participants gave their informed consent to participate in this study.

3. Results

One hundred fifty-six patients were consecutively recruited for this study. All patients had SSc (mean age 59.0 ± 12.4 years; and mean disease duration 32 ± 10.8 years) and satisfied the inclusion criteria. ILD at chest HR-CT was detected in 99 (63.5%) studied patients, of whom 25 (25%) had the UIP pattern, and 74 (75%) the NSIP pattern (46 with GGO alone, and 28 with GGO and reticulations).

A comparison of the main demographic and clinical characteristics of the study groups is shown in Table 1. A higher presence of anti-Scl-70 antibodies was found in patients with ILD (any CT patterns) compared with No-ILD patients. Lower values of FVC in pulmonary function tests were detected in the UIP (84.8 ± 18.3), and NSIP with GGO and reticulations (84 ± 19.4) groups. Regarding other specific CT lung abnormalities, a higher rate of micro-nodules was found in the NSIP with GGO group (56.5%). More extensive and severe ILD, as assessed by the Warrick score, was recognised in the UIP (10.2 ± 4) and NSIP with GGO and reticulations (10.8 ± 1.7) groups (Table 1). A positive correlation between the Warrick score and the number of B-lines was found (r^2: 0.59; p = 0.0001).

Table 1. Comparison among study groups of the main demographic, clinical and instrumental characteristics.

	No-ILD	UIP	NSIP with GGO	NSIP with GGO and Reticulations	*p*-Value
n (%)	57 (36.5)	25 (16)	46 (29.5)	28 (18)	
F/M *n* (%)	55 (96)/2 (3)	23 (92)/2 (8)	43 (93)/3 (6)	27 (96)/1 (4)	0.788
Age (years) (m ± sd)	57.5 ± 14.4	60.4 ± 11.0	58.2 ± 11.5	62.2 ± 10.1	0.370
Disease duration (years) (m ± sd)	9.7 ± 6.9	10.9 ± 5.9	7.4 ± 5.9	7.8 ± 7.7	0.479
BMI (m ± sd)	25.9 ± 4.1	25.2 ± 6.1	25.2 ± 4.3	26.1 ± 4.5	0.170
ESR (mm/h) (n.r. 0–30) (m ± sd)	19.3 ± 14.8	17.6 ± 14	17.7 ± 13.1	24.5 ± 14.4	0.242
CRP (mg/L) (n.r. 0–10) (m ± sd)	4.8 ± 4.8	5.0 ± 1.1	7.5 ± 1.3	9.6 ± 1.8	0.238
ENA					
Anti CENP-B positivity *n* (%)	42 (74)	3 (12)	6 (13)	6 (23)	0.0001 ^
Anti Scl-70 positivity *n* (%)	10 (17.5)	19 (76)	36 (78)	17 (65)	0.0001 "
Nailfold Videocapillaroscopy					
Early scleroderma pattern *n* (%)	16 (32)	4 (20)	16 (40)	4 (17)	0.440
Active scleroderma pattern *n* (%)	24 (45)	8 (40)	15 (37)	12 (52)	0.786
Late scleroderma pattern *n* (%)	13 (24)	8 (40)	9 (22.5)	7 (30)	0.443
Cutaneous subsets					
Limited *n* (%)	49 (86)	20 (80)	34 (74)	25 (89)	0.292
Diffuse *n* (%)	8 (14)	5 (20)	12 (26)	3 (11)	0.296
Presence of LUS abnormalities					
Sliding sign *n* (%)	0 (0)	5 (20)	2 (4)	5 (18.5)	0.002 §
Pleural line irregularity *n* (%)	15 (26)	24 (96)	35 (76)	27 (96)	0.0001 ^
Pleural thickness *n* (%)	2 (3.5)	18 (72)	16 (36)	18 (64)	0.0001 °
B-lines *n* (%)	10 (17.5)	14 (56)	43 (93.5)	27 (96)	0.0001 #
B-lines number (m ± sd)	3.5 ± 16.1	29.0 ± 10.2	69.5 ± 10.7	56.6 ± 11.8	0.0001 #
Subpleural cystis *n* (%)	5 (9)	14 (58)	10 (24)	13 (48)	0.0001 @
Pleural effusion *n* (%)	2 (3.5)	4 (16)	2 (4)	4 (14)	0.103
PFT					
FVC% (m ± sd)	108.1 ± 14.4	84.8 ± 18.3	99.5 ± 15.4	84 ± 19.4	0.0001 &
FEV1% (m ± sd)	100.8 ± 22.2	89.5 ± 26.1	90.9 ± 18.3	80.6 ± 19.6	0.010 $
DLCO% (m ± sd)	77.1 ± 16.9	77.2 ± 17	77.7 ± 18	69.1 ± 18.5	0.627
TLC% (m ± sd)	101.6 ± 12.9	75.2 ± 11	92.9 ± 16.2	91.8 ± 33	0.059
RV% (m ± sd)	105.8 ± 24.9	75.6 ± 13.3	94.8 ± 19.0	93.2 ± 15.7	0.171
6MWT (meter) (m ± sd)	475.9 ± 140.8	390.8 ± 132.6	440.3 ± 102.3	366.1 ± 156.7	0.109
Other lung CT abnormalities					

Table 1. *Cont.*

	No-ILD	UIP	NSIP with GGO	NSIP with GGO and Reticulations	*p*-Value
Micro nodules *n* (%)	6 (13)	8 (32)	26 (56.5)	9 (32)	0.0001 *
Warrick Score *n* (%)	0.0 ± 0.0	10.2 ± 4.0	8 ± 3.4	10.8 ± 1.7	0.0001 $

6MWT: 6 min walking test; BMI: body mass index; CRP: C-reactive proteins; CT: computed tomography; DLCO: diffusing capacity of the lungs for carbon monoxide; ENA: extractable nuclear antigen; ESR: erythrocyte sedimentation rate; F: female; FEV1: forced expiratory volume in the first second; FVC: forced vital capacity; GGO: ground glass; ILD: interstitial lung disease; LUS: lung ultrasound; M: male; n.r.: normal range; NSIP: non-specific interstitial pneumonia; PFT: pulmonary function tests; RV: residual volume; TLC: total lung capacity; UIP: usual interstitial pneumonia. * NSIP with GGO vs. UIP $p \leq 0.05$; NSIP with GGO vs. No-ILD $p \leq 0.05$; NSIP with GGO and reticulations vs. No-ILD $p \leq 0.05$; UIP vs. No-ILD $p \leq 0.05$. § UIP vs. No-ILD $p \leq 0.05$; NSIP with GGO and reticulations vs. No-ILD $p \leq 0.05$. ^ NSIP with GGO vs. No-ILD $p \leq 0.05$; UIP vs. No-ILD $p \leq 0.05$; NSIP with GGO and reticulations vs. No-ILD $p \leq 0.05$. " No-ILD vs. NSIP with GGO $p \leq 0.05$; No-ILD vs. UIP $p \leq 0.05$; No-ILD vs. NSIP with GGO and reticulations $p \leq 0.05$. ° UIP vs. NSIP with GGO $p \leq 0.05$; NSIP with GGO vs. No-ILD $p \leq 0.05$; NSIP with GGO and reticulations vs. No-ILD $p \leq 0.05$; UIP vs. No-ILD $p \leq 0.05$. # UIP vs. No-ILD $p \leq 0.05$; NSIP with GGO vs. No-ILD $p \leq 0.05$; NSIP with GGO and reticulations vs. No-ILD $p \leq 0.05$; NSIP with GGO vs. UIP $p \leq 0.05$; NSIP with GGO and reticulations vs. UIP $p \leq 0.05$. @ UIP vs. No-ILD $p \leq 0.05$; UIP vs. NSIP with GGO $p \leq 0.05$; NSIP with GGO and reticulations vs. No-ILD $p \leq 0.05$. & UIP vs. No-ILD $p \leq 0.05$; NSIP with GGO vs. No-ILD $p \leq 0.05$; NSIP with GGO and reticulations vs. No-ILD $p \leq 0.05$; UIP vs. NSIP with GGO $p \leq 0.05$; NSIP with GGO and reticulations vs. GGO $p \leq 0.05$. $ UIP vs. No-ILD $p \leq 0.05$; NSIP with GGO vs. No-ILD $p \leq 0.05$; NSIP with GGO and reticulations vs. No-ILD $p \leq 0.05$; NSIP with GGO vs. NSIP with GGO and reticulations $p \leq 0.05$.

3.1. LUS Signs and UIP Chest CT Pattern

Regarding UIP at chest CT, the rate of presence of pleural line irregularity (96% vs. 26%), sliding sign (20% vs. 0%), pleural line thickness (72% vs. 3.5%), B-lines (56% vs. 17.5%), and subpleural cysts (58% vs. 9%) were higher compared to the No-ILD group of patients (Table 1). Furthermore, the frequency of the presence of pleural thickness was higher in the UIP pattern than in NSIP with GGO (72% vs. 36%) (Table 1).

In the ROC curve analysis of the UIP pattern, better performance was found in terms of sensitivity and sensibility, with excellent accuracy and CUI positive and negative values, for pleural line irregularity (with a higher negative predictive value), and pleural line thickness (Table 2, Figure 2a). Poor CUI positive values were found for B-lines and pleural effusion. The subpleural cyst assessment had good accuracy, specificity and CUI negative values, but a low value of sensitivity (Table 2).

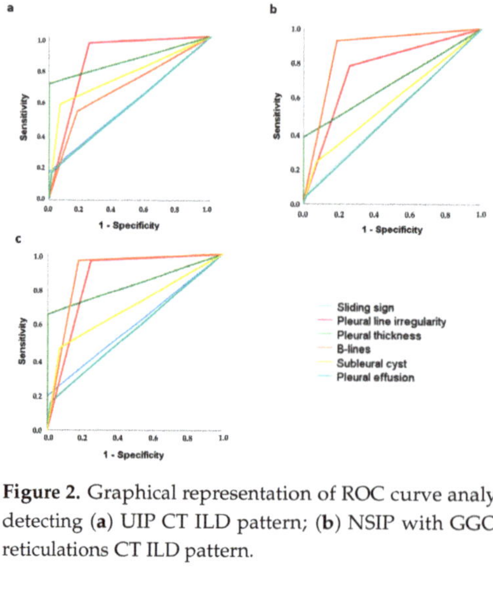

Figure 2. Graphical representation of ROC curve analysis of different transthoracic LUS signs for detecting (**a**) UIP CT ILD pattern; (**b**) NSIP with GGO CT ILD pattern; (**c**) NSIP with GGO and reticulations CT ILD pattern.

Table 2. ROC curve analysis. Sensitivity, specificity, accuracy, positive predictive value, negative predictive value, and clinical utility index values (positive and negative) of different lung ultrasound signs for detecting specific ILD CT lung patterns (UIP, NSIP with GGO, NSIP with GGO and reticulations).

	Sensibility % (95% CI)	Specificity % (95% CI)	PPV %	NPV %	CUI +	CUI −	Accuracy (95% CI)	*p*-Value
UIP								
Sliding sign	20 (6.8–100)	100 (93–100)	100	74	0.2	0.74	0.60 (0.48–0.70)	0.01
Pleural line irregularity	96 (79.6–99.6)	73.6 (60.3–84.5)	61.5	97.7	0.591	0.72	0.848 (0.75–0.91)	<0.0001
Pleural thickness	72 (50.6–87.9)	96.4 (87.9–91.6)	90	88.7	0.648	0.856	0.842 (0.74–0.91)	<0.0001
B-lines	56 (34.9–75.6)	82.4 (70.1–91.3)	58.3	81	0.327	0.668	0.692 (0.58–0.79)	0.007
Subpleural cystis	58.3 (36.6–77.9)	91.2 (80.7–97.1)	73.7	82.5	0.413	0.753	0.748 (0.63–0.83)	<0.0001
Pleural effusion	16 ((4.5–36.1)	96.4 (87.9–91.6)	66.7	72.4	0.107	0.698	0.562 (0.448–0.67)	0.110
NSIP with GGO								
Sliding sign	4.35 (0.5–14.8)	100 (93.5–100)	100	56.4	0.043	0.564	0.552 (0.42–0.62)	0.152
Pleural line irregularity	76.9 (61.2–87.4)	73.6 (60.3–84.5)	71.2	82.4	0.572	0.607	0.749 (0.654–0.829)	<0.0001
Pleural thickness	35.5 (21.9–51.2)	96.4 (87.9–99.6)	88.9	64.7	0.309	0.624	0.660 (0.56–0.75)	<0.0001
B-lines	93.4 (82.1–96.8)	82.4 (70.1–91.3)	81.8	94	0.758	0.775	0.88 (0.80–0.93)	<0.0001
Subpleural cystis	23.1 (12.1–39.5)	91.2 (80.7–97.1)	66.6	59.1	0.145	0.539	0.575 (0.47–0.67)	0.0494
Pleural effusion	4.44 (0.0–7.9)	100 (93.7–100)	3.9	4.3	0.004	0.002	0.505 (0.40–0.60)	0.813
NSIP with GGO and reticulations								
Sliding sign	19.5 (6.3–38.1)	100 (93.5–100)	100	71.3	0.179	0.713	0.593 (0.47–0.7)	0.015
Pleural line irregularity	96.4 (81.7–99.9)	73.6 (60.3–84.5)	64.3	97.7	0.620	0.720	0.851 (0.75–0.91)	<0.0001
Pleural thickness	64.2 (44.1–81.4)	96.4 (87.9–99.6)	90	84.6	0.579	0.816	0.804 (0.70–0.88)	<0.0001
B-lines	96.4 (81.7–99.9)	82.4 (70.1–91.3)	73	97.9	0.704	0.807	0.894 (0.80–0.95)	<0.0001
Subpleural cystis	48.1 (28.7–68.1)	91.2 (80.7–97.1)	72.2	77.6	0.335	0.708	0.697 (0.58–0.79)	0.0002
Pleural effusion	14.2 (4–32.7)	96.4 (87.9–99.6)	66.7	69.6	0.095	0.672	0.554 (0.44–0.66)	0.132

CT: computed tomography; CUI: clinical utility index; GGO: ground glass; ILD: interstitial lung disease; transthoracic LUS: lung ultrasound; NSIP: non-specific interstitial pneumonia; NPV: negative predictive value; PPV: positive predictive value; UIP: usual interstitial pneumonia.

3.2. LUS Signs and NSIP with Ground Glass Lung CT Pattern

Concerning the NSIP with GGO pattern, the transthoracic LUS signs observed with higher frequency were pleural line irregularity (76% vs. 26%) (but significantly lesser than for the UIP pattern), B-lines (93.5% vs. 17.5%), compared with the No-ILD patients. In particular, there was a significant difference in the rate of the presence of B-lines (93.5% vs. 56%) and the total number of B-lines (69.5 ± 10.7 vs. 29 ± 10.2; $p = 0.0001$) between the NSIP with GGO pattern and the UIP pattern (Table 1).

By ROC curve analysis, the best performance among transthoracic LUS signs for the NSIP with GGO pattern was observed just for B-lines, with excellent accuracy, good CUI positive and negative values, high sensitivity, and a high negative predictive value (Table 2, Figure 2b). Regarding the cut-off of total B-line numbers useful for discriminating the presence of NSIP with GGO pattern, a total number of B-lines > 10 maximises the transthoracic LUS performance, with 92.8% sensitivity, 96.4% specificity, and an accuracy of 0.945 (p = 0.0001). Low values of sensitivity, negative predictive value, and acceptable and poor accuracy were found for pleural thickness and pleural line irregularities, respectively (Table 2).

3.2.1. Transthoracic LUS Signs and NSIP with Ground Glass and Reticulations Chest CT Pattern

The NSIP with GGO and reticulations pattern was detected in 28 patients. The principal transthoracic LUS signs assessed with a higher rate in patients with this pattern compared to No-ILD SSc patients were sliding sign (18.5% vs. 0%), pleural line irregularity (96% vs. 26%), pleural line thickness (64% vs. 3.5), B-lines (96% vs. 17.5%), and subpleural cysts (48% vs. 9%). In particular, the presence of B-lines and the number of B-lines were found to be statistically significant between the NSIP with GGO and reticulations group and UIP group, with values of 96% vs. 56%, and 56.6 ± 11.8 vs. 29 ± 10.2, respectively (Table 1).

By ROC curve analysis, the transthoracic LUS signs with higher accuracy, sensitivity, specificity and CUI for NSIP with GGO and reticulations were firstly B-lines, followed by pleural thickness and pleural irregularity, which presented excellent and good CUI negative values, respectively. Concerning the cut-off number of B-lines, the presence of >10 B-lines maintains good sensitivity (91.3%) for the NSIP pattern with GGO and reticulations, but loses some points in terms of specificity (96.4%) and accuracy (0.934; p = 0.0001) compared to the NSIP pattern with GGO.

The true positive, false positive, true negative and false negative of transthoracic LUS signs for each tomographic ILD pattern are reported in Supplementary Table S1.

3.2.2. Transthoracic LUS Signs and NSIP with Ground Glass and Reticulations Lung CT Pattern

In 66 patients, it was possible to perform a comparative evaluation to establish the presence of progressive fibrosing ILD, and 77% of these patients presented the criteria for progressive fibrosing ILD. A higher rate of pleural line irregularities was found in the group of patients with progressive fibrosing ILD compared to those with ILD (71% vs. 40%, p = 0.033). No other significant differences in transthoracic LUS signs were found between the group of patients with ILD and those with progressive fibrosing ILD. A comparison of the principal clinical and instrumental characteristics between the patients with progressive fibrosing ILD and those with non-progressive ILD are shown in Table 3.

Table 3. Comparison between progressive ILD and non-progressive ILD of the main demographic and instrumental characteristics.

	Progressive ILD	Non-Progressive ILD	p-Value
n	51	15	
F/M n (%)	47 (92)/4 (8)	15 (100)/0 (0)	0.347
Age (years) (m ± sd)	60.1 ± 12.4	59.9 ± 10.2	0.936
Disease duration (years) (m ± sd)	13.3 ± 5.8	9.2 ± 4.1	0.013
Presence of LUS abnormalities			
Sliding sign n (%)	4 (8)	0 (0)	0.279

Table 3. *Cont.*

	Progressive ILD	Non-Progressive ILD	*p*-Value
Pleural line irregularity *n* (%)	36 (71)	6 (40)	0.033
Pleural thickness *n* (%)	19 (37)	5 (33)	0.781
B-lines *n* (%)	37 (72)	8 (53)	0.160
B-lines number (m ± sd)	75.3 ± 31.6	48.7 ± 33.7	0.207
Subpleural cystis *n* (%)	11 (21)	4 (26)	0.679
Pleural effusion *n* (%)	2 (4)	0 (0)	0.436
PFT			
FVC% (m ± sd)	87.2 ± 17.6	101.8 ± 19.7	0.008
FEV1 (m ± sd)	86.3 ± 16	91.6 ± 18.7	0.051
DLCO% (m ± sd)	61.5 ± 17.8	76.7 ± 16	0.006
TLC% (m ± sd)	80.4 ± 16.6	101.1 ± 23.7	0.004
RV% (m ± sd)	80.6 ± 22.3	116.6 ± 48.2	0.002
6MWT (meter) (m ± sd)	446.0 ± 82.3	590.8 ± 96.3	0.086
Chest HR-TC			
UIP	6 (12)	1 (7)	0.853
NSIP with GGO	20 (39)	5 (33)	0.849
NSIP with GGO and reticulation	7 (14)	2 (13)	0.592

6MWT: 6 min walking test; CT: computed tomography; DLCO: diffusing capacity of the lungs for carbon monoxide; F: female; FEV1: forced expiratory volume in the first second; FVC: forced vital capacity; GGO: ground glass; ILD: interstitial lung disease; LUS: lung ultrasound; M: male; NSIP: non-specific interstitial pneumonia; PFT: pulmonary function tests; RV: residual volume; TLC: total lung capacity; UIP: usual interstitial pneumonia.

By binary regression analysis, the unique transthoracic LUS sign associated significantly with progressive fibrosing ILD was the presence of pleural line irregularity (OR: 3.6; 95% CI 1.08–11.9; $p = 0.036$). No significance was observed regarding the presence or absence of B-lines and their number.

4. Discussion

This study evidenced principally that the presence of B-lines at transthoracic LUS characterised the chest HR-CT pattern of NSIP with GGO with excellent accuracy (0.88), sensitivity (93.4%) and negative predictive value (94%), as well as a good CUI negative value (0.775) and good CUI positive value (0.758), with a sensitivity of 82.4%. Furthermore, the best indicators of the UIP tomographic pattern were the presence of pleural line thickness (excellent accuracy (0.842) and CUI negative value (0.856), and good CUI positive value (0.648)), and the presence of pleural line irregularities (good accuracy (0.848) and CUI negative value (0.72)). NSIP with GGO and reticulations was characterised in transthoracic LUS principally by B-lines (with good (>0.64) CUI positive and negative values, excellent accuracy (0.894), sensitivity (96.4%), and a high (97.9%) negative predictive value), and by pleural line irregularities (good CUI negative value (0.816), excellent accuracy (0.851), and a high (96.4%) sensitivity). Pleural line thickness presented, for NSIP with the GGO and reticulations pattern, high specificity (96.4%), accuracy (0.804), a positive predictive value (90%) and an excellent (0.816) CUI negative value.

A large number of studies are available on the use of transthoracic LUS in ILD, although a consensus has never been reached on its real clinical value, the coding and scoring system [14]. In SSc, due to the characteristic and orderly progressive involvement of the lung parenchyma, starting from the subpleural basal–posterior lung interstitium, which is easily viewable by LUS [36], the ultrasound examination, in particular the detection of pleural thickness, has been previously proposed as a timely guide for the execution of chest HR-CT [33,37–39]. Our study adds to the other published works, introducing considerations regarding how LUS can differentiate specific lung HR-CT patterns such as NSIP and UIP. Indeed, the present study highlights that transthoracic LUS is a highly sensitive and specific technique in evaluating the presence of typical SSc-ILD tomographic patterns and is also accurate in discriminating its different and characteristic phases. The finding of B-lines on LUS was found to be significantly associated with the presence of active inflammatory phases of SSc-ILD, characterised by GGO, as observed in the NSIP pattern [40–43]. Irregularities of the pleural line and the thickness of the pleural line are, instead, significantly associated with the presence of honeycombing and traction bronchiectasis, as in the UIP pattern, and therefore, of the fibrosis phase. Therefore, principally, the presence or the absence of B-lines significantly differentiates the NSIP pattern from the UIP pattern.

The presence of transthoracic LUS signs, such as the presence of irregularities of the pleural line, thickness of the pleural line, the presence of subpleural cysts, reduced mobility of the pleural line and the presence of B-lines, in patients with pulmonary fibrosis was first described in 1997 [44]. Other studies have evaluated the presence of these same transthoracic LUS signs in ILD secondary to sarcoidosis, rheumatoid arthritis, SSc, mixed connective tissue disease, Sjogren's syndrome and primary biliary cirrhosis, demonstrating their presence also in these secondary forms of ILD [31], and highlighting, in particular, that the presence of B-lines distributed throughout the lung area is associated with more severe pulmonary fibrosis [45]. Concerning the transthoracic LUS evaluation of ILD in connective tissue diseases, in previous studies, it has been observed that a greater number of B-lines could identify SSc-ILD [31,45,46]; and, in particular, that the number of B-lines correlates with the tomographic Warrick score, an index of severity of SSc-ILD [19]. Furthermore, the high sensitivity of B-lines in evaluating the presence of SSc-ILD has been described both in patients with a diagnosis of very early SSc [47] and those with a diagnosis of SSc [33].

The high frequency and sensitivity found in the present study between the presence of B-lines and the tomographic pattern of NSIP with GGO, characteristic of active inflammatory processes [48], could be attributed to the presence of inflammatory infiltrate that would result in variation in the acoustic impedance of the lung parenchyma, creating the air–water interface necessary for the generation of B-lines, as demonstrated in previous studies [49–51].

Concerning the cut-off of B-lines, the limit of >10 B-lines found to be effective for recognising ILD confirms what was previously found for ILD-SSc [52,53]. However, in our study, we found a high sensitivity for the NSIP pattern, especially for the GGO-only pattern, but also for the GGO and reticulations pattern. These findings have not been specifically evaluated in previous studies. However, few data support the presence of B-lines in both GGO and honeycombing patterns [14,37]. Still, no study has ever investigated the possible association between specific chest HR-CT patterns, such as UIP and NSIP, and transthoracic LUS signs. These results could indicate a greater sensitivity for the more inflammatory phases, which would result from the presence of a greater inflammatory infiltrate with a subsequent increase in acoustic impedance.

In a previous study, a low specificity (55%) of B-lines was highlighted for SSc-ILD, despite a high sensitivity (100%); this finding was attributed to the probable presence of pulmonary oedema [31]. Our study improved the previous data with higher sensitivity (93.4% and 96.4%), and higher specificity (82.4% and 82.4%) for both NSIP with GGO and NSIP with GGO and reticulations, respectively, with good positive and negative CUI values.

Regarding the tomographic pattern of UIP, a high sensitivity of transthoracic LUS signs was observed, such as the thickness of the pleural line, the presence of irregularities of the pleural line and the presence of subpleural cysts. This finding could be due to the presence of structural subversion of the lung parenchyma (complete loss of the typical acinar structure of the lung) observed in cases of stabilized pulmonary fibrosis, characterised tomographically by the presence of honeycombing [54].

No data are available on the use of transthoracic LUS in progressive fibrosing ILD; for the first time in this study, we observed a strict association between this ILD pattern and the presence of pleural line irregularities. Probably, the subversion of the parenchymal structure could explain the irregularities of the pleural line visible with transthoracic LUS. Observational studies with larger patient samples would serve to confirm this finding.

In the group of patients enrolled in this study, the percentage of SSc-ILD (63%) is in line with values reported in other studies [1,55]; and the ENA specificity found most frequently in these patients is anti-Scl-70, as observed previously [56].

Therefore, transthoracic LUS, due to its versatility, ease of handling and low maintenance costs, could assume considerable clinical importance for evaluating SSc-ILD. Furthermore, the ease of interpretation of transthoracic LUS images and the rapid execution times (minimum duration < 10 min [33] and maximum 23 min [53]) make transthoracic LUS an efficient instrumental method from a clinical point of view.

The accuracy and completeness of the information provided by the tomographic images, currently not obtainable by other methods, still make the chest HR-CT the better instrumental technique for diagnosing SSc-ILD. Although transthoracic LUS will probably not replace the chest HR-CT for the diagnosis of SSc-ILD, at least in the short term, it could be considered a guide to establish the most appropriate timing to perform the chest HR-CT, minimising the risk of exposure to ionising radiation [33]. In recent decades, this latter risk and the need to contain the costs related to the diagnosis and clinical management of patients has led the scientific community to identify new imaging techniques and new predictors of disease [18–26,57–62]. A transthoracic LUS score with a weighted score for each transthoracic LUS sign would be desirable to improve the diagnostic ability of transthoracic LUS for different HR-CT patterns of ILD. The clinical utility of recognizing the different patterns of ILD-SSc early, and without excessive exposure to ionising radiation, could have important implications in the therapeutic management of patients with SSc. With the advent of artificial intelligence software, it will probably be possible to design multifactorial scores that will meet the needs of clinicians.

Furthermore, considering the good sensitivity values, negative predictive value and negative CUI of the B-lines (for both NSIP patterns) and of the pleural line irregularities (for the UIP and NSIP patterns with GGO and cross-linking), our results strengthen the data on LUS as a promising screening technique for ILD in SSc and as a potential diagnostic tool, significantly impacting the clinical management of SSc patients [63].

There are some limitations to our study. Firstly, the distance between two consecutive B-lines was not reported. This parameter has been shown to correlate with the Warrick score in patients diagnosed with pulmonary fibrosis [64], but its evaluation was not considered useful for the purposes of our study. It would have been, furthermore, interesting to compare the ultrasound data of patients with SSc-ILD with the data of healthy subjects and other groups of patients with pathologies complicated with ILD. However, in other studies conducted on this topic, the same transthoracic LUS signs highlighted in this study were also found for other forms of ILD [31,42,65]. We did not use the scoring system of transthoracic LUS, as it has not yet been validated. Regarding progressive fibrosing ILD, it would have been useful to evaluate the transthoracic LUS in an observational study at two time points in conjunction with the lung HR-CT. Instead, we only have one transthoracic LUS detection in conjunction with the last lung HR-CT performed.

Considering the generalizability of our results, it is important to underline that transthoracic LUS could represent a screening technique, and probably in the future a valid diagnostic tool, for ILD characterised by lesions located in subpleural areas. However, it would be less reliable in ILD characterised by lesions localised along the peribronchiolar, peribronchio–vascular and perilymphatic regions, as in sarcoidosis, bronchiolitis-associated interstitial lung disease, and lymphoid interstitial pneumonia.

5. Conclusions

The present study, comparing two instrumental methods (transthoracic LUS and lung HR-CT) demonstrates that there is a good association between these two methods in evaluating the presence of specific patterns of ILD. In particular, transthoracic LUS has shown high sensitivity in distinguishing different tomographic ILD patterns, such as UIP with pleural line irregularities, pleural line thickness and the presence of subpleural cells; NSIP with GGO, characterised by B-lines; and NSIP with GGO and reticulations, characterised by pleural thickness, pleural irregularities and B-lines. Therefore, the hypothesis that transthoracic LUS is an effective screening method for the evaluation of the presence of SSc-ILD and can be used to establish, therefore, the correct timing of a chest HR-CT, in order to avoid patients from excessive exposure to ionising radiation, is supported. Our results also could open new avenues to the possibility of using LUS as a future recognised diagnostic technique in ILD-SSc. Observational studies should be performed to support our results and confirm LUS findings to further characterise the pattern of progressive fibrosing ILD.

Supplementary Materials: The following supporting information can be downloaded at https://www.mdpi.com/article/10.3390/diagnostics15040488/s1, Table S1. True positive, true negative, false positive, false negative.

Author Contributions: C.R. and A.C. designed the study; wrote, and critically revised the final draft of the manuscript; and approved the submitted version. C.R. performed the statistical analysis and interpreted the data. G.B., R.B., V.R., F.C., M.F., F.I., C.M.I.Q., D.L., A.T. and F.P.C. collected the data and revised the final draft of the manuscript. All authors have read and agreed to the published version of the manuscript.

Funding: This research received no external funding.

Institutional Review Board Statement: This study was approved by the local ethics committee (Ethics Review Board of Policlinico of Foggia, protocol number 115/C.E., 21 September 2021).

Informed Consent Statement: Written informed consent has been obtained from the patient(s) to publish this paper.

Data Availability Statement: The data presented in this study are available on request from the corresponding author.

Acknowledgments: Special thanks to C. Bonali for his contribution.

Conflicts of Interest: The authors declare no conflicts of interest.

Abbreviations

The following abbreviations are used in this manuscript:

BMI	body mass index
CRP	C-reactive proteins
ESR	erythrocyte sedimentation rate
GGO	ground glass opacities
LUS	lung ultrasound
ILD	interstitial lung disease
Lung HR-CT	chest high-resolution computed tomography

NSIP	non-specific interstitial pneumonia
PFT	pulmonary function tests
ROC	receiver operating characteristic
SSc	systemic sclerosis
UIP	usual interstitial pneumonia

References

1. Walker, U.A.; Tyndall, A.; Czirják, L.; Denton, C.P.; Farge-Bancel, D.; Kowal-Bielecka, O.; Müller-Ladner, U.; Bocelli-Tyndall, C.; Matucci-Cerinic, M. Clinical risk assessment of organ manifestations in systemic sclerosis: A report from the EULAR Scleroderma Trials and Research group database. *Ann. Rheum. Dis.* **2007**, *66*, 754–763. [CrossRef]
2. Steen, V.D.; Medsger, T.A. Changes in causes of death in systemic sclerosis, 1972–2002. *Ann. Rheum. Dis.* **2007**, *66*, 940–944. [CrossRef] [PubMed]
3. Distler, O.; Assassi, S.; Cottin, V.; Cutolo, M.; Danoff, S.K.; Denton, C.P.; Distler, J.H.; Hoffmann-Vold, A.-M.; Johnson, S.R.; Ladner, U.M.; et al. Predictors of progression in systemic sclerosis patients with interstitial lung disease. *Eur. Respir. J.* **2020**, *55*, 1902026. [CrossRef] [PubMed]
4. Elhai, M.; Meune, C.; Avouac, J.; Kahan, A.; Allanore, Y. Trends in mortality in patients with systemic sclerosis over 40 years: A systematic review and meta-analysis of cohort studies. *Rheumatology* **2012**, *51*, 1017–1026. [CrossRef]
5. Tyndall, A.J.; Bannert, B.; Vonk, M.; Airò, P.; Cozzi, F.; Carreira, P.E.; Bancel, D.F.; Allanore, Y.; Müller-Ladner, U.; Distler, O.; et al. Causes and risk factors for death in systemic sclerosis: A study from the EULAR Scleroderma Trials and Research (EUSTAR) database. *Ann. Rheum. Dis.* **2010**, *69*, 1809–1815. [CrossRef] [PubMed]
6. Volkmann, E.R.; Tashkin, D.P.; Sim, M.; Li, N.; Goldmuntz, E.; Keyes-Elstein, L.; Pinckney, A.; Furst, D.E.; Clements, P.J.; Khanna, D.; et al. Short-term progression of interstitial lung disease in systemic sclerosis predicts long-term survival in two independent clinical trial cohorts. *Ann. Rheum. Dis.* **2018**, *78*, 122–130. [CrossRef] [PubMed]
7. Johnson, S.R.; Bernstein, E.J.; Bolster, M.B.; Chung, J.H.; Danoff, S.K.; George, M.D.; Khanna, D.; Guyatt, G.; Mirza, R.D.; Aggarwal, R.; et al. 2023 American College of Rheumatology (ACR)/American College of Chest Physicians (CHEST) Guideline for the Screening and Monitoring of Interstitial Lung Disease in People with Systemic Autoimmune Rheumatic Diseases. *Arthritis Rheumatol.* **2024**, *76*, 1201–1213. [CrossRef]
8. Diot, E.; Boissinot, E.; Asquier, E.; Guilmot, J.-L.; Lemarié, E.; Valat, C.; Diot, P. Relationship between abnormalities on high-resolution CT and pulmonary function in systemic sclerosis. *Chest* **1998**, *114*, 1623–1629. [CrossRef] [PubMed]
9. Orlandi, M.; Landini, N.; Sambataro, G.; Nardi, C.; Tofani, L.; Bruni, C.; Randone, S.B.; Blagojevic, J.; Melchiorre, D.; Hughes, M.; et al. The role of chest CT in deciphering interstitial lung involvement: Systemic sclerosis versus COVID-19. *Rheumatology* **2021**, *28*, 615.
10. Wells, A.U.; Hansell, D.M.; Rubens, M.B.; King, A.D.; Cramer, D.; Black, C.M.; Du Bois, R.M. Fibrosing alveolitis in systemic sclerosis. Indices of lung function in relation to extent of disease on computed tomography. *Arthritis Rheum.* **1997**, *40*, 1229–1236. [PubMed]
11. Clukers, J.; Lukers, J.; Lanclus, M.; Belmans, D.; van Holsbeke, C.; de Backer, W.; Vummidi, D.; Cronin, P.; Lavon, B.R.; de Backer, J.; et al. Interstitial lung disease in systemic sclerosis quantification of disease classification and progression with high-resolution computed tomography: An observational study. *J. Scleroderma Relat. Disord.* **2021**, *6*, 154–164. [CrossRef]
12. Desai, S.R.; Veeraraghavan, S.; Hansell, D.M.; Nikolakopolou, A.; Goh, N.S.; Nicholson, A.G.; Colby, T.V.; Denton, C.P.; Black, C.M.; Du Bois, R.M.; et al. CT features of lung disease in patients with systemic sclerosis: Comparison with idiopathic pulmonary fibrosis and non-specific interstitial pneumonia. *Radiology* **2004**, *232*, 560–567. [CrossRef] [PubMed]
13. Launay, D.; Remy-Jardin, M.; Michon-Pasturel, U.; Mastora, I.; Hachulla, E.; Lambert, M.; Delannoy, V.; Queyrel, V.; Duhamel, A.; Matran, R.; et al. High resolution computed tomography in fibrosing alveolitis associated with systemic sclerosis. *J. Rheumatol.* **2006**, *33*, 1789–1801. [PubMed]
14. Ruaro, B.; Baratella, E.; Confalonieri, P.; Wade, B.; Marrocchio, C.; Geri, P.; Busca, A.; Pozzan, R.; Andrisano, A.G.; Cova, M.A.; et al. High-Resolution Computed Tomography: Lights and Shadows in Improving Care for SSc-ILD Patients. *Diagnostics* **2021**, *11*, 1960. [CrossRef] [PubMed]
15. Bouros, D.; Wells, A.U.; Nicholson, A.G.; Colby, T.V.; Polychronopoulos, V.; Pantelidis, P.; Haslam, P.L.; Vassilakis, D.A.; Black, C.M.; du Bois, R.M. Histopathologic subsets of fibrosing alveolitis in patients with systemic sclerosis and their relationship to outcome. *Am. J. Respir. Crit. Care Med.* **2002**, *165*, 1581–1586. [CrossRef]
16. Damiani, A.; Orlandi, M.; Bruni, C.; Bandini, G.; Lepri, G.; Scaletti, C.; Ravaglia, C.; Frassanito, F.; Guiducci, S.; Moggi-Pignone, A.; et al. The role of lung biopsy for diagnosis and prognosis of interstitial lung disease in systemic sclerosis: A systematic literature review. *Respir. Res.* **2024**, *25*, 138. [CrossRef] [PubMed]

17. Lu, T.Y.T.; Hill, C.L.; Pontifex, E.K.; Roberts-Thomson, P.J. Breast cancer and systemic sclerosis: A clinical description of 21 patients in a population-based cohort study. *Rheumatol. Int.* **2008**, *28*, 895–899. [CrossRef]
18. Demizio, D.J.; Bernstein, E. Detection and classification of systemic sclerosis-related interstitial lung disease: A review. *Curr. Opin. Rheumatol.* **2019**, *31*, 553–560. [CrossRef] [PubMed]
19. Delle Sedie, A.; Doveri, M.; Frassi, F.; Gargani, L.; D'Errico, G.; Pepe, P.; Bazzichi, L.; Riente, L.; Caramella, D.; Bombardieri, S. Ultrasound Lung Comets in Systemic Sclerosis: A Useful Tool to Detect Lung Interstitial Fibrosis. *Clin. Exp. Rheumatol.* **2010**, *28*, S54. [PubMed]
20. Wang, Y.; Gargani, L.; Barskova, T.; Furst, D.E.; Cerinic, M.M. Usefulness of lung ultrasound B-lines in connective tissue disease-associated interstitial lung disease: A literature review. *Arthritis Res. Ther.* **2017**, *19*, 206. [CrossRef]
21. Gargani, L.; Doveri, M.; D'Errico, L.; Frassi, F.; Bazzichi, M.L.; Delle Sedie, A.; Scali, M.C.; Monti, S.; Mondillo, S.; Bombardieri, S.; et al. Ultrasound Lung Comets in Systemic Sclerosis: A Chest Sonography Hallmark of Pulmonary Interstitial Fibrosis. *Rheumatology* **2009**, *48*, 1382–1387. [CrossRef] [PubMed]
22. Vicente-Rabaneda, E.F.; Bong, D.A.; Castañeda, S.; Möller, I. Use of ultrasound to diagnose and monitor interstitial lung disease in rheumatic diseases. *Clin. Rheumatol.* **2021**, *40*, 3547–3564. [CrossRef] [PubMed]
23. Fairchild, R.; Chung, M.; Yang, D.; Sharpless, L.; Li, S.; Chung, L. Development and Assessment of Novel Lung Ultrasound Interpretation Criteria for the Detection of Interstitial Lung Disease in Systemic Sclerosis. *Arthritis Care Res.* **2021**, *73*, 1338–1342. [CrossRef] [PubMed]
24. Hoffmann-Vold, A.M.; Molberg, Ø. Detection, screening, and classification of interstitial lung disease in patients with systemic sclerosis. *Curr. Opin. Rheumatol.* **2020**, *32*, 497–504. [CrossRef] [PubMed]
25. Trombetta, A.C.; Smith, V.; Gotelli, E.; Ghio, M.; Paolino, S.; Pizzorni, C.; Vanhaecke, A.; Ruaro, B.; Sulli, A.; Cutolo, M. Vitamin D deficiency and clinical correlations in systemic sclerosis patients: A retrospective analysis for possible future developments. *PLoS ONE* **2017**, *12*, e0179062. [CrossRef] [PubMed]
26. Volpicelli, G. Lung Ultrasound B-Lines in Interstitial Lung Disease Moving from Diagnosis to Prognostic Stratification. *Chest* **2020**, *158*, 1323–1324. [CrossRef]
27. van den Hoogen, F.; Khanna, D.; Fransen, J.; Johnson, S.R.; Baron, M.; Tyndall, A.; Matucci-Cerinic, M.; Naden, R.P.; Medsger, T.A., Jr.; Carreira, P.E.; et al. 2013 classification criteria for systemic sclerosis: An American college of rheumatology/European league against rheumatism collaborative initiative. *Ann. Rheum. Dis.* **2013**, *72*, 1747–1755. [CrossRef]
28. Goh, N.S.; Desai, S.R.; Veeraraghavan, S.; Hansell, D.M.; Copley, S.J.; Maher, T.M.; Corte, T.J.; Sander, C.R.; Ratoff, J.; Devaraj, A.; et al. Interstitial lung disease in systemic sclerosis: A simple staging system. *Am. J. Respir. Crit. Care Med.* **2008**, *177*, 1248–1254. [CrossRef]
29. Raghu, G.; Remy-Jardin, M.; Richeldi, L.; Thomson, C.C.; Inoue, Y.; Johkoh, T.; Kreuter, M.; Lynch, D.A.; Maher, T.M.; Martinez, F.J.; et al. Idiopathic Pulmonary Fibrosis (an Update) and Progressive Pulmonary Fibrosis in Adults: An Official ATS/ERS/JRS/ALAT Clinical Practice Guideline. *Am. J. Respir. Crit. Care Med.* **2022**, *205*, e18–e47. [CrossRef]
30. Volpicelli, G.; Elbarbary, M.; Blaivas, M.; Lichtenstein, D.A.; Mathis, G.; Kirkpatrick, A.W.; Melniker, L.; Gargani, L.; Noble, V.E.; Via, G.; et al. International evidence-based recommendations for point-of-care lung ultrasound. *Intensive Care Med.* **2012**, *38*, 577–591. [CrossRef] [PubMed]
31. Sperandeo, M.; Varriale, A.; Sperandeo, G.; Filabozzi, P.; Piattelli, M.L.; Carnevale, V.; Decuzzi, M.; Vendemiale, G. Transthoracic ultrasound in the evaluation of pulmonary fibrosis: Our experience. *Ultrasound Med. Biol.* **2009**, *35*, 723–729. [CrossRef]
32. Moazedi-Fuerst, F.C.; Zechner, P.M.; Tripolt, N.J.; Kielhauser, S.M.; Brickmann, K.; Scheidl, S.; Lutfi, A.; Graninger, W.G. Pulmonary echography in systemic sclerosis. *Clin. Rheumatol.* **2012**, *31*, 1621–1625. [CrossRef] [PubMed]
33. Barskova, T.; Gargani, L.; Guiducci, S.; Randone, S.B.; Bruni, C.; Carnesecchi, G.; Conforti, M.L.; Porta, F.; Pignone, A.; Caramella, D.; et al. Lung ultrasound for the screening of interstitial lung disease in very early systemic sclerosis. *Ann. Rheum. Dis.* **2013**, *72*, 390–395. [CrossRef] [PubMed]
34. Swets, J.A. Measuring accuracy of diagnostic systems. *Science* **1988**, *240*, 1285–1293. [CrossRef] [PubMed]
35. Mitchell, A.J. Sensitivity $\times$ PPV is a recognized test called the clinical utility index (CUI+). *Eur. J. Epidemiol.* **2011**, *26*, 251–252. [CrossRef]
36. Falcetta, A.; Leccardi, S.; Testa, E.; Papaleo, F.; Fenoglio, L.; Melchio, R. The role of lung ultrasound in the diagnosis of interstitial lung disease. *Shanghai Chest* **2018**, *2*, 41. [CrossRef]
37. Lacedonia, D.; Scioscia, G.; Giardinelli, A.; Quarato, C.M.I.; Sassani, E.V.; Foschino Barbaro, M.P.; Maci, F.; Sperandeo, M. The Role of Transthoracic Ultrasound in the Study of Interstitial Lung Diseases: High-Resolution Computed Tomography Versus Ultrasound Patterns: Our Preliminary Experience. *Diagnostics* **2021**, *11*, 439. [CrossRef] [PubMed]
38. Sperandeo, M.; De Cata, A.; Molinaro, F.; Trovato, F.M.; Catalano, D.; Simeone, A.; Var-Riale, A.; Martines, G.F.; Trovato, G. Ultrasound signs of pulmonary fibrosis in systemic sclerosis as timely indicators for chest computed tomography. *Scand. J. Rheumatol.* **2015**, *44*, 389–398. [CrossRef]

39. Buda, N.; Piskunowicz, M.; Porzezinska, M.; Kosiak, W.; Zdrojewski, Z. Lung Ultraso-nography in the Evaluation of Interstitial Lung Disease in Systemic Connective Tissue Diseases: Criteria and Severity of Pulmonary Fibrosis-Analysis of 52 Patients. *Ultras-Chall Med.* **2016**, *37*, 379–385.
40. Lee, J.; Im, J.; Ahn, J.; Kim, Y.; Han, M. Fibrosing alveolitis: Prognostic implication of ground-glass attenuation at high-resolution CT. *Radiology* **1992**, *184*, 451–454. [CrossRef] [PubMed]
41. Muller, N.; Staples, C.A.; Miller, R.R.; Vedal, S.; Thurlbeck, W.M.; Ostrow, D.N. Disease activity in idiopathic pulmonary fibrosis: CT and pathologic correlation. *Radiology* **1987**, *165*, 731–734. [CrossRef] [PubMed]
42. Shahin, A.A.; Sabri, Y.Y.; Mostafa, H.A.; Sabry, E.Y.; Hamid, M.A.; Gamal, H.; Shahin, H.A. Pulmonary function tests, high-resolution computerized tomography, α1-antitrypsin measurement, and early detection of pulmonary involvement in patients with systemic sclerosis. *Rheumatol. Int.* **2001**, *20*, 95–100. [CrossRef] [PubMed]
43. Terriff, B.A.; Kwan, S.Y.; Chan-Yeung, M.; Müller, N. Fibrosing alveolitis: Chest radiography and CT as predictors of clinical and functional impairment at follow-up in 26 patients. *Radiology* **1992**, *184*, 445–449. [CrossRef] [PubMed]
44. Lichtenstein, D.; Meziere, G.; Biderman, P.; Gepner, A.; Barre, O. The comet-tail artifact: An ultrasound sign of alveolar-interstitial syndrome. *Am. J. Respir. Crit. Care Med.* **1997**, *156*, 1640–1646. [CrossRef]
45. Targhetta, R.; Chavagneux, R.; Balmes, P.; Lemerre, C.; Mauboussin, J.M.; Bourgeois, J.M.; Pourcelot, L. Sonographic lung surface evaluation in pulmonary sarcoidosis: Preliminary results. *J. Ultrasound Med.* **1994**, *13*, 381–388. [CrossRef]
46. Reissig, A.; Kroegel, C. Transthoracic Sonography of Diffuse Parenchymal Lung Disease the Role of Comet Tail Artifacts. *J. Ultrasound Med.* **2003**, *22*, 173–180. [CrossRef]
47. Avouac, J.; Fransen, J.; Walker, U.; Riccieri, V.; Smith, V.; Muller, C.; Miniati, I.; Tarner, I.; Randone, S.B.; Cutolo, M.; et al. Preliminary criteria for the very early diagnosis of systemic sclerosis: Results of a Delphi Consensus Study from EULAR Scleroderma Trials and Research Group. *Ann. Rheum. Dis.* **2011**, *70*, 476–481. [CrossRef] [PubMed]
48. Warrick, J.H.; Bhalla, M.; Schabel, S.I.; Silver, R.M. High resolution computed tomography in early scleroderma lung disease. *J. Rheumatol.* **1991**, *18*, 1520–1528. [PubMed]
49. Jambrik, Z.; Monti, S.; Coppola, V.; Agricola, E.; Mottola, G.; Miniati, M.; Picano, E. Usefulness of ultrasound lung comets as a nonradiologic sign of extravascular lung water. *Am. J. Cardiol.* **2004**, *93*, 1265–1270. [CrossRef] [PubMed]
50. Doveri, M.; Frassi, F.; Consensi, A.; Vesprini, E.; Gargani, L.; Tafuri, M.; Picano, E.; Della Rossa, A.; Sedie, A.D.; D'Ascanio, A.; et al. Ultrasound lung comets: New echographic sign of lung interstitial fibrosis in systemic sclerosis. *Reumatismo* **2008**, *60*, 180–184. [CrossRef]
51. Picano, E.; Frassi, F.; Agricola, E.; Gligorova, S.; Gargani, L.; Mottola, G. Ultrasound lung comets: A clinically useful sign of extravascular lung water. *J. Am. Soc. Echocardiogr.* **2006**, *19*, 356–363. [CrossRef] [PubMed]
52. Gargani, L.; Romei, C.; Bruni, C.; Lepri, G.; El-Aoufy, K.; Orlandi, M.; D'errico, L.; Bandini, G.; D'angelo, G.; Guiducci, S.; et al. Lung ultrasound B-lines in systemic sclerosis: Cut-off values and methodological indications for interstitial lung disease screening. *Rheumatology* **2022**, *61*, SI56–SI64. [CrossRef]
53. Tardella, M.; Di Carlo, M.; Carotti, M.; Filippucci, E.; Grassi, W.; Salaffi, F. Ultrasound B-lines in the evaluation of interstitial lung disease in patients with systemic sclerosis: Cut-off point definition for the presence of significant pulmonary fibrosis. *Medicine* **2018**, *97*, e0566. [CrossRef] [PubMed]
54. Hansell, D.M.; Bankier, A.A.; MacMahon, H.; McLoud, T.C.; Muller, N.L.; Remy, J. Fleischner Society: Glossary of terms for thoracic imaging. *Radiology* **2008**, *246*, 697–722. [CrossRef]
55. Bussone, G.; Mouthon, L. Interstitial lung disease in systemic sclerosis. *Autoimmun. Rev.* **2011**, *10*, 248–255. [CrossRef] [PubMed]
56. Greidinger, E.L.; Flaherty, K.T.; White, B.; Rosen, A.; Wigley, F.M.; Wise, R.A. African-American race and antibodies to topoisomerase I are associated with increased severity of scleroderma lung disease. *Chest* **1998**, *114*, 801–807. [CrossRef] [PubMed]
57. Assassi, S.; Sharif, R.; Lasky, R.E.; McNearney, T.A.; Estrada-Y-Martin, R.M.; Draeger, H.; Nair, D.K.; Fritzler, M.J.; Reveille, J.D.; Arnett, F.C.; et al. Predictors of interstitial lung disease in early systemic sclerosis: A prospective longitudinal study of the GENISOS cohort. *Arthritis Res. Ther.* **2010**, *12*, R166. [CrossRef] [PubMed]
58. Wu, W.; Jordan, S.; Becker, M.O.; Dobrota, R.; Maurer, B.; Fretheim, H.; Ye, S.; Siegert, E.; Allanore, Y.; Hoffmann-Vold, A.-M.; et al. Prediction of progression of interstitial lung disease in patients with systemic sclerosis: The SPAR model. *Ann. Rheum. Dis.* **2018**, *77*, 1326–1332. [CrossRef] [PubMed]
59. Le Gouellec, N.; Duhamel, A.; Perez, T.; Hachulla, A.L.; Sobanski, V.; Faivre, J.B.; Morell-Dubois, S.; Lambert, M.; Hatron, P.Y.; Hachulla, E.; et al. Predictors of lung function test severity and outcome in systemic sclerosis-associated interstitial lung disease. *PLoS ONE* **2017**, *12*, e0181692. [CrossRef] [PubMed]
60. Landini, N.; Orlandi, M.; Bruni, C.; Carlesi, E.; Nardi, C.; Calistri, L.; Morana, G.; Tomassetti, S.; Colagrande, S.; Matucci-Cerinic, M. Computed tomography predictors of mortality or disease progression in systemic sclerosis—Interstitial lung disease: A Systematic Review. *Front. Med.* **2022**, *8*, 807982. [CrossRef]

61. Rotondo, C.; Urso, L.; Praino, E.; Cacciapaglia, F.; Corrado, A.; Cantatore, F.P.; Iannone, F. Thoracic lymphoadenopathy as possible predictor of the onset of interstitial lung disease associated to systemic sclerosis. *J. Scleroderma Relat. Disord.* **2020**, *5*, 210–218. [CrossRef] [PubMed]
62. Hoffmann-Vold, A.-M.; Maher, T.M.; Philpot, E.E.; Ashrafzadeh, A.; Barake, R.; Barsotti, S.; Bruni, C.; Carducci, P.; Carreira, P.E.; Castellví, I.; et al. The identification and management of interstitial lung disease in systemic sclerosis: Evidence-based European consensus statements. *Lancet Rheumatol.* **2020**, *2*, e71–e83. [CrossRef] [PubMed]
63. Hughes, M.; Bruni, C.; Cuomo, G.; Sedie, A.D.; Gargani, L.; Gutierrez, M.; Lepri, G.; Ruaro, B.; Santiago, T.; Suliman, Y.; et al. The role of ultrasound in systemic sclerosis: On the cutting edge to foster clinical and research advancement. *J. Scleroderma Relat. Disord.* **2021**, *6*, 123–132. [CrossRef]
64. Makhlouf, H.; Hasan, A. B-lines: Transthoracic chest ultrasound signs useful in assessment of interstitial lung diseases. *Ann. Thorac. Med.* **2014**, *9*, 99. [CrossRef] [PubMed]
65. Tardella, M.; Gutierrez, M.; Salaffi, F.; Carotti, M.; Ariani, A.; Bertolazzi, C.; Filippucci, E.; Grassi, W. Ultrasound in the assessment of pulmonary fibrosis in connective tissue disorders: Correlation with high-resolution computed tomography. *J. Rheumatol.* **2012**, *39*, 1641–1647. [CrossRef] [PubMed]

diagnostics

Article

Flow-Field Inference for Turbulent Exhale Flow Measurement

Shane Transue [1], Do-kyeong Lee [2], Jae-Sung Choi [3], Seongjun Choi [4], Min Hong [5,*] and Min-Hyung Choi [1]

1 Department of Computer Science and Engineering, University of Colorado Denver, Denver, CO 80204, USA; shane.transue@ucdenver.edu (S.T.); min.choi@ucdenver.edu (M.-H.C.)
2 Department of Software Convergence, Soonchunhyang University, Asan 31538, Republic of Korea
3 Department of Internal Medicine, Cheonan Hospital, College of Medicine, Soonchunhyang University, Cheonan 31151, Republic of Korea
4 Department of Otolaryngology-Head and Neck Surgery, Cheonan Hospital, College of Medicine, Soonchunhyang University, Cheonan 31151, Republic of Korea
5 Department of Computer Software Engineering, Soonchunhyang University, Asan 31538, Republic of Korea
* Correspondence: mhong@sch.ac.kr

Abstract: Background: Vision-based pulmonary diagnostics present a unique approach for tracking and measuring natural breathing behaviors through remote imaging. While many existing methods correlate chest and diaphragm movements to respiratory behavior, we look at how the direct visualization of thermal CO_2 exhale flow patterns can be tracked to directly measure expiratory flow. Methods: In this work, we present a novel method for isolating and extracting turbulent exhale flow signals from thermal image sequences through flow-field prediction and optical flow measurement. The objective of this work is to introduce a respiratory diagnostic tool that can be used to capture and quantify natural breathing, to identify and measure respiratory metrics such as breathing rate, flow, and volume. One of the primary contributions of this work is a method for capturing and measuring natural exhale behaviors that describe individualized pulmonary traits. By monitoring subtle individualized respiratory traits, we can perform secondary analysis to identify unique personalized signatures and abnormalities to gain insight into pulmonary function. In our study, we perform data acquisition within a clinical setting to train an inference model (FieldNet) that predicts flow-fields to quantify observed exhale behaviors over time. Results: Expiratory flow measurements capturing individualized flow signatures from our initial cohort demonstrate how the proposed flow field model can be used to isolate and analyze turbulent exhale behaviors and measure anomalous behavior. Conclusions: Our results illustrate that detailed spatial flow analysis can contribute to unique signatures for identifying patient specific natural breathing behaviors and abnormality detection. This provides the first-step towards a non-contact respiratory technology that directly captures effort-independent behaviors based on the direct measurement of imaged CO_2 exhaled airflow patterns.

Keywords: CO_2 pulmonary evaluation; non-contact pulmonary measurement; flow field interpolation; UNET flow prediction; pulmonary diagnostics

1. Introduction

Non-contact pulmonary evaluation presents a promising method for capturing natural breathing behaviors that are normally altered when using conventional diagnostic methods such as spirometry [1–3], plethysmography [4,5], and capnography (end-tidal CO_2) [6–8]. While many of these pulmonary diagnostics provide highly accurate results, they also modify the natural breathing behavior by forcing unnatural posture, breathing through tubes that restrict flow, alter posture, and increase physiological effort during monitoring. This leads to effort-dependent evaluation of pulmonary behavior, which can alter the interpretation of subtle condition traits that can be detected through continuous non-contact evaluation. Unlike many new approaches to pulmonary monitoring that attempt to utilize wearable devices to infer breathing behaviors, the direct analysis of exhaled airflow cannot be captured through these methods. Rather, they rely on indirect measures of related physiological signals that provide a basis for evaluating pulmonary function. Additionally, advanced forms of respiratory diagnostics include X-ray Computed Tomography (CT) [9,10], Magnetic Resonance Imaging (MRI) [11], and Positron emission tomography (PET) scans [12]; however, these advanced scans have limited availability and incur procedures that are warranted once chronic conditions become serious enough to require detailed evaluation. As an alternative to these existing methods, we present a non-contact approach to pulmonary behavioral evaluation aimed at identifying unique characteristics directly observed expiratory behaviors through thermal CO_2 imaging [13,14]. The proposed method sits in between existing wearable, tube-based, and advanced scan solutions, presenting an approach that can provide a basis for pre-screening and anomalous behavior detection.

In this work, we present a method for utilizing imaged exhale behaviors captured through thermal CO_2 flow fields to directly measure exhale flow and form unique characteristic waveforms that uniquely describe individualized pulmonary signatures (Figure 1). By capturing the unique behaviors from 20 different subjects, we identify how a flow inference model can be utilized to extract, filter, and interpolate exhale flow characteristics. This enables us to define and characterize how individual posture and effort contribute to the exhale behaviors that give insight into natural breathing. Our goal is to provide a basis for identifying how these unique traits can be used to create insight into how natural breathing reflects conditions including Chronic Obstructed Pulmonary Disease (COPD) [15,16], asthma, Acute Respiratory Distress Syndrome (ARDS) [17–19], Obstructive Sleep Apnea [20,21], and Respiratory Syncytial Virus (RSV) [22]. In our approach, we provide a pipeline that translates recorded or real-time sequences of thermal images into flow-field estimates that can be used to measure and uniquely identify individualistic traits of expiratory behavior. This includes: (1) using facial tracking to (2) performing dense, detailed flow analysis, enabling us to (3) quantify and model flow behaviors within an open clinical environment. We introduce a new Convolutional Neural Network architecture: *FieldNet*, based on architectures similar to U-Net [23] and FlowNet [24] to predict exhale flow fields. The proposed model operates on vector fields to perform flow-field inference to isolate and extract exhale flow behaviors that can define individualized pulmonary traits.

While capturing detailed exhale behaviors through a non-contact method presents unique advantages, there are several technical challenges that make this approach competitive within the existing field of pulmonary diagnostic technologies. The challenges of using a vision-based approach to measuring exhale behaviors are greatly influenced by three primary factors: (1) subject movement must be accounted for to provide continuous and reliable monitoring, (2) capturing flow behaviors of the thermal signature requires sensitive imaging hardware and detailed flow tracking, and (3) airflow behaviors are mixed with general air movements as well as backgrounds that interfere with flow isolation. To address the first requirement, we adopt facial tracking and minimize the impact on computed flow fields. Second, we leverage Mid Wavelength Infrared Camera (MWIR, 640 × 512@25 Hz) thermography paired with a spectral band-pass filter (3–5 μm) and parallel optical flow tracking [25] for generating dense 2D flow-fields. Finally, we leverage the ability of a flow-field inference model to extract and isolate flow behaviors to create exhale measurements. The proposed flow model provides three primary functions: (1) the ability to segment and separate airflow behaviors for measurement, (2) the ability to predict interpolated flow-fields, and (3) isolate exhale behaviors from flow fields to improve signal extraction.

The aim of this study is to capture and quantify individualized traits of natural exhale behaviors to measure pulmonary function and isolate instances of anomalous breathing. To accomplish this, we provide three primary contributions as the objectives of the study: (1) track and capture exhale behaviors from various subjects to identify inter-subject exhale behavioral variance, (2) train a convolutional deep neural network to isolate exhale flow behaviors from open-air recordings, and (3) generate and analyze expiratory waveforms regarding the unique signatures of individualized pulmonary behaviors to identify possible anomalies within exhale episodes using a model trained on expected versus abnormal exhale patterns.

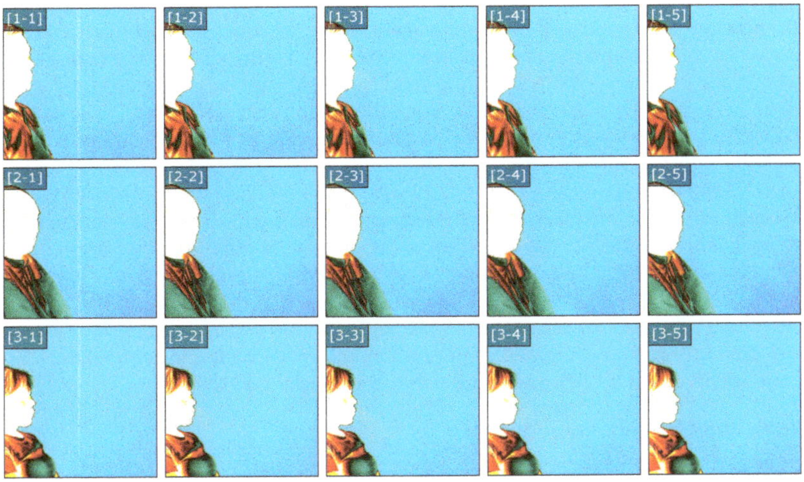

Figure 1. Flow sequences from three selected subjects illustrating unique exhale patterns. Each sequence presents static states of the captured infrared thermal CO_2 flow for a single exhale episode for each of the selected subjects. Each sequence presents five ($h = 512$, $w = 640$) frames for each subject that exemplify the flow of each patient's exhale pattern.

2. Related Work

Numerous respiratory monitoring devices and pulmonary diagnostic technologies are prominent in modern healthcare [26–30]. These typically fall within four primary modalities: (1) tube-based solutions that provide accurate airflow measurements, (2) advanced scanning technologies that include X-ray and CT scans [19,31], (3) wearable solutions that are worn continuously such as transducer or accelerometer belts [32,33] and health monitors [34], and (4) non-contact technologies such as imaging [35–39], wireless [40,41], and smartphone technologies [40]. Tube-based solutions provide the most accurate measurements that include respiratory rate (BPM), flow (L/s), volume (L) which are formulated into Pulmonary Function (PFT) Testing measures (VC FVC, ERV, TV, etc.) [15,42]. This segment contains many of the primary pulmonary diagnostics utilized in clinical settings including spirometry [1–3], plethysmography [4,5], and capnography [6–8]. If monitored pulmonary conditions indicate severe progression, advanced scanning technologies such as X-ray and CT scans may be used to identify interstitial lung abnormalities (ILAs), regional ventilation, perfusion, and gas exchange [43,44]. While these scans provide some of the most accurate physiological measures, they are also costly and largely restricted to clinical use. Looking beyond clinical pulmonary diagnostics, wearable devices focus on the translation of chest movements or other physiological signals into pulmonary metrics. The most common forms of wearable devices that measure breathing rate and volume are transducer and accelerometer belts [32,33]. These devices translate chest movements and expansion as a correlated function that describes respiratory behaviors. Improvements in these devices have lead to accurate measurements of behaviors and are commonly used in exercise and sleep studies [20,21]. The benefit of wearable devices is that they provide continuous signals, but accuracy is generally reduced by sensor placement and movement. Smart watches and fitness trackers also include biometric readings related to respiration; however, while these devices are accurate for heart rate, they still have to infer respiratory characteristics through other physiological signals. Recent methods have utilized machine learning to infer these measures from reliable sources such as heart rate [45]. The majority of vision-based methods for respiratory analysis have leveraged the relationship between observable chest movements with the cyclical breathing patterns [46]. Many of these approaches utilize depth imaging or color images to identify shifts in movement that correspond to the observable respiratory movements [47,48]. The weakness of these approaches is in the correlation between the observed movement and how the measurement is translated into a quantitative signal representing the inspiratory and expiratory behaviors. While this correlation is closely related to the premise used for transducer and accelerometer respiratory belts, vision-based methods are typically less accurate due to the unbound relationship between camera and subject [49,50]. The second common form of vision-based approaches to respiratory monitoring utilize infrared imaging; however, these approaches focus on the detection of temperature differentials on the face (nostrils) to identify breathing behaviors [49,51–53]. These approaches operate in a similar fashion to thermistors that are used in sleep studies, but are largely limited to measuring respiratory rate. Vision-based models have been proposed for identifying, segmenting, and extracting gaseous flow behaviors from images and image sequences for other applications. The detection of plume behaviors for automated smoke segmentation [54] and vapour formations have been approached for a wide variety of applications including forest fire tracking [55], automated fire detection [56,57], and cloud movement tracking [58,59].

3. Materials and Methods

The objective of vision-based thermal flow expiratory modeling is to identify, track, and extract meaningful respiratory traits that uniquely describe pulmonary function. The unique contribution to this approach is that imaging expiratory behaviors can provide a detailed visual of natural breathing which, characteristic traits that model individualized traits. By directly capturing exhaled airflow, we can identify characteristic traits of an individual's subconscious breathing pattern that is not directly altered by the monitoring method. To achieve this, we present an automated pipeline that allows the monitored subject to relax and breathe naturally for a short period of time while capturing their unique breathing signature. During this monitoring phase, we record the stream of thermal images that capture exhale behavior. These image sequences are then provided to a processing pipeline that: (1) tracks the subjects movement and localizes the Region-of-Interest (ROI) to the nose-mouth sub-image, (2) computes a fine-grained dense optical flow to generate flow-fields that capture expiratory behavior, (3) encodes this information to train a generative flow-field model to predict flow behaviors, that can then be used to (4) generate waveform representations of the observed exhale behaviors. An overview of the implemented data processing pipeline is shown in Figure 2.

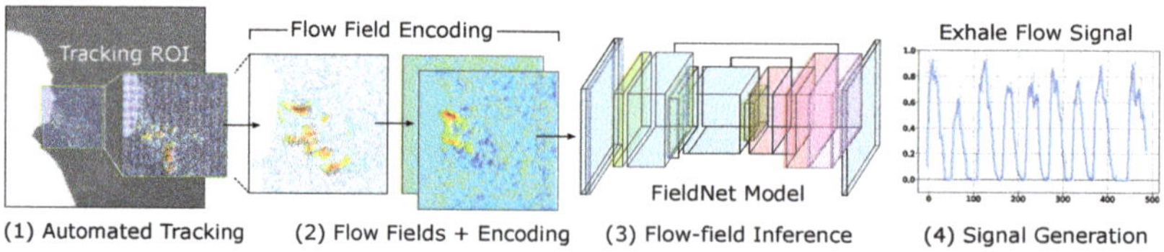

Figure 2. Approach overview. Facial tracking within thermal sequences capturing exhale behaviors is used to localize the ROI for which 2D flow fields are generated. These fields are then encoded and used to train the proposed *FieldNet* architecture for predicting flow fields. These fields and then used to generate signals representing captured exhale behaviors.

The foundation of the approach is based on the creation of image datasets that contain exhale behaviors from various subjects that are used to train and evaluate the proposed FieldNet model and generate individualized respiratory measurements. This process is divided into four primary steps that include: data acquisition, ROI selection and tracking, flow field generation, and training data collation. The resulting trained model is then used to predict flow fields with isolated exhale vectors that can be used as a measurement to evaluate airflow. This provides the ability to use the model on each of the 20 subjects included in the initial cohort to generate non-contact respiratory signals that exhibit physiological differences in the highly individualistic measurement waveforms. This provides a basis for observing natural breathing behaviors as a pulmonary diagnostic that can be used to identify unique characteristics of respiratory traits.

3.1. Data Acquisition

The data collection procedure was performed within an outpatient clinical setting where each subject was recorded for approximately five minutes. Data acquisition was performed at Soonchunghyang University Cheonan Hospital, IRB (no. 2023-10-012). The basic arrangement provided: (1) tripod mounted camera approximately 1.5 [m] from the subject, (2) seating position perpendicular to the camera, parallel to the imaging plane, and (3) the clinical tech and recording software running on a locally connected computer. Given this setup, turbulent exhale flows are captured and projected from the 3D space between the camera and background and projected to a 2D flow representation that is imaged by the thermal camera. Thermal images are captured and stored as raw 16-bit (0-2^{16}-1) *count* (sensor activations to pixel intensity) image sequences stored in contiguous binary field sequence files. These pixel values typically range between 5000–9000 counts an interior room at 22 °C. The content of the image can be heavily modified by the selected range of raw pixel values that are normalized ($p \in [0,1]$); therefore, we form two algorithms for: (1) a normalization that prioritizes tracking and (2) a normalization range that emphasizes observable exhale flow. For tracking, the normalization is set to constant at the high and low values of the image to provide clear feature-points that can be tracked to enable the automated facial tracking: $[F_c, I_{max}] \rightarrow [0,1]$, where F_c is the count corresponding to the face temperature. For the exhale normalization we sample the background values and identify this as the general ambient count (A_c) within an ROI of the image and identify high bound as the face temperature F_c. Since the exhale intensities will fall between these values, this normalization is defined as: $[A_c, F_c] \rightarrow [0,1]$. This creates two parallel uses of each image contained within the recording allowing for subject movements to be tracked to dynamically adjust the ROI used to capture exhale flow behaviors.

3.2. Automated Tracking

The imaging process captures the subject as well as the surrounding area, with a portion of the image reserved to capture exhale behaviors within the tracked ROI. The immediate challenge for accurately capturing a subject's breathing behavior is closely tracking the nose-mouth region from the profile view while minimizing the impact of subject movement on the process of generating the flow field. As the ROI moves, this can introduce uniform shifts within the flow field that contribute to error within the measurement of the exhale behavior. To address this, we employ automated tracking that: (1) accurately tracks the nose-mouth region and (2) minimizes the per-frame jump required to maintain a constant tracking region that best captures the exhale behavior. Since the outcome of the tracking is an ROI shift, the offset for the computation of the flow field can be addressed through an inverse of the transformation. This is utilized to eliminate the shift of the ROI that would be exhibited within the flow field for subsequent frames. While facial tracking in infrared images has been studied [60], these methods are not directly applicable to profile tracking restricted to the nose-mouth region. Similarly, Haar cascades can be adopted to infrared imaging [61]; however, this approach is not well suited for profile facial tracking in this application due to translational jumps during tracking.

To achieve accurate tracking within the recorded sequence, we utilize a fixed sub-region of interest of fixed size that correlates to the the input dimensionality of the flow model. This sub region window will be used to capture the imaged exhale flow as the thermal distribution of the flow over time in contrast to the static background. First, we identify the nose/mouth region of the subject and then define the fixed size sub-region within the initial frame. As each frame is processed, key-points within the sub-region corresponding to features on the subject's face are identified. Lucas-Kanade optical flow [62] is then used to estimate the translation vectors of these key-points to update or shift the sub-region during the subsequent frame. To provide real-time feedback, the proposed parallel dense optical flow is used to compute the flow vectors within the sub-region corresponding to the exhale behavior. The image sequence in Figure 3 illustrates the captured image, the sub-region, and the magnitude of the flow vectors within the region.

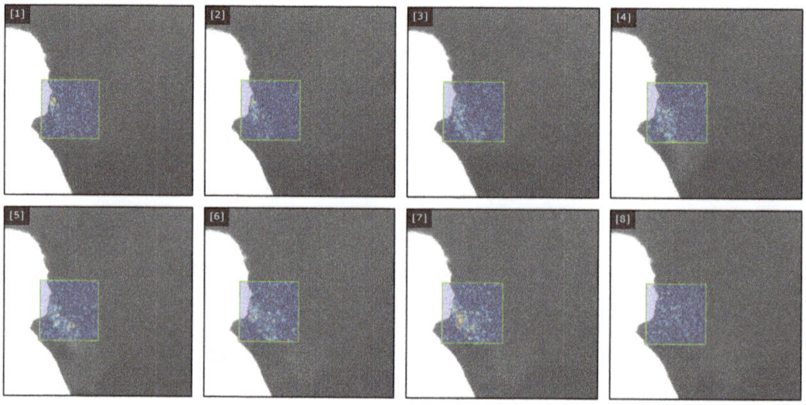

Figure 3. Tracking region for select frames 1–8. Optical flow is utilized to identify the correspondence and movement between subsequent frames of the tracked region. This fixed window is tracked throughout all frames of the sequence. The flow vectors are displayed using the real-time approximation (stride = 2 + line segment rendering) for images of size ($h = 512$, $w = 640$).

Within the tracked sub-region, the flow is computed using a parallel implementation of a noise-sensitive dense optical flow and superimposed on the image used to provide feedback during clinical recording. The rendering is based on drawing scaled line segments representing the flow field with a stride of $s = 2$ and color-mapped flow vector. This provides an approximation of the vector field being computed from the recorded frames. The image sequence in Figure 4 illustrates the states of the vector field representing the apparent exhale flow for eight consecutive images. Based on the progression of the observable exhale flow, the resulting flow field captures the magnitude of the localized flow as the image intensities shift between frames.

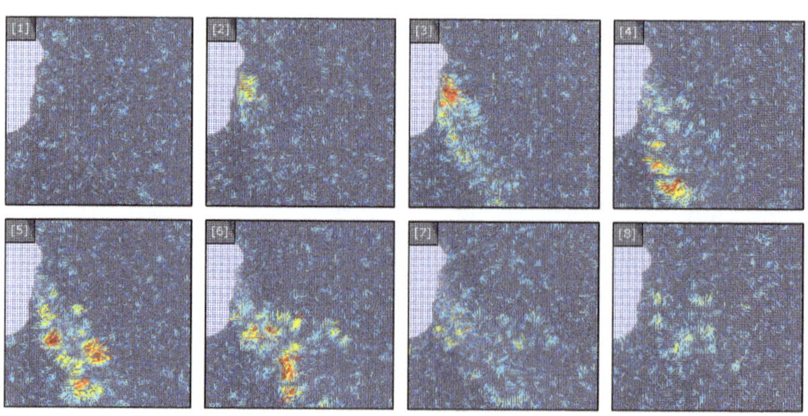

Figure 4. Tracked flow sequence for select frames 1–8. The tracked sub-region of the exhale is recorded for both training and inference datasets. Flow vector magnitude and direction is displayed using line segments (stride = 2) for the tracked sub-region. Color indicates flow vector magnitude.

The apparent flow identified within the tracked ROI provides the basis of the measurements used to quantify and analyze exhale behavior. Exhale flow fields are computed between subsequent frames within the collected sequences and are then used to train the FieldFlow model. The process of generating the fields is based on performing dense optical flow [25] that creates a 2D vector field of the contribution of the gradient shift in image intensity represented by vector direction and magnitude. The iterative scheme of the optical flow algorithm is then used to detect minute directional flow changes observed within the imaged exhale pattern to both separate exhale flow from background air movements and provide the basis for quantifying the changes in observable flow patterns between subsequent frames.

3.3. Flow Field Generation

Observable flow contained within the recorded infrared image sequences is computed as the optical flow of the shifts of intensity obtained from image pairs [25,63]. To generate flow sequences from each recording, we define the flow as the *apparent* shifts in intensity obtained from iteratively solving for the brightness and smoothness constraints of the optical flow formulation [25]. This computes the flow field $\vec{F}_t(i,j)$ between consecutive normalized image frames I_t and I_{t+1} based on the minimization form of the brightness and smoothness constraints as shown in Equation (1).

$$\min \iint (\nabla I \cdot v + \frac{\partial I}{\partial t})^2 + \alpha(\|\nabla v_x\|^2 + \|\nabla v_y\|^2)\ dxdy \tag{1}$$

where $\vec{v} = (v_x, v_y)$ represents the flow vector at each pixel correlated between the input images I_t and I_{t+1} and the scalar α specifies the contribution of the gradient magnitudes. To ensure that small fluctuations in flow are maintained within the generated flow field, α is reduced to small values ($\alpha = 0.15$). This ensures that intensity shifts close to the noise floor are counted as valid contributions to the observable exhale. The iterative implementation is based on the computation of two updated flow images U_i and V_i representing the x and y derivative images respectively at i iterations. The following iterative scheme presented in Equations (2) and (3) are used to update the derivative images for n-iterations.

$$v_x^{i+1} = \bar{v}_x^i - \left[\frac{\frac{\partial I}{\partial x}\bar{v}_x^i + \frac{\partial I}{\partial y}\bar{v}_y^i + \frac{\partial I}{\partial t}}{\alpha + \frac{\partial I}{\partial x}^2 + \frac{\partial I}{\partial y}^2}\right]\frac{\partial I}{\partial x} = U_i \tag{2}$$

$$v_y^{i+1} = \bar{v}_y^i - \left[\frac{\frac{\partial I}{\partial x}\bar{v}_x^i + \frac{\partial I}{\partial y}\bar{v}_y^i + \frac{\partial I}{\partial t}}{\alpha + \frac{\partial I}{\partial x}^2 + \frac{\partial I}{\partial y}^2}\right]\frac{\partial I}{\partial y} = V_i \tag{3}$$

The spatial derivatives $\partial I/\partial x$ and $\partial I \partial y$ are approximated using a Sobel kernel [64]. This results in the computation of the two channel flow field with $\partial I/\partial x$ and $\partial I/\partial y$ derivatives stored as a $(h, w, 2)$ tensor that defines the flow field $\vec{F}_t(i,j)$ after n iterations. The resulting form of this tensor is shown in Equation (4). These flow frame instances will then be used to create the training dataset used to train the flow interference model.

$$\vec{F}_t(i,j,0) = v_x^n \qquad \vec{F}_t(i,j,1) = v_y^n \tag{4}$$

While numerous advances have been proposed for the prediction of optical flow fields, the implementation is designed for real-time feedback within the clinical setting. Therefore, the our approach utilizes an optimized parallel C++/CUDA implementation compiled into a compiled Python module using pybind11 [65]. This implementation does not utilize the hierarchy from the LK method [62]. The implementation can be interchanged with numerous other modern optical flow implementations [66] as well as recent advances in adopted transformer model implementations [45]. From the implemented flow field generation, we visualize the flow sequence as shown in Figure 5.

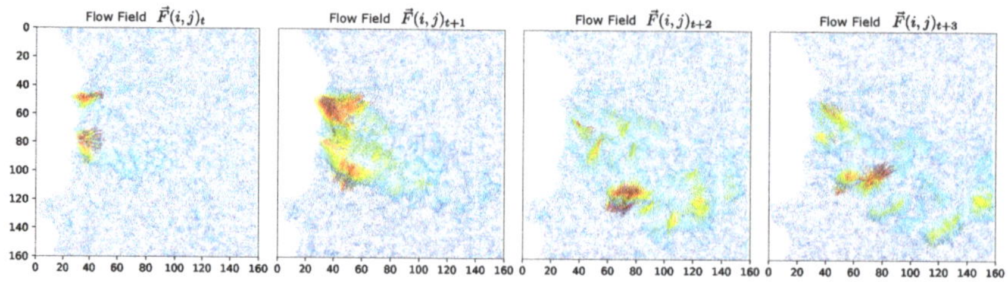

Figure 5. Illustration of the flow sequence vector fields for one exhale sequence. The sequence is illustrated by the flow fields for $i\%n$ frames where $n = 3$. The flow is characterized by the initial release and dissipation of the exhale, illustrated as a function of magnitude.

3.4. Signal Generation

The objective of a respiratory diagnostic is to identify the key traits and signatures that illustrate pulmonary health and how various conditions are exhibited within recorded breathing behaviors. Typically, pulmonary behavior must be reduced to a clinically meaningful metric that adheres to existing measurements that correspond to the behavioral traits of various conditions. Since the direct presentation of 2D flow behaviors can be difficult to discern for pulmonary diagnostics, we simplify the computed flow behaviors to a one-dimensional exhale waveform that represents the overall flow over time. To do this, we assume that the gradient of flow vector contributions gradually tappers from a maximal thermal signature to zero as it dissipates. Therefore, the contributions of each flow vector within the flow field are added together to provide a one-dimensional time-series that describes the overall behavior of the captured natural breathing. To compute this for a given flow field at time t for the field $\vec{F}_t$, we sum the contributions from each vector within the ROI as shown in Equation (5).

$$f(t) = \sum_{i=1}^{h} \sum_{j=1}^{w} |\vec{F}_t(i,j)| \tag{5}$$

where $f(t)$ provides a reduction from 2D field space to a 1D time-series signal based on the sum of all contributing vector magnitudes. Each flow field provides an individual sample within this reduction, allowing us to generate a 1D time-series per sequence. While spatial information is lost, the overall characteristics of the exhale traits result in considerable differences between subjects. However, the problem with directly utilizing flow fields computed from the optical flow is that since exhaled air rapidly dissipates, vector magnitudes quickly diminish to the noise floor of the thermal cameras detection of background fluctuations. This presents a problem for obtaining clean flow behaviors that can be measured. This provides the motivation for the creation of a model that is capable of isolating the contributions of these flow vectors as they dissipate to create accurate estimates of exhale flow over time. Therefore, we present the creation of a FieldFlow model that aids in both isolation and extraction of exhale flow measurements as part of an interpolation network that enables flow-field inference to create new flow fields that: (1) eliminate background noise, (2) characterise flow behaviors, and (3) enable quantitative evaluation of respiratory behaviors.

3.5. Flow Field Modeling

There are numerous approaches to isolating and segmenting important information from images to obtain clinically meaningful information. U-Net architectures have been widely used in medical imaging for segmentation [23]; this architecture can also be used as a foundation for the process of extracting minute exhale behaviors that can be lost as they dissipate to background noise-levels. Additionally, FlowNet [24] based network architectures can be utilized as generative models, providing estimates of flow fields based on observed training behaviors. Based on this we present a hybrid between these network architectures that captures the intermediate behaviors of observable flow states to estimate exhale flow behaviors. To implement this, we construct a training dataset based on a subset of the recorded data formulated as contiguous flow fields that can be used to predict intermediate flow fields to: (1) isolate exhale behaviors and (2) predict filtered flow fields.

3.6. Flow Field Encoding

Capturing spatially contiguous fluid flow behavioral characteristics is the primary objective of the model. This requires encoding the relative spatial distribution of angles with adjacent flow vectors to represent the change of flow over time. Projected flow fields are represented by a two-dimensional coordinate space populated by a dense set of vectors defined as $(x, y) \in \mathbb{R}^2$. There are two approaches to encode this information within the input tensor of the model: (1) encode the angle-magnitude representation or (2) normalize the x-y components of the field. The angle encoding provides an intuitive approach; however, immediate problem with encoding angles directly into the spatial representation of the 2D field is that the encoding will include discontinuous angle values ($0 \leq \theta \leq 2\pi$), require magnitude scaling, and still need to be normalized. A solution to this problem is to define the encoding as follows:

$$\begin{aligned} \theta &\rightarrow \big(y_1 = \sin(\theta),\ y_2 = \cos(\theta)\big) \\ (y_1, y_2) &\rightarrow arctan2(y_2,\ y_1) \end{aligned} \tag{6}$$

where we encode the angle to a two part representation where $y_1 = \sin(\theta) \in [-1, 1]$ and $y_2 = \cos(\theta) \in [-1, 1]$. While the angle is normalized in this representation, the values still must be scaled to include the magnitude of each vector. By obtaining the maximum vector magnitude (m) of the field (or field sequence), a linear transformation can be used to normalize the encoding that includes both angle and norm of each vector as (θ, n) into the scaled pair $(y1,\ y2) \cdot (n/m)$ where the quantity $(n/m) \in [0, 1]$. The inverse order of operations can be used to obtain the decoding of these vector where $\theta = arctan2(y_2,\ y_1)$ and the original $(x, y) = \big(\sin(\theta),\ \cos(\theta)\big) \cdot (n \cdot m)$. The range of $\sin(\theta)$ is based on the angle values contained within the input; therefore, to maximize the distribution of the values within the training domain, the angle-encoded values still need to be normalized. Based on the numerical domain of this encoding, we can compare this to the direct normalization of the (x, y) pair. Since each component can be scaled through a linear transformation given the maximum observed value to the range $[0, 1]$, we obtain an encoding equivalency between these representations. Encoding an asymmetrical vector field, we obtain the results shown in Figure 6 where the outcome from approach (1) becomes equivalent to (2) with the exception of the slices of the tensor being swapped.

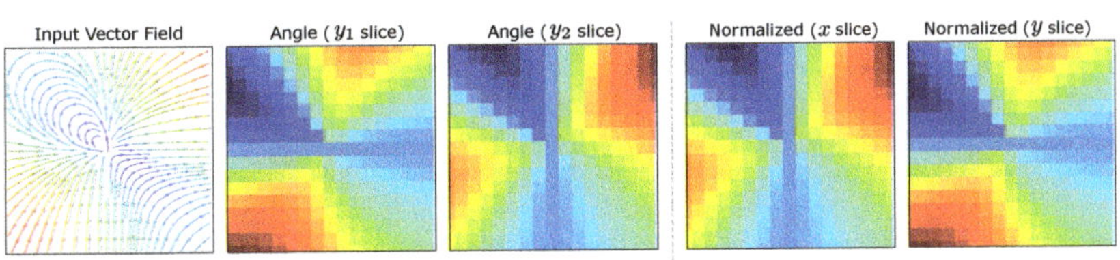

Figure 6. Encoding equivalency. Angular encoding (1) is equivalent to direct normalization on the bounds [0, 1] when magnitude is included. Therefore the optimal encoding is the direct parallel normalization of the vector field by component. Note: Angle (y_1 slice) is equivalent to Normalized (y slice) and Angle (y_2 slice) is equivalent to Normalized (x slice).

This illustrates that the angular encoding is functionally interchangeable with the direct normalization with the exception of a basic flip and interchange between the two encoded channels (slices). The standard linear transformation for this normalization can be computed through an element-wise parallel process, greatly reducing the computational complexity of the encoding. This reduction is due to removing numerous trigonometric functions and required normalization if the angles within the field do not correspond to $\sin(\theta) = 1$, leading to an under-utilization of the training domain range. Therefore, the encoding of the network is reduced to the parallel normalization of the field components to the interval $[0, n_{max}] \to [-1, 1]$ based on the maximum vector magnitude n_{max}. This results in two channels of the x and y slices of the flow field as shown in Figure 7. Due to the masking introduced within tracking to remove the flow components of the face, the encoded fields will contain zero regions. While the zero regions within the encoded training dataset represents instance bias, the augmentation is used to alleviate this problem.

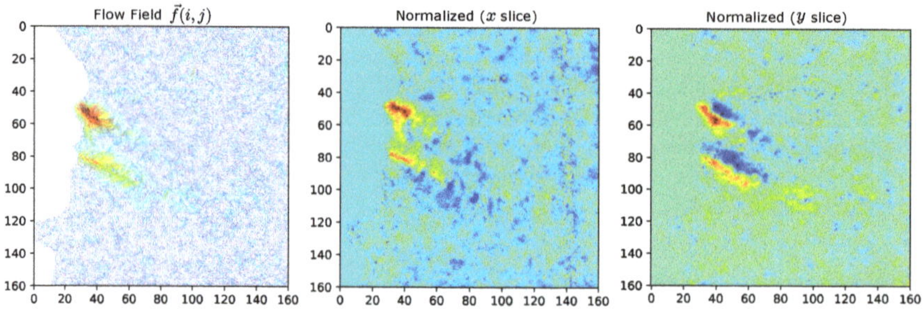

Figure 7. Flow field encoding visualization. The flow field (left) containing $(h \cdot w)$ vector pairs is encoded into the $(h, w, 2)$ tensor representing the flow between subsequent frames. The masked region of the subject contains no flow as compared to the background noise floor. The normalized slices x and y are stacked and provided as input to the FieldNet model.

3.7. Dataset Creation

Recorded sequences from 20 subjects have been used to generate flow sequences that comprise the foundation of the collected dataset that is used for both training and validation. A subset (5) of the subjects where selected to define the flow fields of the training dataset. Of each of the five training subjects, a (90/10) training/validation split was defined. The generation of a flow field sequence from a provided recording creates a collection of vector fields representing the apparent exhale flow from time t_0 to t_{n-1}. To provide an expected value for the intermediate flow that will be predicted by the model, there are two possible solutions: (1) the dataset can be structured to utilize intermediate frames as the expected value or (2) generate frames from a 2D fluid simulation to provide flow characteristics. Both approaches have several benefits and challenges. For one, utilizing existing data provides training data that closely matches the behavioral characteristics of expected behavior and is easily utilized. Whereas, utilizing 2D fluid flow dynamics provides correct physical behavior, but can present challenges in determining the initial conditions and progression of the simulation that closely match recorded data. In our approach, we utilize solution (1) where we select the intermediate frame t_{it+1} between frames t_i and t_{i+2} as the expected ground truth that represents the intermediate flow frame. This formulates an input x_{train} containing two flow frames $(\vec{F}_t, \vec{F}_{f+2})$ with a shape of $(h, w, 4)$. Based on this scheme, we define the x_{train} set as every flow field at t_i and t_{i+2} and the y_{train} set as the field at t_{i+1}. This creates an interleaved structure that defines the input and expected values of the input for training the model as shown in Figure 8.

The input of subsequent flow fields and output of the intermediate flow field creates an interpolation network; however, since the model also extracts the primary behavioral characteristics of the observed flow behavior, we utilize the model to segment flow behaviors from the surrounding environment and airflow. The dependence on subsequent frames is not a limitation; for continuous recordings of length n, the model can be used to predict the filtered flow fields for $n - 1$ frames, providing a method for isolating and measuring flow patterns of the training dataset. This can be utilized to greatly reduce noise in the resulting waveforms generated for each subject.

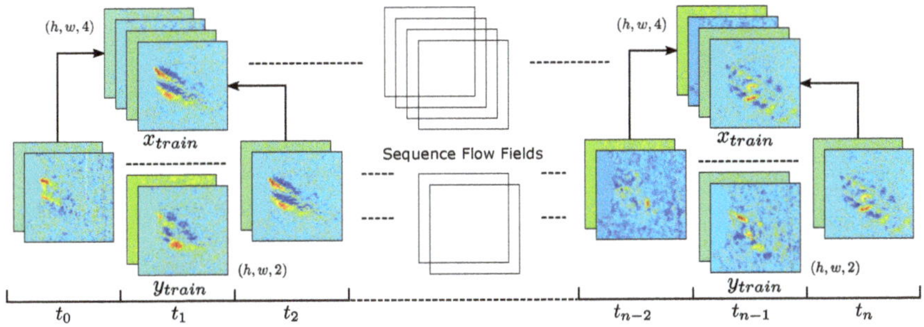

Figure 8. Interleaved training dataset generation. The model input x_{train} is composed of two encoded flow frames from times t_i and t_{i+2}, forming the input tensor of shape $(h, w, 4)$. The y_{train} expected value is defined as the intermediate frame at time t_{i+1} with shape $(h, w, 2)$.

3.8. Augmentation

Encoded flow fields can be directly augmented using standard image-based augmentation methods. Based on primitive transformations, these include rotation, translative shifting, and flipping. Due to the 2-channel representation of the encoded flow fields, standard library augmentation functions for images can be applied to expand the training dataset. The experimental setup within the clinical setting remained constant throughout the data collection process. Therefore, each recorded sequence places the subject to left of the each image, breathing to the right. This creates an inherent bias within the training dataset that has to be corrected. Augmentation of the training dataset includes both rotation and mirrored instances of the flow fields to account for this bias. The resulting augmentation extends the training dataset significantly; therefore, we limit the inclusion of subject training data to sub-sections of the recorded data for the subject data that is included within the training dataset. This subset of data will then be augmented prior to forming the training (x_{train}, y_{train}) and validation (x_{test}, y_{test}) datasets where the x set is composed of paired input fields $x_i = (\vec{F}_t, \vec{F}_{t+2})$ and the expected value composing the y set represents the intermediary flow field $y_i = \vec{F}_{t+1}$.

3.9. Model Architecture

The objective of the FieldNet model is to isolate and extract flow behaviors and minimize the contributions from background airflow and artifacts that may be introduced from background irregularities that contribute to noise. This is achieved by utilizing a convolutional network that operates on the flow fields that are generated from the parallel optical flow obtained from sequential frames. For the architecture, we adopt a U-Net convolutional model [23] that takes as input two encoded flow-fields, each of shape $(h, w, 2)$, resulting in an input tensor shape $(h, w, 4)$. The predicted output represents an encoded flow field of shape $(h, w, 2)$ which is decoded to represent the Cartesian vector field of the predicted intermediate flow state between the input fields. The implemented model architecture is shown in Figure 9 with simplified layer dimension labels. The architecture has been implemented in Keras [67] using the TensorFlow [68] back-end and trained on Google Colab Tesla T4 GPU. The model is composed of the input + 18 total layers and contains a total of 18.7 million trainable params.

Since the objective is to capture the spatial relationship between the input flow field states and the resulting prediction, *skip connections* are introduced at two levels within the architecture. This aids the model in preserving the spatial relationship between flow vectors which is critical for representing both the consistency in vector direction, but also allows the model to capture flow phenomena including vortex behaviors. Due to the range of the encoded fields $[-1, 1]$, the activation function used for all layers is the hyperbolic tangent function $tanh(x) = e^x - e^{-x}/e^x + e^{-x}$. This correlates well with the representation of the directional components of the predicted flow vectors.

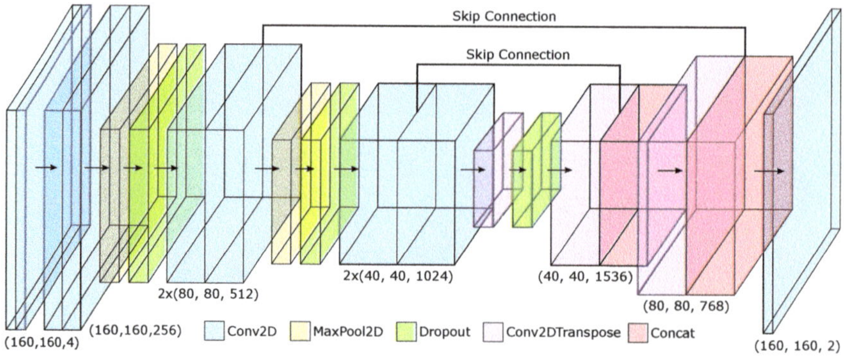

Figure 9. Model Architecture. The model is responsible for predicting the intermediate flow state of two flow fields stacked as an input tensor of shape $(h, w, 4)$. This is achieved by defining a UNET [23] inspired architecture that takes two encoded flow fields and predicts the encoded output flow field. Skip connections are utilized to preserve the spatial distribution of the flow throughout the network. The output defines a single intermediate flow field of shape $(h, w, 2)$.

3.10. Training

The training dataset is composed of the sequence of flow fields that are computed from the recorded thermal images from 20 subjects. To create the dataset used to train the model, we selected five of the subjects randomly and pulled sub-sequences of 500 flow frames from each of the selected subjects. This creates an initial set of 2500 flow images that are used as the training dataset that is was then interleaved following the process defined in Figure 8. This creates the input x_{train} and y_{train} expected values for model fit. For model loss we use standard Mean Squared Error (MSE). The selected model optimizer is Adam [69] with a learning rate of 1.0×10^{-3}. For training, we use a batch size of 64 for 16 epochs. The dropout rate is 0.025 = 2.5% for all dropout layers.

3.11. Flow-Field Interpolation

The trained model will be used to segment the observable flow movements and generate the resulting exhale flow signal. This is based on running the model on each subject's recorded sequence. The model will provide the segmented flow fields as a function of the current and next flow fields computed using optical flow. To obtain the interpolation prediction provided by the trained model, we provide the model subsequent flow frames as the current field $\vec{F}_t$ and the next field $\vec{F}_{t+2}$ to estimate the flow between these two captured states as the predicted flow field $\vec{F}_{t+1}$.

By selecting alternate frames, we can now compare the predicted flow field $\vec{F}_{t+1}$ at time $t + 1$ with the ground truth (optical flow result) at time $t + 1$. A visualization of the model predicted flow field and its corresponding ground-truth are shown in Figure 10.

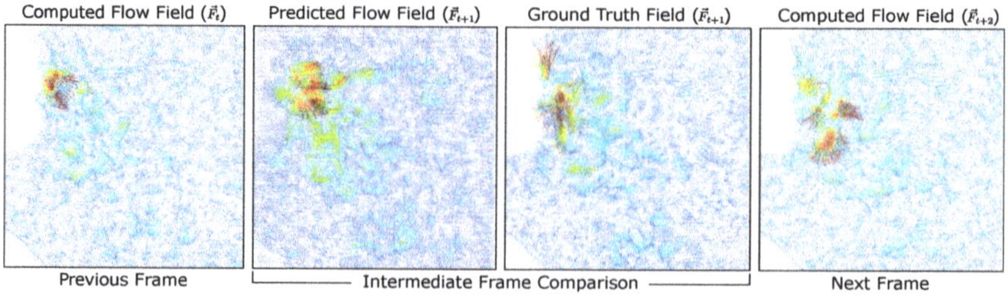

Figure 10. Flow field interpolation. Given the current and next frames at times $(t, t + 2)$, the predicted frame is compared to the ground truth, both of which represent the intermediate flow at time $(t + 1)$.

3.12. Exhale Measures and Unique Signatures

A of the prominent aim of this work was to provide an accurate method for isolating and measuring exhale behaviors to identify behavioral differences in natural breathing between subjects. To validate the results generated from the model predicted flow fields, we generate exhale waveforms that represent the magnitude of observable flow for the recorded sequence. First, we illustrate the result of using the model to isolate and extract exhale behaviors compared to the result provided by using optical flow. From this, we obtain exhale signals with less noise. For all signals, the results are processed using three primary filters: (1) outlier removal outside of 1.5 standard deviation, (2) Savitzky–Golay with a window size of 9 samples, and (3) min-max normalization to the interval $[0, 1]$.

Due to exhale behaviors rapidly dissipating within the thermal image sequences, the optical flow must be capable of identifying these flow characteristics compared to the noise floor. This results in the optical flow fields containing a significant amount of noise. We address this by utilizing FieldNet to predict flow images based on the original signal generated from the flow fields computed using optical flow. When the newly generated flow fields are predicted from the model, the resulting signal is significantly provides a better representation of the exhale behaviors with a better noise tolerance. Figure 11 illustrates the difference between the raw optical flow signal versus the signal generated from the FieldNet field prediction and demonstrates the resulting higher signal-to-noise ratio produced by the predictions of the model.

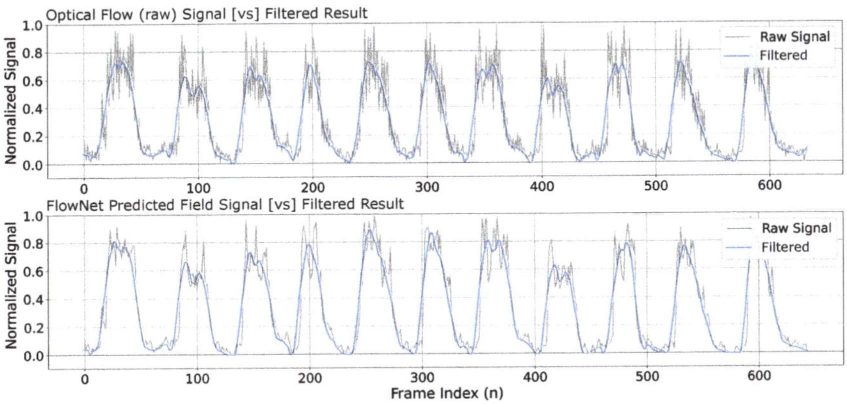

Figure 11. Optical flow (raw) versus model predicted flow field signal. FieldNet predicted flow fields contain a high signal-to-noise ratio that provides an implicit filtering of flow fields. The direct optical flow result (**top**) is compared with the model generated flow result (**bottom**).

3.13. Exhale Episode Anomaly Detection

Capturing open-air behaviors as the foundation of a pulmonary diagnostic requires pre-processing and dataset normalization that can standardize observed respiratory behaviors. This is challenging due to the natural variance observed from turbulent exhale behaviors. However, while open-air turbulent flows contain per-exhale variance, they also exhibit clear patterns unique to each subject. These unique patterns demonstrate how natural breathing behaviors can be captured and exhibit the basis for furthering our understanding of pulmonary behavior. Therefore, we present a method for processing observed exhale waveforms to identify the unique characteristics of individualized behaviors and generate an anomaly score model to measure anomalous respiratory behaviors. We implement this by introducing two new models (1) an autoencoder model to reduce turbulent flow noise and (2) an anomaly prediction model. By creating a standardized set of normal or healthy classified behaviors used to train an anomaly score model, subsequent subject respiratory patterns can be evaluated through a trained convolutional model.

3.14. Exhale Segmentation

To evaluate anomalies within recorded exhale flow sequences predicted by FieldNet, we present an automated analysis for segmenting exhales and scoring each based on a trained anomaly model. Flow sequences are translated into continuous time-series sequences to isolate individual exhale episodes through peak detection. This converts each sequence from a continuous waveform into a collection of individual exhales as shown in Figure 12. Each exhale segment is then resampled to match the input of the pre-processing autoencoder and anomaly networks with input size of $n = 256$.

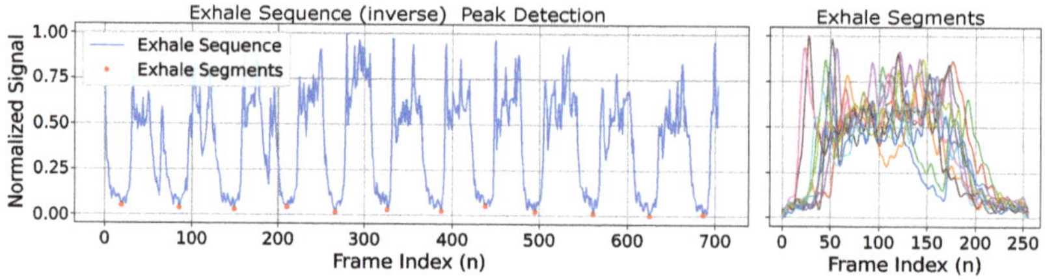

Figure 12. Exhale sequence peak detection and exhale segment extraction. Each recorded sequence is subdivided into individual exhales based on the minimal of each exhale. These are then consolidated into exhale segments (uniquely colored) used to train the filtering and anomaly detection models.

3.15. Filtering Model

Turbulent open-air flow captured within each exhale exhibits a significant variance. While analytical noise removal can be applied to reduce these behaviors, they typically maintain anomalous spikes due to flow field variance. To address this, we use a one-dimensional filtering autoencoder to capture the key behavioral characteristics of the recorded exhale datasets. This requires a training dataset composed of individual exhale segments of length (256, 1), where the expected value output is 1-to-1 with the provided input. The architecture of the autoencoder is presented in Figure 13. The model is composed of four layers (Conv1D, Conv1DTranspose) with rectified linear (relu) activation, kernel size $k = 7$, trained using the Adam [69] optimizer with learning rate of 1.0×10^{-3}. This model generates a filtered segment used for the anomaly detection model.

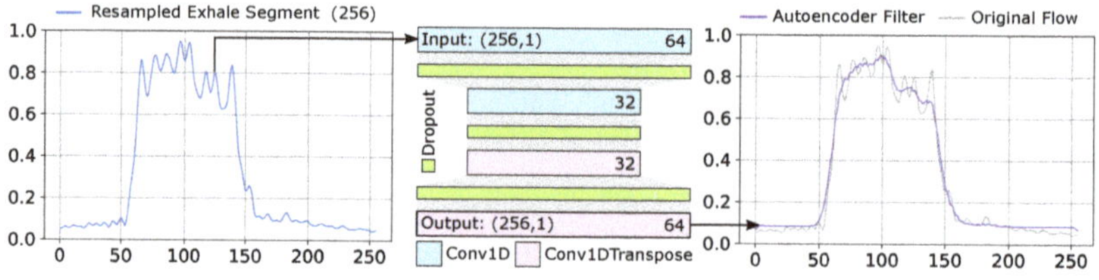

Figure 13. Segment filtering model. An autoencoder is employed as a noise reduction method to eliminate the high variance within each exhale segment caused by turbulent flow field magnitudes.

3.16. Anomaly Model

The aim of the proposed diagnostic is to contribute to the existing set of tests that can be performed to identify unique traits and possible anomalous behaviors within recorded breathing patterns. To provide the basis for model that can estimate anomalous behavior [70] within each exhale segment, we employ an *idealized* reference training set composed of instances from standardized exhale behaviors. These are exhales that exhibit the most regular behavior captured from subjects. This distinction is made on a simple health versus subject with pulmonary condition basis. To construct this the training set, we select ($n = 61$) individual exhale segments $f(t)_{Ref}$ and a cross-mixed set of ($m = 103$) exhale segments $f(t)_{Input}$ that contain both regularized and anomalous signatures for a total training set of $n \cdot m = 6283$. From these two sets, we form an error set by computing the absolute difference from each reference and input test sequence:

$e(t) = |f(t)_{Ref} - f(t)_{Input}|$. This results in an $(n \cdot m)$ training dataset size, with each instance is defined as $(x_{train} = f(t)_{Input}, y_{train} = e(t))$ and the predicted value $e(t)$ represents the anomaly error of the sequence. This provides a model that can predict the relative error between new exhale sequences and the reference set used to train the model. This model operates on the fixed length sequences of shape (256, 1) and is shown in Figure 14. The model uses rectified linear (relu) activation, kernel size $k = 7$, trained using the Adam [69] optimizer with learning rate of 1.0×10^{-3}. Predicted error metrics are then presented as an anomaly scale that identifies unique traits of input exhale segments.

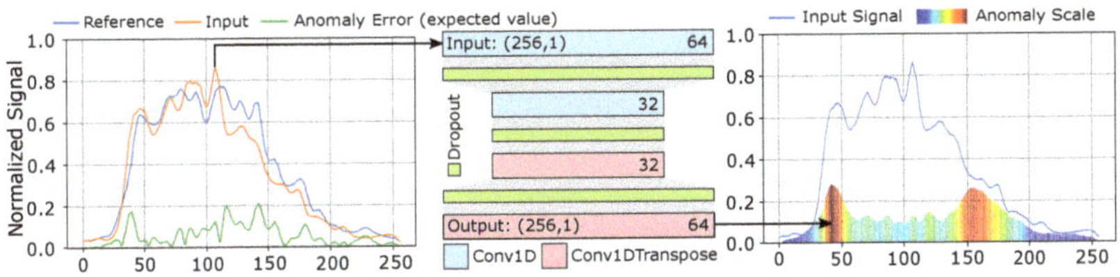

Figure 14. Anomaly model. Exhale instances are used to generate anomaly error waveforms by computing the absolute difference between the instance and the provided reference. This anomaly waveform is then used as the expected training value for the output of the network.

4. Results

Validation of the proposed method was performed using the data collected from 20 patients which provided twenty recordings of observable exhale flow. Each sequence contains 6–18 individual exhale episodes. These image sequences are provided to the processing pipeline that includes: (1) automated tracking, (2) localized flow field generation, (3) generation of the 1D exhale waveforms, and (4) anomaly detection. These results demonstrate the capability of the proposed method for capturing detailed natural exhale behaviors. This provides insight into how the natural breathing patterns of each subject result in different waveform characteristics that can be used to assist in the diagnostic process. By further evaluating the differences in these generated waveforms, anomalous behavioral analysis can be applied to identify potential variance in pulmonary function.

4.1. Localized Exhale Flow Prediction

The prediction of the trained model obtains flow characteristics within the spatial distribution, magnitude, and localized behaviors. In various instances, including those visualized in Figure 15, the spatial distribution of the flow vectors represents an estimated approximation of the ground truth; however, the distribution of the vectors with the largest magnitudes may not directly correlate. Additionally, due to the zero regions introduced by the facial masking, the model has residual non-zero vector values that are estimated. While these vectors have a relatively small magnitude, they still contribute to error within the measurement of the exhale flow behavior.

4.2. Individualized Exhale Behaviors

One of the primary objectives of this diagnostic is to provide a method for directly capturing exhale airflow behaviors without imposing any constraints on the subject. By eliminating the need for tubes and wearable solutions, the patient can resume subconscious breathing behaviors that are typically modified by exerting effort trying to maintain the constraints of the diagnostic system. This allows for the natural behaviors of each subject to be captured within the generated exhale waveform. This results in unique waveform characteristics for each subject including breathing rate, breath duration, and secondary metrics that can be obtained such as flow rate and volume. The generated exhale waveforms for select subjects have been presented within Figure 16. Each plot represents the recorded sequence waveform after processing. Each recording was performed for approximately 25 [s] for a total of 650 frames (samples) per waveform.

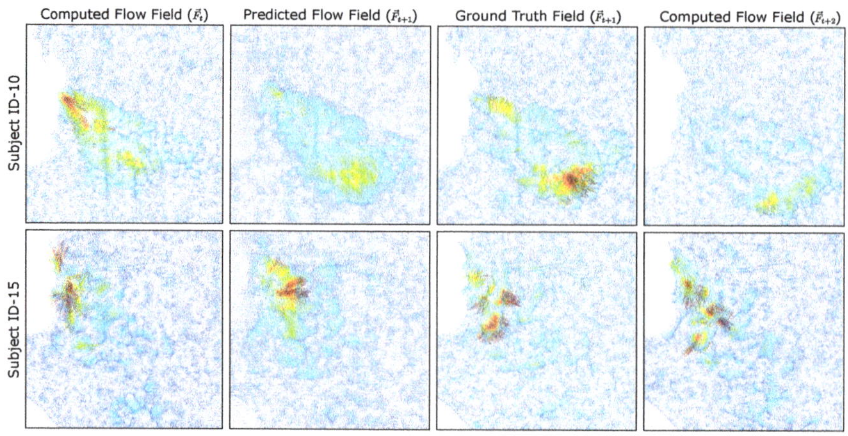

Figure 15. Multi-subject flow field interpolation. The interpolated result for arbitrary selected frames containing exhale flows are illustrated for three subjects. The ground truth field $\vec{F}_{t+1}$ is provided by the computed optical flow and compared to the predicted field $\vec{F}_{t+1}$. Color represents flow magnitude.

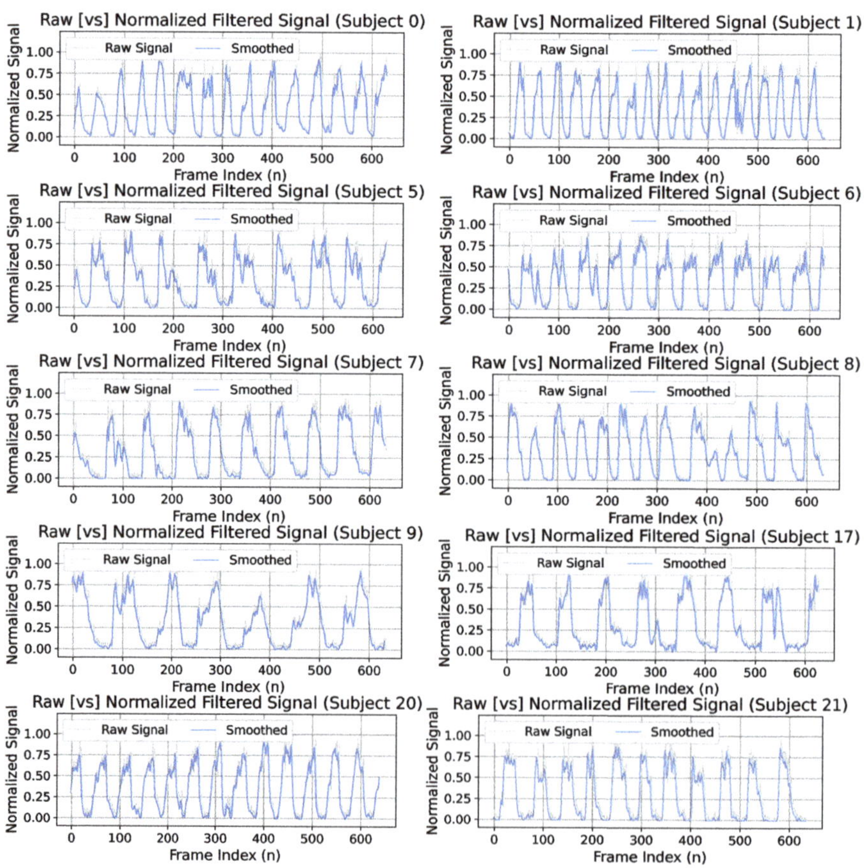

Figure 16. Select subject sequences (1–10). Each plot represents a collection of 500 flow frames that are used to generate expiatory waveforms. Each provides a unique subject exhale behaviors.

4.3. Anomaly Model Predictions

The anomaly model predicts an objective estimation of how much the provided exhale segment diverges from the training set reference. This provides two useful purposes: (1) we can automate the process of scoring and isolating exhale segments that reach a given anomaly threshold and (2) we establish the basis for performing secondary diagnostics related to condition specific exhale behavioral traits. The anomaly measurements predicted for six randomly selected exhale segments are shown in Figure 17. This creates a unique form of behavioral analysis that provides an door opening opportunity to explore how the direct visualisation of exhale behaviors can contribute to the unique signatures of specific pulmonary conditions. While this approach provides this unique opportunity, there are still several challenges associated with visualized exhale flow imaging for pulmonary function analysis. These are covered in our discussion.

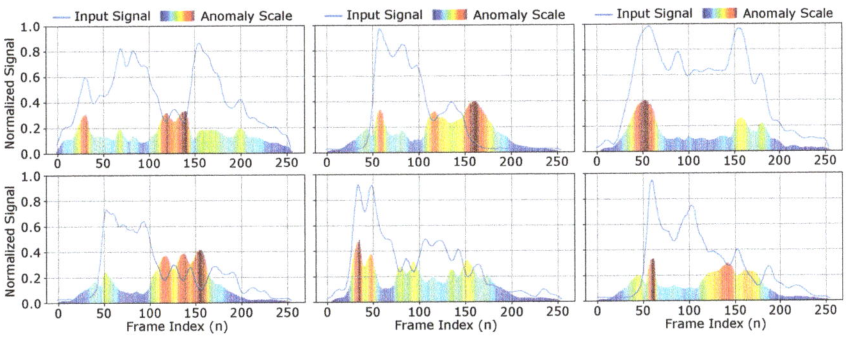

Figure 17. Anomaly results for select exhale segments. The input waveform is illustrated with the anomaly scale color mapped by magnitude. The regions of the waveform that illustrate anomalous behaviors result in higher predicted variance from the reference model.

5. Discussion

The primary goal of the presented non-contact method is to capture, isolate, and measure the unique exhale traits that can be used to aid respiratory diagnostics. Similar to many of the different pulmonary diagnostic technologies, this approach has a mixture of advantages and disadvantages. As with all vision-based techniques for evaluating pulmonary behavior, there are a set of immediate challenges including, resolution, frame rate, occlusion, and overall accuracy. While these aspects of vision-based approaches present challenges in capturing respiratory behavior, it is one of the only methods for directly capturing exhaled CO_2 flow at a distance. However, there are still factors that contribute to continued challenges with this approach. The first challenge is that there are physiological factors that may influence how recorded data reflects observable breathing patterns. While non-contact breathing can provide a unique look at pulmonary behavior, additional factors such as posture, mouth shape, and nasal blockages can contribute to variance within recorded observations. These factors partially contribute to the visualization obtained from the thermal camera, resulting in airflow differences. Experimental constraints can be used to minimize these factors, but they still may contribute to signal variance. The second is flow field estimation errors. There are instances where the face profile contributes to error within the generated flow sequences due to how the model will estimate flow contributions for null regions within the input flow. Other challenges are related to the formulation of the flow field model and signal generation. The FieldNet model can be used for two purposes: (1) interpolating between frames to increase the amount of 2D information translated into generated exhale waveforms and (2) to filter and isolate the exhale behavior to improve signal quality and reduce noise. The input of the model dictates that there must exist two flow states to predict an intermediary form. While this limits projective forecasting (or future state estimation) estimates, recordings of exhale behaviors are continuous within the collected data.

While this approach presents challenges associated with vision-based respiratory analysis, it also provides unique contributions towards the objective of obtaining natural, unrestricted breathing signals for diagnostic evaluation. This introduces the capability of identifying personalized characteristics of exhale behaviors by removing the conscious effort needed to maintain tube-based devices but also provides direct exhale flow observations that can be used to identify anomalies in regular breathing. By directly measuring the open airflow without impeding the subject's breathing, we can obtain distinct patterns that can be linked to physiological effort, types of breathing, and traits belonging to specific pulmonary conditions. As part of our result evaluation, we identified unique waveform signatures from each subject, indicating that this could potentially be the case.

As a secondary extension of this approach, the proposed metho d can combined with existing respiratory monitoring devices that monitor deformations of the chest or diaphragm to capture inhale/exhale behaviors. This combination could be used to infer pulmonological traits from chest movements by correlating movement signals with airflow obtained from the direct CO_2 flow visualization. By correlating visual flow data with existing methods for evaluating chest movement (ex. depth imaging, transducer belts), we can evaluate the relationship between chest/abdomen movement and exhaled CO_2 airflow. This could enable the ability to aid in identifying the level of muscular activation indirectly from observable flow patterns, providing diagnostic synergy with existing solutions.

6. Future Developments

The presented work provides a basis for performing behavioral analysis of expiratory airflow to identify key traits that contribute to the unique pulmonary signatures. The objective of this approach was to illustrate how this novel method for capturing airflow through a non-contact method to provide effort-independent behavioral patterns. To further the utility of this approach, additional studies conducted with healthy subject versus subjects with known conditions must be performed to isolate how these behaviors contribute to unique identifiers that could be used to classify specific conditions. This requires the extension of the study to a larger subject sample size to identify the key trends that vary between breathing behaviors of one subject versus cross-subject patterns. Additionally, the models presented within this work provide a functional proof-of-concept but would benefit from larger training datasets to identify if condition-specific behavioral traits can be isolated. These additional factors would then provide the basis for comparing healthy subjects to subjects with specific pulmonary conditions.

7. Conclusions

In this work, we presented a method for pulmonary evaluation aimed at extracting the unique features of exhale behaviors from thermal image CO_2 flow sequences. This provides a non-contact diagnostic for extracting exhale behaviors to obtain a natural and unrestricted measurement of expiratory airflow. Through the implementation of a convolutional neural network *FieldNet*, we evaluated how flow fields can be used to capture and isolate exhale behaviors from open-air recordings. Secondary convolutional networks were introduced and trained to reduce signal noise reduction and enable basic anomaly detection. This work provides the foundation for the classification of different pulmonary behaviors that can be identified through vision-based flow analysis. As part of ongoing research in direct exhale visualization and measurement, this approach provides an initial attempt at capturing individualized expiratory traits of natural breathing through CO_2 flow analysis. While this approach does not directly distinguish between biological traces or provide direct validation of an exact pulmonary condition, it can be utilized to identify key individualistic behaviors that can assist in the evaluation of long-term trends of natural breathing and suggest additional screening based on detected anomalous behaviors. This provides alternative screening diagnostic that can be performed in between the larger context of pulmonary monitoring devices including wearables, existing PFT devices, and advanced scanning technologies to model natural breathing behaviors.

Author Contributions: Conceptualization, S.T. and M.-H.C.; methodology S.T.; validation S.T., J.-S.C. and S.C.; investigation, J.-S.C., S.C. and D.-k.L.; data curation, S.T. and D.-k.L.; writing—original draft presentation, S.T. and M.-H.C.; visualization, S.T.; project administration M.H. and M.-H.C. All authors have read and agreed to the published version of the manuscript.

Funding: This research was partially supported by the Brain Pool program funded by the Ministry of Science and ICT through the National Research Foundation of Korea (NRF-2022H1D3A2A01093538) and was partially supported by Soonchunhyang University Research.

Institutional Review Board Statement: The study was conducted in accordance with the Institutional Review Board (IRB no. 2023-10-012) conducted at Soonchunghyang University Cheonan Hospital.

Informed Consent Statement: Informed consent was obtained from all subjects involved in the study.

Data Availability Statement: The data presented in this study are available on request from the corresponding author.

Acknowledgments: We would like to acknowledge the contributions of Ryan Van Valkenburg for research and development of tracking and image processing.

Conflicts of Interest: The founders had no role in the design of the study; in the collection, analyses, or interpretation of data; in the writing of the manuscript; or in the decision to publish the results.

References

1. Lopes, A.J. Advances in spirometry testing for lung function analysis. *Expert Rev. Respir. Med.* **2019**, *13*, 559–569. [CrossRef] [PubMed]
2. Pierce, R. Spirometry: An essential clinical measurement. *Aust. Fam. Physician* **2005**, *34*, 535–539. [PubMed]
3. Barreiro, T.J.; Perillo, I. An approach to interpreting spirometry. *Am. Fam. Physician* **2004**, *69*, 1107–1114. [PubMed]
4. Criée, C.; Sorichter, S.; Smith, H.; Kardos, P.; Merget, R.; Heise, D.; Berdel, D.; Köhler, D.; Magnussen, H.; Marek, W.; et al. Body plethysmography—Its principles and clinical use. *Respir. Med.* **2011**, *105*, 959–971. [CrossRef] [PubMed]
5. Coates, A.; Peslin, R.; Rodenstein, D.; Stocks, J. Measurement of lung volumes by plethysmography. *Eur. Respir. J.* **1997**, *10*, 1415–1427. [CrossRef] [PubMed]
6. Sun, X.; Yang, W.; Gong, S.; Liang, S.; Gu, S.; Lu, H.; Liu, J.; Xu, J. Diagnostic value of volumetric capnography in patients with chronic cough variant asthma. *Clinics* **2020**, *75*, e1662. [CrossRef] [PubMed]
7. Abid, A.; Mieloszyk, R.J.; Verghese, G.C.; Krauss, B.S.; Heldt, T. Model-Based Estimation of Respiratory Parameters from Capnography, with Application to Diagnosing Obstructive Lung Disease. *IEEE Trans. Biomed. Eng.* **2017**, *64*, 2957–2967. [PubMed]
8. John, R.E.S. End-Tidal Carbon Dioxide Monitoring. *Crit. Care Nurse* **2003**, *23*, 83–88. [CrossRef]
9. Bhandary, A.; Prabhu, G.A.; Rajinikanth, V.; Thanaraj, K.P.; Satapathy, S.C.; Robbins, D.E.; Shasky, C.; Zhang, Y.D.; Tavares, J.M.R.; Raja, N.S.M. Deep-learning framework to detect lung abnormality—A study with chest X-Ray and lung CT scan images. *Pattern Recognit. Lett.* **2020**, *129*, 271–278. [CrossRef]
10. Hu, S.; Hoffman, E.; Reinhardt, J. Automatic lung segmentation for accurate quantitation of volumetric X-ray CT images. *IEEE Trans. Med. Imaging* **2001**, *20*, 490–498. [CrossRef]
11. Romei, C.; Turturici, L.; Tavanti, L.; Miedema, J.; Fiorini, S.; Marletta, M.; Wielopolski, P.; Tiddens, H.; Falaschi, F.; Ciet, P. The use of chest magnetic resonance imaging in interstitial lung disease: A systematic review. *Eur. Respir. Rev.* **2018**, *27*, 180062. [CrossRef]
12. Frood, R.; Prestwich, R.; Tsoumpas, C.; Murray, P.; Franks, K.; Scarsbrook, A. Effectiveness of Respiratory-gated Positron Emission Tomography/Computed Tomography for Radiotherapy Planning in Patients with Lung Carcinoma—A Systematic Review. *Clin. Oncol.* **2018**, *30*, 225–232. [CrossRef]
13. Zhu, Z.; Fei, J.; Pavlidis, I. Tracking human breath in infrared imaging. In Proceedings of the Fifth IEEE Symposium on Bioinformatics and Bioengineering (BIBE 2005), Minneapolis, MN, USA, 19–21 October 2005; IEEE: Piscataway, NJ, USA, 2005; pp. 227–231.
14. Murthy, R.; Pavlidis, I.; Tsiamyrtzis, P. Touchless monitoring of breathing function. In Proceedings of the 26th Annual International Conference of the IEEE Engineering in Medicine and Biology Society, San Francisco, CA, USA, 1–5 September 2004; pp. 1196–1199.
15. Kakavas, S.; Kotsiou, O.S.; Perlikos, F.; Mermiri, M.; Mavrovounis, G.; Gourgoulianis, K.; Pantazopoulos, I. Pulmonary function testing in COPD: Looking beyond the curtain of FEV1. *NPJ Prim. Care Respir. Med.* **2021**, *31*, 23. [CrossRef] [PubMed]
16. Tomasic, I.; Tomasic, N.; Trobec, R.; Krpan, M.; Kelava, T. Continuous remote monitoring of COPD patients—Justification and explanation of the requirements and a survey of the available technologies. *Med. Biol. Eng. Comput.* **2018**, *56*, 547–569. [CrossRef]
17. Lecompte-Osorio, P.; Pearson, S.D.; Pieroni, C.H.; Stutz, M.R.; Pohlman, A.S.; Lin, J.; Hall, J.B.; Htwe, Y.M.; Belvitch, P.G.; Dudek, S.M.; et al. Bedside estimates of dead space using end-tidal CO_2 are independently associated with mortality in ARDS. *Crit. Care* **2021**, *25*, 333. [CrossRef]
18. Li, X.; Ma, X. Acute respiratory failure in COVID-19: Is it "typical" ARDS? *Crit. Care* **2020**, *24*, 198. [CrossRef]

19. Masclans, J.R.; Roca, O.; Muñoz, X.; Pallisa, E.; Torres, F.; Rello, J.; Morell, F. Quality of Life, Pulmonary Function, and Tomographic Scan Abnormalities After ARDS. *Chest* **2011**, *139*, 1340–1346. [CrossRef] [PubMed]
20. An, J.Y.; Shin, H.J.; Yang, M.; Park, D.-Y.; Yang, J.; Kim, H.J. Non-contact diagnosis of sleep breathing disorders using infrared optical gas imaging: A prospective observational study. *Sci. Rep.* **2022**, *12*, 21052. [CrossRef] [PubMed]
21. Jakkaew, P.; Onoye, T. Non-Contact Respiration Monitoring and Body Movements Detection for Sleep Using Thermal Imaging. *Sensors* **2020**, *20*, 6307. [CrossRef]
22. Verwey, C.; Nunes, M.C.; Dangor, Z.; Madhi, S.A. Pulmonary function sequelae after respiratory syncytial virus lower respiratory tract infection in children: A systematic review. *Pediatr. Pulmonol.* **2020**, *55*, 1567–1583. [CrossRef]
23. Ronneberger, O.; Fischer, P.; Brox, T. U-Net: Convolutional Networks for Biomedical Image Segmentation. *arXiv* **2015**, arXiv:1505.04597.
24. Fischer, P.; Dosovitskiy, A.; Ilg, E.; Häusser, P.; Hazirbas, C.; Golkov, V.; van der Smagt, P.; Cremers, D.; Brox, T. FlowNet: Learning Optical Flow with Convolutional Networks. *arXiv* **2015**, arXiv:1504.06852,
25. Horn, B.K.; Schunck, B.G. Determining optical flow. *Artif. Intell.* **1981**, *17*, 185–203. [CrossRef]
26. Costanzo, I.; Sen, D.; Rhein, L.; Guler, U. Respiratory Monitoring: Current State of the Art and Future Roads. *IEEE Rev. Biomed. Eng.* **2022**, *15*, 103–121. [CrossRef] [PubMed]
27. Massaroni, C.; Nicolò, A.; Sacchetti, M.; Schena, E. Contactless Methods For Measuring Respiratory Rate: A Review. *IEEE Sens. J.* **2021**, *21*, 12821–12839. [CrossRef]
28. Vanegas, E.; Igual, R.; Plaza, I. Sensing Systems for Respiration Monitoring: A Technical Systematic Review. *Sensors* **2020**, *20*, 5446. [CrossRef] [PubMed]
29. Fekr, A.R.; Radecka, K.; Zilic, Z. Design and Evaluation of an Intelligent Remote Tidal Volume Variability Monitoring System in e-Health Applications. *IEEE J. Biomed. Health Informatics* **2015**, *19*, 1532–1548. [CrossRef] [PubMed]
30. Folke, M.; Cernerud, L.; Ekström, M.; Hök, B. Critical review of non-invasive respiratory monitoring in medical care. *Med Biol. Eng. Comput.* **2003**, *41*, 377–383. [CrossRef] [PubMed]
31. Lucidarme, O.; Coche, E.; Cluzel, P.; Mourey-Gerosa, I.; Howarth, N.; Grenier, P. Expiratory CT scans for chronic airway disease: Correlation with pulmonary function test results. *Am. J. Roentgenol.* **1998**, *170*, 301–307. [CrossRef]
32. Sharma, P.; Hui, X.; Zhou, J.; Conroy, T.B.; Kan, E.C. Wearable radio-frequency sensing of respiratory rate, respiratory volume, and heart rate. *npj Digit. Med.* **2020**, *3*, 98. [CrossRef]
33. Piuzzi, E.; Pisa, S.; Pittella, E.; Podestà, L.; Sangiovanni, S. Wearable Belt With Built-In Textile Electrodes for Cardio—Respiratory Monitoring. *Sensors* **2020**, *20*, 4500. [CrossRef] [PubMed]
34. Vitazkova, D.; Foltan, E.; Kosnacova, H.; Micjan, M.; Donoval, M.; Kuzma, A.; Kopani, M.; Vavrinsky, E. Advances in Respiratory Monitoring: A Comprehensive Review of Wearable and Remote Technologies. *Biosensors* **2024**, *14*, 90. [CrossRef] [PubMed]
35. Deepika, C.; Kandaswamy, A.; Pradeepa, G. An efficient method for detection of inspiration phase of respiration in thermal imaging. *J. Sci. Ind. Res.* **2016**, *75*, 40–44.
36. Hanawa, D.; Ohguchi, T.; Oguchi, K. Basic study on non-contact measurement of human oral breathing by using far infra-red imaging. In Proceedings of the International Conference on Telecommunications and Signal Processing (TSP'16), Vienna, Austria, 27–29 June **2016**; pp. 681–684.
37. Abbas, A.K.; Leonhardt, S. Intelligent neonatal monitoring based on a virtual thermal sensor. *BMC Med. Imaging* **2014**, *14*, 9. [CrossRef] [PubMed]
38. Murthy, J.; Van Jaarsveld, J.; Fei, J.; Pavlidis, I.; Harrykissoon, R.I.; Lucke, J.F.; Faiz, S.; Castriotta, R.J. Thermal Infrared Imaging: A Novel Method to Monitor Airflow during Polysomnography. *Sleep* **2009**, *32*, 1521–1527. [CrossRef]
39. Chekmenev, S.; Rara, H.; Farag, A.A. Non-contact, wavelet-based measurement of vital signs using thermal imaging. In Proceedings of the First International Conference on Graphics, Vision, and Image Processing (GVIP), Cairo, Egypt, 19 December 2005; pp. 107–112.
40. Adhikari, A.; Hetherington, A.; Sur, S. mmFlow: Facilitating At-Home Spirometry with 5G Smart Devices. In Proceedings of the IEEE (SECON'18), Rome, Italy, 6–9 July 2021; pp. 1–9.
41. Hillyard, P.; Luong, A.; Abrar, A.S.; Patwari, N.; Sundar, K.; Farney, R.; Burch, J.; Porucznik, C.A.; Pollard, S.H. Comparing Respiratory Monitoring Performance of Commercial Wireless Devices. *arXiv* **2018**, arXiv:1807.06767.
42. Ponce, M.C.; Sankari, A.; Sharma, S. *Pulmonary Function Tests*; StatPearls Publishing: Treasure Island, FL, USA, 2023.
43. Ohno, Y.; Seo, J.B.; Parraga, G.; Lee, K.S.; Gefter, W.B.; Fain, S.B.; Schiebler, M.L.; Hatabu, H. Pulmonary Functional Imaging: Part 1—State-of-the-Art Technical and Physiologic Underpinnings. *Radiology* **2021**, *299*, 508–523. [CrossRef]
44. Gefter, W.B.; Lee, K.S.; Schiebler, M.L.; Parraga, G.; Seo, J.B.; Ohno, Y.; Hatabu, H. Pulmonary Functional Imaging: Part 2—State-of-the-Art Clinical Applications and Opportunities for Improved Patient Care. *Radiology* **2021**, *299*, 524–538. [CrossRef] [PubMed]
45. Huang, Z.; Shi, X.; Zhang, C.; Wang, Q.; Cheung, K.C.; Qin, H.; Dai, J.; Li, H. FlowFormer: A Transformer Architecture for Optical Flow. In *Proceedings of the Computer Vision—ECCV 2022*; Avidan, S., Brostow, G., Cissé, M., Farinella, G.M., Hassner, T., Eds.; Springer Nature: Cham, Switzerland, 2022; pp. 668–685.
46. Lin, K.Y.; Chen, D.Y.; Tsai, W.J. Image-based motion-tolerant remote respiratory rate evaluation. *IEEE Sens. J.* **2016**, *16*, 3263–3271. [CrossRef]
47. He, S.; Han, Z.; Iglesias, C.; Mehta, V.; Bolic, M. A Real-Time Respiration Monitoring and Classification System Using a Depth Camera and Radars. *Front. Physiol.* **2022**, *13*, 799621. [CrossRef]

48. Addison, A.P.; Addison, P.S.; Smit, P.; Jacquel, D.; Borg, U.R. Noncontact Respiratory Monitoring Using Depth Sensing Cameras: A Review of Current Literature. *Sensors* **2021**, *21*, 1135. [CrossRef] [PubMed]
49. Maurya, L.; Zwiggelaar, R.; Chawla, D.; Mahapatra, P. Non-contact respiratory rate monitoring using thermal and visible imaging: A pilot study on neonates. *J. Clin. Monit. Comput.* **2023**, *37*, 815–828. [CrossRef] [PubMed]
50. Manullang, M.C.T.; Lin, Y.H.; Lai, S.J.; Chou, N.K. Implementation of Thermal Camera for Non-Contact Physiological Measurement: A Systematic Review. *Sensors* **2021**, *21*, 7777. [CrossRef]
51. Shu, S.; Liang, H.; Zhang, Y.; Zhang, Y.; Yang, Z. Non-contact measurement of human respiration using an infrared thermal camera and the deep learning method. *Meas. Sci. Technol.* **2022**, *33*, 075202. [CrossRef]
52. Yang, F.; He, S.; Sadanand, S.; Yusuf, A.; Bolic, M. Contactless Measurement of Vital Signs Using Thermal and RGB Cameras: A Study of COVID 19-Related Health Monitoring. *Sensors* **2022**, *22*, 627. [CrossRef] [PubMed]
53. Takahashi, Y.; Gu, Y.; Nakada, T.; Abe, R.; Nakaguchi, T. Estimation of Respiratory Rate from Thermography Using Respiratory Likelihood Index. *Sensors* **2021**, *21*, 4406. [CrossRef] [PubMed]
54. Jia, Y.; Du, H.; Wang, H.; Yu, R.; Fan, L.; Xu, G.; Zhang, Q. Automatic early Smoke Segmentation Based on Conditional Generative Adversarial Networks. *Optik* **2019**, *193*, 162879. [CrossRef]
55. Martins, L.; Guede-Fernández, F.; Valente de Almeida, R.; Gamboa, H.; Vieira, P. Real-Time Integration of Segmentation Techniques for Reduction of False Positive Rates in Fire Plume Detection Systems during Forest Fires. *Remote Sens.* **2022**, *14*, 2701. [CrossRef]
56. Nguyen, A.Q.; Nguyen, H.T.; Tran, V.C.; Pham, H.X.; Pestana, J. A Visual Real-Time Fire Detection Using Single Shot MultiBox Detector for UAV-Based Fire Surveillance. In Proceedings of the 2020 IEEE Eighth International Conference on Communications and Electronics (ICCE), Phu Quoc, Vietnam, 13–15 January 2021; pp. 338–343.
57. Saponara, S.; Elhanashi, A.; Gagliardi, A. Real-time video fire/smoke detection based on CNN in antifire surveillance systems. *J. Real-Time Image Process.* **2021**, *18*, 889–900. [CrossRef]
58. Guo, Y.; Cao, X.; Liu, B.; Gao, M. Cloud Detection for Satellite Imagery Using Attention-Based U-Net Convolutional Neural Network. *Symmetry* **2020**, *12*, 1056. [CrossRef]
59. Xie, W.; Liu, D.; Yang, M.; Chen, S.; Wang, B.; Wang, Z.; Xia, Y.; Liu, Y.; Wang, Y.; Zhang, C. SegCloud: A novel cloud image segmentation model using a deep convolutional neural network for ground-based all-sky-view camera observation. *Atmos. Meas. Tech.* **2020**, *13*, 1953–1961. [CrossRef]
60. Kopaczka, M.; Breuer, L.; Schock, J.; Merhof, D. A Modular System for Detection, Tracking and Analysis of Human Faces in Thermal Infrared Recordings. *Sensors* **2019**, *19*, 4135. [CrossRef]
61. Setjo, C.H.; Achmad, B. Thermal image human detection using Haar-cascade classifier. In Proceedings of the 2017 7th International Annual Engineering Seminar (InAES), Yogyakarta, Indonesia, 1–2 August 2017; pp. 1–6.
62. Lucas, B.D.; Kanade, T. An iterative image registration technique with an application to stereo vision. In Proceedings of the 7th International Joint Conference on Artificial Intelligence, San Francisco, CA, USA, 28 August 1981; IJCAI'81; Volume 2, pp. 674–679.
63. Alvarez, L.; Weickert, J.; Sánchez, J. Reliable Estimation of Dense Optical Flow Fields with Large Displacements. *Int. J. Comput. Vis.* **2000**, *39*, 41–56. [CrossRef]
64. Sobel, I. An Isotropic 3 × 3 Image Gradient Operator. In *Presentation at Stanford A.I. Project 1968*; Academic Press: Cambridge, MA, USA, 2014.
65. Jakob, W.; Rhinelander, J.; Moldovan, D. Github: PyBind11. Available online: https://github.com/pybind/pybind11 (accessed on 31 May 2024).
66. Fortun, D.; Bouthemy, P.; Kervrann, C. Optical flow modeling and computation: A survey. *Comput. Vis. Image Underst.* **2015**, *134*, 1–21. [CrossRef]
67. Chollet, F. Keras: Deep Learning for Humans. Software. 2015. Available online: https://keras.io (accessed on 31 May 2024).
68. Abadi, M.; Agarwal, A.; Barham, P.; Brevdo, E.; Chen, Z.; Citro, C.; Corrado, G.S.; Davis, A.; Dean, J.; Devin, M.; et al. TensorFlow: Large-Scale Machine Learning on Heterogeneous Systems. Software. 2015. Available online: https://tensorflow.org (accessed on 31 May 2024).
69. Kingma, D.; Ba, J. Adam: A Method for Stochastic Optimization. *arXiv* **2014**, arXiv:1412.6980.
70. Golgowski, M.; Osowski, S. Anomaly detection in ECG using wavelet transformation. In Proceedings of the 2020 IEEE 21st International Conference on Computational Problems of Electrical Engineering (CPEE'20), Online, 16–19 September **2020**; pp. 1–4.

diagnostics

MDPI

Article

Topotecan in a Real-World Small-Cell Lung Cancer Cohort: Prognostic Biomarkers Improve Selection of Patients for Second-Line Treatment

Laura Lambrecht [1,2], Paola Arnold [1,2], Jürgen Behr [1,2], Pontus Mertsch [1,2], Amanda Tufman [1,2] and Diego Kauffmann-Guerrero [1,2,*]

1 Department of Medicine V, University Hospital, LMU Munich, 81337 Munich, Germany; laura.lambrecht@med.uni-muenchen.de (L.L.); paola.arnold@med.uni-muenchen.de (P.A.); juergen.behr@med.uni-muenchen.de (J.B.); pontus.mertsch@med.uni-muenchen.de (P.M.); amanda.tufman@med.uni-muenchen.de (A.T.)

2 Comprehensive Pneumology Center (CPC), Member of the German Center for Lung Research (DZL), University of Munich (LMU), 81337 Munich, Germany

* Correspondence: diego.kauffmannguerreo@med.uni-muenchen.de

Abstract: Background: Small-cell lung cancer (SCLC) is a highly aggressive tumor, and overall survival (OS) remains poor despite intensive efforts to develop new treatment strategies. In second line, topotecan is the only approved drug, with a median OS of 5.9 months. However, real-world SCLC patients are often in worse condition and harbor more comorbidities than study populations. Therefore, the real-world performance of topotecan may differ from that seen in studies. Here, we analyzed outcomes of SCLC patients receiving topotecan and identified predictive and prognostic markers. Patients and Methods: We retrospectively analyzed 44 consecutive SCLC patients receiving topotecan between 2015 and 2022. We analyzed baseline characteristics (age, ECOG-PS, topotecan cycles, and dosage) and pre-treatment blood values (LDH, CRP, sodium) as well as prognostic scores (neutrophil/lymphocyte ratio (NLR), thrombocyte/lymphocyte ratio (TLR), Glasgow Prognostic Score, prognostic nutritional score, systemic inflammation index (SII), and the prognostic index) extracted from electronic patients' charts to identify predictive and prognostic markers. Results: In our cohort, mPFS and mOS were only 1.9 and 5.6 months, respectively. Gender, ECOG-PS, active brain metastases, NLR, GPS, PNI, and SII significantly influenced PFS and OS in univariate analysis. ECOG-PS ($p > 0.001$), active brain metastases ($p = 0.001$), and SII ($p = 0.008$) were significant independent prognostic variables in a multivariate COX regression model. Selecting patients by these three markers achieved an mPFS of 5.7 months and thus increased the mPFS three-fold. Patients not meeting all criteria had an mPFS of 1.8 months ($p = 0.006$). Patients identified by prognostic markers had an mOS of 9.1 months ($p = 0.002$). Conclusions: The efficacy of topotecan in SCLC real-world patients is poor, indicating that many patients were treated without any benefit. Easy-to-obtain markers can predict response and treatment efficacy and should therefore be validated in larger cohorts to identify patients who are more likely to benefit from topotecan.

Keywords: SCLC; topotecan; chemotherapy; prognostic markers

1. Introduction

Small-cell lung cancer (SCLC) accounts for about 15% of all diagnosed lung cancers and is characterized by rapid and aggressive growth [1]. Despite good initial response to first-line chemotherapy, median survival of advanced SCLC patients remains poor, ranging from 7 to 10 months. The 1-year overall survival is only about 20–40% [1,2]. This is mainly due to early relapse after first-line therapy and lack of effective therapies in further treatment lines [3].

Topotecan is the only FDA- and EMA-approved second-line treatment in SCLC based on the observation that topotecan versus cyclophosphamide, doxorubicin, and vincristine (CAV) in patients relapsing at least 60 days after first-line treatment resulted in a numerical higher response rate of 24.3% (topotecan) compared to 18.3% (CAV). Median overall survival (OS) was 5.7 months in both groups, but the topotecan group achieved a more favorable toxicity profile [4]. O'Brien et al. compared oral topotecan versus best supportive care (BSC) in patients with relapsed SCLC not considered for standard intravenous therapy and found a significantly longer overall survival in the topotecan group with median survival times of 25.9 weeks compared to 13.9 weeks in the BSC group [5]. Since this time, despite many clinical trials evaluating promising agents, no further second-line treatment achieved approval for SCLC patients in Europe.

However, topotecan is associated with relevant, especially hematological side effects, and most patients do no benefit from this treatment [6]. Furthermore, real-world SCLC patients are mostly in worse condition and harbor more comorbidities than study populations. Therefore, one has to fear that the small survival advantage of topotecan might vanish in a real-world cohort. Furthermore, although topotecan is the only approved drug for the second-line treatment of patients with SCLC, we still have no verified markers to forecast the tolerance and effectiveness of the treatment in individual patients.

Reliable predictive and prognostic markers are therefore needed to select patients with higher probability for response to topotecan. For this reason, this study aims at investigating outcomes of real-world SCLC patients receiving topotecan and tries to identify predictive and prognostic markers.

2. Materials and Methods

2.1. Study Population

We identified all patients with SCLC treated with topotecan between 2015 and 2022 at our tertiary care lung cancer center. Patients were identified using our electronic cancer documentation system and our pharmacy delivery repository searching for all patients documented with SCLC of the lung receiving topotecan. Results have been verified by pathological reports and clinical documentation. Patient characteristics in terms of gender, age, performance status, tumor stage, smoking status, information on treatment modalities, and dosage were collected from patient charts (paper-based and electronic records). Age was calculated at the time of diagnosis. Treatment response was evaluated by CT- or PET-CT-based monitoring using the Response Evaluation Criteria in Solid Tumors 1.1 (RECIST) [7]. We did not use iRECIST criteria in patients receiving immunotherapy treatment before topotecan because of relevant varying times since immuno-oncology teratments (IO). Progression-free survival (PFS) was calculated from the date of first topotecan dose until confirmed disease progression or death. Overall survival (OS) was calculated from the date of first topotecan dose until death. Pre treatment blood values were extracted from the electronic patients' charts. The time frame in which the values have to be recorded was between 0–10 days before treatment start.

2.2. Statistical Analyses

For data description, mean values and standard deviations were used. Comparisons between groups were performed by the Mann–Whitney-U test, or by chi-square-tests in case of categorical variables. Estimation of progression-free survival (PFS) and overall survival (OS) was carried out using the Kaplan–Meier method with the logrank test. For multivariate survival analysis, COX-regression models were used, including variables with significant results in the univariate analysis. Cut-offs for metric valiables are set by the upper or lower reference value of our laboratory for clinical chemistry (blood values) or by cut-off from the literature or our own publications (calculated values) [8,9]. The level of statistical significance was determined at $p < 0.05$. All statistical analyses were performed using SPSS 29 statistical software (IBM Corp., Armonk, NY, USA).

3. Results

We analyzed all patients with SCLC receiving topotecan between 2015 und 2022. The cohort showed a light male predominance (56.8%), and the mean age was 64.3 years. All patients had stage IV disease confirmed by CT or PET-CT scans before treatment start with topotecan and were active or former smokers. A total of 43.2% harbored reduced performance status (ECOG-PS 2 or higher), and 34.1% had active brain metastases at the start of topotecan. About a third were pretreated with chemo-immunotherapy (atezolizumab) and 29.5% received radio-chemotherapy (RCT) in first-line treatment. These patients received either concurrent or sequential RCT in a limited stage as a first-line treatment. Only two patients received thoracic consolidative radiotherapy after systemic treatment in stage IV. Those are listed under chemo(-immune) therapy. Table 1 summarizes baseline characteristics of the study cohort.

Table 1. Baseline characteristics of the study cohort. ECOG-PS (Eastern Cooperative Oncology Group Performance Status), IO (immune oncology), WBRT (whole-brain radiation therapy), Sdev (standard deviation), PR (partial response), SD (stable disease), PD (progressive disease).

		***n* = 44**
Male (*n* (%))		25 (56.8%)
Female (*n* (%))		19 (43.2%)
Age (mean ± Sdev)		64.3 ± 9.9
	Males	65.52 ± 10.4
	Females	62.6 ± 9.2
ECOG-PS		
	0	6 (13.6%)
	1	19 (43.2%)
	2	14 (31.8%)
	3	5 (11.4%)
Smoking status (*n* (%))		
	Current smoker	9 (20.5%)
	Former smoker	35 (79.5%)
Pharmaceutical form (*n* (%))		
	Intravenous	40 (90.9%)
	Oral	4 (9.1%)
Median of applied cycles [range]		3 (1–17)
Mean applied dosage [range]		1.22 mg/m^2 [0.625–2.4 mg/m^2]
Median time from first diagnosis until topotecan		8.3 months
Median time from last treatment until topotecant (treatment free intervall, TFI)		1.7 months
	<3 months (*n* (%))	32 (72.7)
	≥3 months	12 (27.3)
Prior IO treatment (*n* (%))		
	Yes	15 (34.1%)
	No	29 (65.9%)
Active brain metastases (*n* (%))		
	Yes	15 (34.1%)
	No	29 (64.9%)

Table 1. *Cont.*

	n = 44
WBRT before topotecan (n (%))	
Yes	20 (45.5%)
No	24 (54.5%)
1st-line treatment (n (%))	
Chemo(-immuno) therapy	31 (70.5%)
Radiochemotherapy	13 (29.5%)
Best response to topotecan treatment (n (%))	
PR	6 (13.6%)
SD	8 (18.2%)
PD	23 (52.3%)
Death	7 (15.9%)

3.1. Survival Analysis of the Whole Cohort

The median PFS of topotecan treatment in the whole cohort was only 1.9 months [CI 95% 1.5–2.3] (Figure 1A), while overall survival was 5.6 months [CI 95% 2.4–8.8] (Figure 1B). Overall response rate (ORR) was 13.6% with a disease control rate (DCR) of 31.8%. The median duration of response (mDoR) was 5.1 months [CI 95% 3.7–6.7].

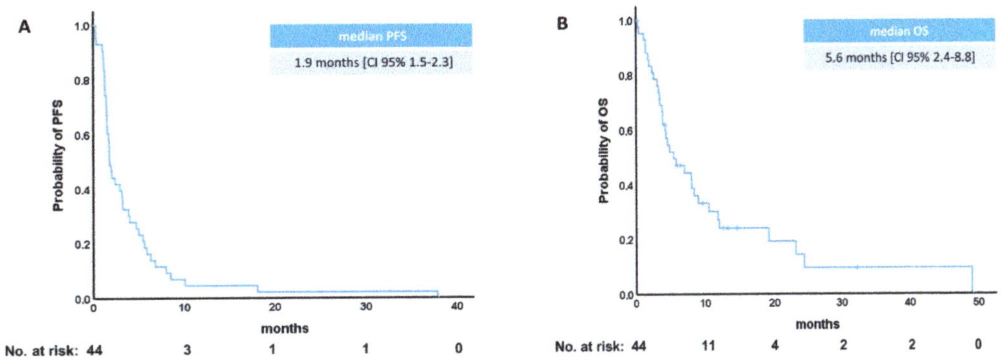

Figure 1. Survival curves of the study cohort. (**A**) Progression-free survival (PFS). (**B**) Overall survival (OS).

Interestingly, women showed significantly improved median PFS (3.3 vs. 1.9 months) and median OS (8.5 vs. 4.0 months) compared to male patients (Figure 2A,B).

As expected, ECOG performance status (PS) did significantly influence PFS and OS, showing that patients with ECOG-PS of 2 or higher had no meaningful benefit from topotecan treatment (Figure 3A,B). Moreover, older patients (≥70 years) had no reduced survival per se.

Time from first diagnosis of SCLC until topotecan treatment and treatment-free interval (TFI) before topotecan did not affect survival after topotecan treatment as well as prior treatment with atezolizumab. Furthermore, mean topotecan dosage did not change PFS, but, interestingly, patients with lower mean dosages (<1.25 mg/m^2) had numerically longer OS. Patients without active brain metastases at the beginning of topotecan treatment had significantly longer PFS, which did not translate into an OS advantage. On the other hand, patients with and without prior WBRT did not differ regarding their survival. First-line treatment, sole systemic therapy, or radio-chemotherapy (RCT) had no significant impact on survival as well as type of systemic first-line treatment (with or without IO). However, patients with initial RCT had a meaningful longer OS with 12.0 months compared to patients with systemic treatment (Table 2).

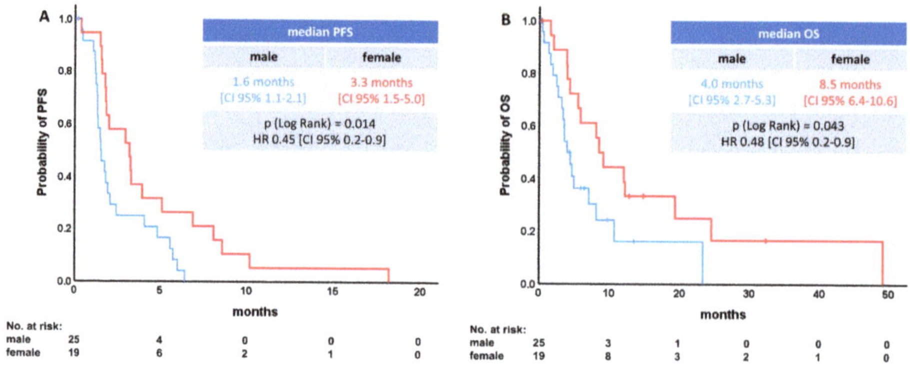

Figure 2. Survival differences between male and female patients. (**A**) Progression-free survival (PFS). (**B**) Overall survival (OS).

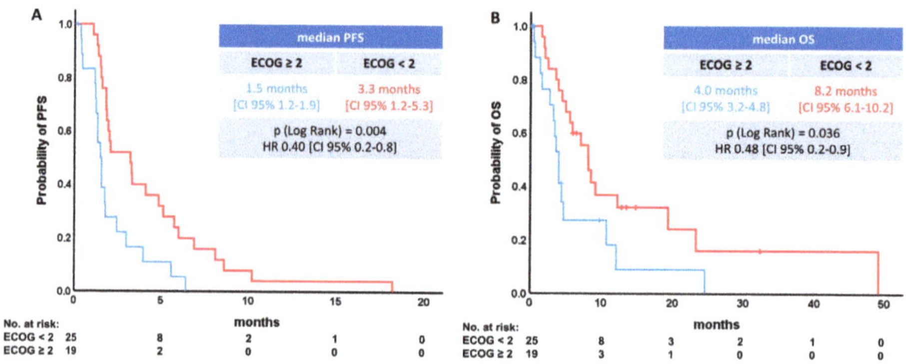

Figure 3. Survival differences stratified by ECOG-Performance status. (**A**) Progression-free survival (PFS). (**B**) Overall survival (OS).

Table 2. Univariate survival analyses of the study cohort stratified by baseline characteristics. PFS (progression-free survival) in months, OS (overall survival) in months, HR (hazard ratio), CI (confidence interval).

	Median PFS [CI 95%]	HR [CI 95%]	*p*-Value [Log Rank]	Median OS [CI 95%]	HR [CI 95%]	*p*-Value [Log Rank]
Gender						
Male	1.6 [1.1–2.1]		$p = 0.014$	4.0 [2.7–5.3]		$p = 0.043$
Female	3.3 [1.5–5.0]	0.45 [0.2–0.9]		8.5 [6.4–10.6]	0.48 [0.2–0.9]	
ECOG						
<2	3.3 [1.2–5.3]	0.40 [0.2–0.8]	$p = 0.004$	8.2 [6.1–10.2]	0.48 [0.2–0.9]	$p = 0.036$
≥2	1.5 [1.2–1.9]			4.0 [3.2–4.8]		
Age						
<70 years	1.9 [0.4–3.5]		$p = 0.433$	5.9 [2.0–9.8]		$p = 0.713$
≥70 years	1.9 [1.5–2.3]	1.2 [0.6–2.6]		4.0 [3.4–4.6]	1.1 [0.5–2.6]	

Table 2. *Cont.*

	Median PFS [CI 95%]	HR [CI 95%]	*p*-Value [Log Rank]	Median OS [CI 95%]	HR [CI 95%]	*p*-Value [Log Rank]
Time until topotecan						
≤8.3 months	1.9 [1.5–2.3]		$p = 0.719$	4.4 [3.6–5.1]		$p = 0.484$
>8.3 months	1.9 [0.3–3.5]	0.66 [0.3–1.3]		8.6 [4.4–12.7]	0.78 [0.4–1.6]	
Median time from last treatment until topotecant (treatment free interval, TFI)						
<3 months (*n* (%))	1.9 [1.6–2.2]	1.9 [0.6–5.8]	$p = 0.205$	5.0 [3.4–6.5]	1.5 [0.1–2.5]	$p = 0.711$
≥3 months	1.9 [0.0–5.8]			9.1 [1.7–16.6]		
Mean topotecan dosage						
<1.25 mg/m^2	3.3 [1.2–5.3]		$p = 0.547$	8.1 [3.9–12.3]		$p = 0.111$
≥1.25 mg/m^2	1.6 [1.2–2.0]	1.4 [0.8–2.7]		4.3 [3.1–5.7]	1,7 [0.9–3.5]	
Prior IO treatment						
Yes	1.8 [1.2–2.4]		$p = 0.903$	5.0 [0.8–9.1]		$p = 0.855$
No	2.1 [0.4–3.9]	0.9 [0.5–1.6]		5.9 [0.5–11.4]	2.1 [0.5–8.8]	
Active brain metastases						
No	3.0 [1.5–4.6]		$p = 0.028$	7.1 [2.5–11.7]		$p = 0.633$
Yes	1.6 [1.3–1.9]	1.9 [1.0–3.6]		4.7 [3.7–5.7]	1.2 [0.6–2.6]	
WBRT before topotecan						
No	2.1 [0.0–4.4]		$p = 0.475$	7.1 [2.5–11.7]		$p = 0.571$
Yes	1.8 [1.7–2.0]	1.1 [0.6–2.0]		5.0 [2.7–7.2]	0.8 [0.4–1.6]	
1st-line treatment						
Chemo(-immuno) therapy	1.8 [0.9–2.8]		$p = 0.638$	4.4 [3.6–5.2]		$p = 0.089$
Radiochemotherapy	1.9 [1.5–2.4]	0.7 [0.3–1.4]		12.0 [6.8–17.2]	0.5 [0.2–1.1]	

3.2. *Toxicity*

As expected, hematological side effects were common. A total of 88% of all patients had any grade of anemia, with grade 3 and 4 accounting for 45.5% and 4.5%, respectively. A total of 65.9% had thrombocytopenia (grade 3: 22.7% and grade 4: 27.3%). Leucopenia was observed in 66% of the patients (grade 3: 22.7% and grade 4: 22.7%).

3.3. *Prediction of Response*

Higher topotecan dosage (≥1.25 mg/m^2) was significantly associated with better response to treatment ($p = 0.010$). Besides this, we found no further predictive markers for treatment response for either categorial or for continuous variables in our cohort (Tables 3 and 4).

Table 3. Categorial and continuous variables tested for impact on response of topotecan treatment.

Categorial Variables			
Variable	***p*-Value**	**Variable**	***p*-Value**
Gender (male vs. women)	$p = 0.054$	Active brain metastases (yes vs. no)	$p = 0.598$
ECOG (0 vs. 1 vs. 2 vs. 3)	$p = 0.085$	WBRT (yes vs. no)	$p = 0.813$
Application (p.o. vs. i.v.)	$p = 0.152$	1st-line treatment (CX vs. RCT)	$p = 0.420$
GPS (0 vs. 1 vs. 2)	$p = 0.207$	Age (<70 vs. ≥70)	$p = 0.709$

Table 3. *Cont.*

Categorial Variables			
Variable	***p*-Value**	**Variable**	***p*-Value**
Prior IO treatment (yes vs. no)	$p = 0.128$	Time till Topotecan (≤8.3 vs. >8.3 months)	$p = 0.195$
Dosage (<1.25 mg/m^2 vs. ≥1.25 mg/m^2)	$p = 0.010$	LDH (≤250 vs. >250 U/l)	$p = 0.195$
CRP (≤0.5 vs. >0.5 mg/dL	$p = 0.32$	NLR (≤6 vs. >6)	$p = 0.598$
Sodium (<135 vs. ≥135 mmol/L)	$p = 0.092$	TLR (≤200 vs. >200)	$p = 1$
PNI (<40 vs. ≥40)	$p = 0.262$	SII (<1500 vs. ≥1500)	$p = 0.349$
PI (1 + 2 vs. 3 + 4)	$p = 0.287$		
Continuous Variables			
Variable	***p*-Value**	**Variable**	***p*-Value**
Age	$p = 0.830$	NLR	$p = 0.257$
Mean topotecan dosage	$p = 0.009$	TLR	$p = 0.860$
LDH	$p = 0.059$	PNI	$p = 0.174$
CRP	$p = 0.007$	SII	$p = 0.166$
Sodium	$p = 0.419$		

Table 4. Prognostic impact on survival of examined markers and scores. PFS (progression-free survival) in months, OS (overall survival) in months, HR (hazard ratio), CI (confidence interval).

	Median PFS [CI 95%]	HR [CI 95%]	*p*-Value [Log Rank]	Median OS [CI 95%]	HR [CI 95%]	*p*-Value [Log Rank]
LDH						
≤250 U/L	3.0 [0.9–5.1]		$p = 0.006$	8.5 [4.0–13.1]		$p = 0.370$
>250 U/L	1.6 [1.2–2.1]	2.2 [1.2–4.3]		3.6 [1.8–5.4]	1.4 [0.7–2.8]	
CRP						
≤0.5 mg/dL	3.4 [0.7–6.0]		$p = 0.089$	9.1 [5.7–12.6]		$p = 0.102$
>0.5 mg/dL	1.6 [1.3–2.0]	1.6 [0.8–3.1]		4.4 [2.7–6.0]	1.6. [0.8–3.1]	
NLR						
≤6	3.0 [0.9–5.1]		$p = 0.002$	8.2 [4.1–12.2]		$p = 0.01$
>6	1.3 [1.1–1.6]	2.5 [1.3–4.9]		4.4 [1.8–7.1]	2.6 [1.2–5.5]	
Sodium						
<135 mmol/L	1.6 [1.5–1.7]		$p = 0.066$	4.4 [2.9–5.8]		$p = 0.029$
≥135 mmol/L	3.0 [1.6–4.4]	0.5 [0.3–1.1]		8.1 [4.1–12.2]	0.4 [0.2–0.9]	
TLR						
≤200	3.0 [0.4–5.7]		$p = 0.020$	8.1 [3.0–13.2]		$p = 0.734$
>200	1.9 [1.5–2.3]	1.9 [1.0–3.7]		4.7 [3.5–5.8]	1.1 [0.6–2.2]	
GPS						
0	3.0 [0.4–5.6]		$p = 0.011$	9.1 [5.6–12.66]		$p = 0.008$
≥1	1.6 [1.2–2.1]	2.3 [1.2–4.6]		4.0 [2.2–5.7]	2.6 [1.3–5.4]	
PNI						
<40	1.4 [1.2–1.7]		$p = 0.015$	2.4 [1.2–3.6]		$p < 0.001$
≥40	2.5 [0.9–4.1]	0.4 [0.2–0.9]		8.6 [2.4–8.8]	0.2 [0.1–0.4]	

Table 4. *Cont.*

		Median PFS [CI 95%]	HR [CI 95%]	*p*-Value [Log Rank]	Median OS [CI 95%]	HR [CI 95%]	*p*-Value [Log Rank]
SII							
	<1500	3.0 [0.9–5.1]		$p = 0.014$	8.1 [3.0–13.3]		$p = 0.431$
	≥1500	1.6 [1.3–1.8]	2.1 [1.1–3.9]		4.7 [2.9–6.5]	1.3 [0.7–2.7]	
PI							
	1–2	3.3 [2.6–4.0]		$p = 0.003$	9.1 [2.4–15.9]		$p = 0.001$
	3–4	1.6 [1.3–1.9]	2.4 [1.3–4.7]		3.6 [2.9–4.2]	3.2 [1.5–6.6]	

3.4. *Prognostic Value of Different Markers and Scores*

Elevated pre-therapeutic LDH levels were significantly associated with poorer PFS, OS was prolonged by trend, but not statistically significant in patients with normal LDH values. CRP values did not significantly correlate with PFS or OS. However, we found a neutrophil-to-lymphocyte ratio (NLR) with a cut-off of 6 significantly affecting PFS and OS of topotecan-treated patients. Low sodium values were associated with reduced OS but showed no significant impact on PFS. Low thrombocyte-to-lymphocyte ratio (TLR) values were revealed to be associated with significantly longer PFS but did not translate to longer OS. A Glasgow Prognostic Score (GPS) score of 0 was strongly correlated with better OS and PFS. Elevated SII was significantly associated with worse PFS. The prognostic index (PI) differed highly significantly in patients regarding both PFS and OS. Results of univariate survival analyses are shown in Table 4.

3.5. *Multivariate Analysis*

Integrating all significant variables from the univariate analysis in a multivariate COX regression model for PFS revealed ECOG-PS ($p < 0.001$), active brain metastases ($p = 0.001$), and SII ($p = 0.008$) as significant independent prognostic markers. Selecting patients by these three markers achieved a PFS of 5.7 months [CI 95% 2.3–10.4] and thus increased the PFS of the total cohort three-fold. Patients not meeting all criteria had a PFS of 1.8 months [CI 95% 1.5–2.1] (Figure 4A). For OS, multivariate analysis also revealed ECOG-PS ($p < 0.001$) and PNI ($p < 0.001$) as independent prognostic markers. Selecting patients by these markers achieved a median OS of 9.1 months [CI 95% 3.8.–14.5] and thus increased the OS of the total cohort 1.6-fold. Patients not meeting all criteria had a median PFS of 3.6 months [CI 95% 2.7–4.5] (Figure 4B).

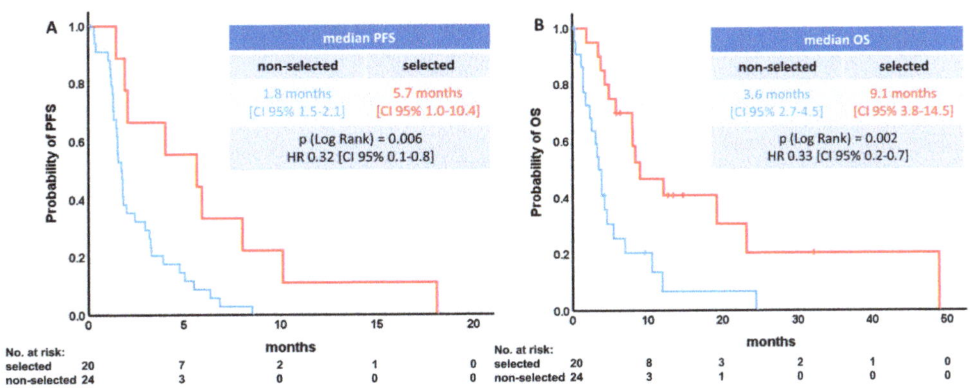

Figure 4. Survival analysis after patient selection based on the examined prediction models. (**A**) Progression-free survival (PFS). (**B**) Overall survival (OS).

4. Discussion

In this retrospective single-center analysis, we sought to analyze efficacy of topotecan second-line treatment in unselected SCLC patients and identify predictive and prognostic markers for improved patient selection.

Efficacy of topotecan second-line treatment in SCLC have been examined in some phase II and phase III studies. In a first phase II trial, Perez-Soler et al. showed a modest effect of topotecan in pretreated SCLC patients with a response rate of 11%. The duration of response recached from 7 to 19 weeks, and the median survival was 4.6 months [10]. A larger phase II trial of second-line topotecan with 92 patients was published one year later. The authors found a response rate of only 6.4% in platinum refractory patients, but of 37.8% in sensitive disease, respectively. The median survival was 4.7. months in refractory and 6.9 months in sensitive patients [11]. The first randomized second-line trial published by von Pawel et al. tested topotecan versus cyclophosphamide, doxorubicin, and vincristine (CAV) in patients relapsing at least 60 days after first-line treatment. Response rates were 24.3% (Topotecan) and 18.3% (CAV). Median PFS and median OS (5.7 months in both groups) were not significantly different, but the topotecan group achieved a more favorable toxicity profile [4]. A Japanese phase II trial of second-line topotecan in relapsed SCLC showed a response rate of 26% an OS of 8.6 months [12]. In our cohort, the response rate was 13.6%, median PFS was 1.9 months, and median OS was 5.6 months and thus are in line with survival results found in the pivotal studies.

We found a significantly improved PFS and OS in female patients. Looking deeper into baseline characteristics, it showed that women received more WBRT often and are often initially treated with chemoradiotherapy.

Nearly 10% of our patients have been treated with oral topotecan. We found no difference of efficacy compared to intravenous therapy. This confirmed the results of two earlier studies and indicates that a more frequent use of oral topotecan should be considered to improve patients' comfort [13,14]. Besides this, in our cohort, lower topotecan dosage seemed not to impact PFS with numerically increased OS, although response rates were increased in higher topotecan dosages. This might be attributed to the relevant increased toxicity with higher doses of topotecan. Moreover, the drug leads to higher response rates, but limits survival by treatment-associated complications. Furthermore, we found that age per se does not impair effectiveness of topotecan treatment. This is in line with a previous study showing that topotecan monotherapy resulted in more toxicity in elderly patients, but was as effective as in younger patients [15].

We identified the ECOG-PS, active brain metastases, and the systemic inflammation index (SII) as potent selection parameters to identify patients with better PFS. By selecting patients with these three markers, the PFS could be increased three-fold in our cohort.

For OS selecting patients by ECOG-PS and prognostic nutritional index (PNI), we found a significant increase of 1.6-fold in survival compared with unselected patients.

Other studies support these markers. Treat and colleagues demonstrated in a pooled analysis of 480 SCLC patients treated with second-line topotecan that patients' performance status is associated with overall survival [16]. Another meta-analysis showed that improvement of PS is a reliable predictor of response to topotecan treatment [17].

Topotecan can cross the blood–brain barrier. But although a phase II study showed an intracranial response rate of 33% in pretreated patients, the OS remains at only 3.6 months in this analysis [18]. Another small study did not show a meaningful intracranial activity of topotecan in SCLC [19], underlying the negative effect of brain metastases we also found in our cohort.

Taking these results together, you should assume that topotecan as second-line treatment in an unselected real-world population might not be beneficial, especially in light of high toxicity rates. Therefore, patient selection, for example using the easy-to-obtain markers we examined in this study, might increase treatment efficacy and prevent unnecessary toxicity. Furthermore, other more effective second-line therapy options are urgently needed as more aggressive treatment regimens by combining topotecan with other chemotherapeutic drugs as cisplatin or etoposide resulted in no or only minimal increases in overall survival but relevant more toxicity [20–25]. Also, the combination of biologicals with topotecan as Bcl2-Inhibotors or VEGF-Inhibitors did not improve efficacy in this setting [26–28]. Amrubicin demonstrated possible superiority over topotecan in the second-line setting in phase II studies [29,30]. Approved in Japan, amrubicin is not approved in Europe and the US due to a negative phase III study [31]. Also, the combination of amrubicin and topotecan showed promising efficacy, but was not investigated further [32].

Our study harbors some limitations. The first is the small size of the cohort and the retrospective character of the study. However, as we included all consecutive SCLC patients treated with second-line topotecan and did not exclude patients from the analysis, we analyzed a true real-world cohort and our results may help to guide treatment decisions in second-line to identify patients who are more likely to benefit from topotecan. Furthermore, given the low efficacy of the topotecan treatment in combination with high toxicity, patients should, whenever possible, be enrolled in clinical studies helping to identify more potent therapy options in this setting.

5. Conclusions

The efficacy of topotecan in SCLC real-world patients is poor, indicating that many patients were treated without any benefit. The analysis of our cohort revealed easy-to-obtain markers (ECOG-PS, active brain metastases, and the systemic inflammation index (SII) for PFS as well as ECOG-PS and prognostic nutritional index (PNI) for OPS) to predict response and treatment efficacy. These markers should therefore be validated in larger cohorts to identify patients more likely to benefit from topotecan.

Author Contributions: Conceptualization, L.L. and D.K.-G.; methodology, L.L. and D.K.-G.; investigation, L.L., P.A., J.B., P.M., A.T. and D.K.-G.; data curation, L.L., P.A., J.B., P.M., A.T. and D.K.-G.; writing—original draft preparation, L.L. and D.K.-G.; writing—review and editing, L.L., P.A., J.B., P.M., A.T. and D.K.-G.; visualization, L.L. and D.K.-G.; supervision, D.K.-G. All authors have read and agreed to the published version of the manuscript.

Funding: This research received no external funding.

Institutional Review Board Statement: The study was conducted in accordance with the Declaration of Helsinki, and approved by the Institutional Ethics Committee of the University Hospital of Munich (LMU), (Approval-code: 476-16 UE, approval date 5 August 2016).

Informed Consent Statement: Patient consent was waived due to due to the retrospective nature of the data. The ethics board consented to the analysis without requirements beyond the anonymization of the data prior to the analysis.

Data Availability Statement: The authors state to share the original data on reasonable request.

Conflicts of Interest: The authors declare no conflicts of interest.

References

1. Wang, Q.; Gümüş, Z.H.; Colarossi, C.; Memeo, L.; Wang, X.; Kong, C.Y.; Boffetta, P. SCLC: Epidemiology, Risk Factors, Genetic Susceptibility, Molecular Pathology, Screening, and Early Detection. *J. Thorac. Oncol.* **2023**, *18*, 31–46. [CrossRef] [PubMed]
2. Kahnert, K.; Kauffmann-Guerrero, D.; Huber, R.M. SCLC-State of the Art and What Does the Future Have in Store? *Clin. Lung Cancer* **2016**, *17*, 325–333. [CrossRef] [PubMed]
3. Saida, Y.; Watanabe, S.; Kikuchi, T. Extensive-Stage Small-Cell Lung Cancer: Current Landscape and Future Prospects. *OncoTargets Ther.* **2023**, *16*, 657–671. [CrossRef] [PubMed]
4. von Pawel, J.; Schiller, J.H.; Shepherd, F.A.; Fields, S.Z.; Kleisbauer, J.P.; Chrysson, N.G.; Stewart, D.J.; Clark, P.I.; Palmer, M.C.; Depierre, A.; et al. Topotecan versus cyclophosphamide, doxorubicin, and vincristine for the treatment of recurrent small-cell lung cancer. *J. Clin. Oncol.* **1999**, *17*, 658–667. [CrossRef] [PubMed]
5. O'Brien, M.E.R.; Ciuleanu, T.-E.; Tsekov, H.; Shparyk, Y.; Čučeviá, B.; Juhasz, G.; Thatcher, N.; Ross, G.A.; Dane, G.C.; Crofts, T. Phase III Trial Comparing Supportive Care Alone With Supportive Care With Oral Topotecan in Patients With Relapsed Small-Cell Lung Cancer. *J. Clin. Oncol.* **2006**, *24*, 5441–5447. [CrossRef] [PubMed]

6. Horita, N.; Yamamoto, M.; Sato, T.; Tsukahara, T.; Nagakura, H.; Tashiro, K.; Shibata, Y.; Watanabe, H.; Nagai, K.; Inoue, M.; et al. Topotecan for Relapsed Small-cell Lung Cancer: Systematic Review and Meta-Analysis of 1347 Patients. *Sci. Rep.* **2015**, *5*, 15437. [CrossRef] [PubMed]
7. Eisenhauer, E.A.; Therasse, P.; Bogaerts, J.; Schwartz, L.H.; Sargent, D.; Ford, R.; Dancey, J.; Arbuck, S.; Gwyther, S.; Mooney, M.; et al. New response evaluation criteria in solid tumours: Revised RECIST guideline (version 1.1). *Eur. J. Cancer* **2009**, *45*, 228–247. [CrossRef] [PubMed]
8. Kauffmann-Guerrero, D.; Kahnert, K.; Kiefl, R.; Sellmer, L.; Walter, J.; Behr, J.; Tufman, A. Systemic inflammation and pro-inflammatory cytokine profile predict response to checkpoint inhibitor treatment in NSCLC: A prospective study. *Sci. Rep.* **2021**, *11*, 10919. [CrossRef] [PubMed]
9. Kauffmann-Guerrero, D.; Kahnert, K.; Syunyaeva, Z.; Tufman, A.; Huber, R.M. Pretherapeutic Inflammation Predicts Febrile Neutropenia and Reduced Progression-Free Survival after First-Line Chemotherapy in SCLC. *Oncol. Res. Treat* **2018**, *41*, 506–512. [CrossRef] [PubMed]
10. Perez-Soler, R.; Glisson, B.S.; Lee, J.S.; Fossella, F.V.; Murphy, W.K.; Shin, D.M.; Hong, W.K. Treatment of patients with small-cell lung cancer refractory to etoposide and cisplatin with the topoisomerase I poison topotecan. *J. Clin. Oncol.* **1996**, *14*, 2785–2790. [CrossRef]
11. Ardizzoni, A.; Hansen, H.; Dombernowsky, P.; Gamucci, T.; Kaplan, S.; Postmus, P.; Giaccone, G.; Schaefer, B.; Wanders, J.; Verweij, J. Topotecan, a new active drug in the second-line treatment of small-cell lung cancer: A phase II study in patients with refractory and sensitive disease. The European Organization for Research and Treatment of Cancer Early Clinical Studies Group and New Drug Development Office, and the Lung Cancer Cooperative Group. *J. Clin. Oncol.* **1997**, *15*, 2090–2096.
12. Takeda, K.; Negoro, S.; Sawa, T.; Nakagawa, K.; Kawahara, M.; Isobe, T.; Kudoh, S.; Masuda, N.; Niitani, H.; Fukuoka, M. A phase II study of topotecan in patients with relapsed small-cell lung cancer. *Clin. Lung Cancer* **2003**, *4*, 224–228. [CrossRef]
13. Von Pawel, J.; Gatzemeier, U.; Pujol, J.-L.; Moreau, L.; Bildat, S.; Ranson, M.; Richardson, G.; Steppert, C.; Rivière, A.; Camlett, I.; et al. Phase II Comparator Study of Oral Versus Intravenous Topotecan in Patients With Chemosensitive Small-Cell Lung Cancer. *J. Clin. Oncol.* **2001**, *19*, 1743–1749. [CrossRef]
14. Eckardt, J.R.; von Pawel, J.; Pujol, J.-L.; Papai, Z.; Quoix, E.; Ardizzoni, A.; Poulin, R.; Preston, A.J.; Dane, G.; Ross, G. Phase III study of oral compared with intravenous topotecan as second-line therapy in small-cell lung cancer. *J. Clin. Oncol.* **2007**, *25*, 2086–2092. [CrossRef]
15. Imai, H.; Yamada, Y.; Minemura, H.; Sugiyama, T.; Kotake, M.; Kaira, K.; Kanazawa, K.; Nakamura, Y.; Kasai, T.; Shibata, Y.; et al. Topotecan monotherapy for the treatment of relapsed small cell lung cancer in elderly patients: A retrospective analysis. *Thorac. Cancer* **2018**, *9*, 1699–1706. [CrossRef] [PubMed]
16. Treat, J.; Huang, C.H.; Lane, S.R.; Levin, J. Topotecan in the treatment of relapsed small cell lung cancer patients with poor performance status. *Oncologist* **2004**, *9*, 173–181. [CrossRef] [PubMed]
17. Lilenbaum, R.C.; Huber, R.M.; Treat, J.; Masters, G.; Kaubitzsch, S.; Lane, S.; Wissel, P. Topotecan therapy in patients with relapsed small-cell lung cancer and poor performance status. *Clin. Lung Cancer* **2006**, *8*, 130–134. [CrossRef]
18. Korfel, A.; Oehm, C.; von Pawel, J.; Keppler, U.; Deppermann, M.; Kaubitsch, S.; Thiel, E. Response to topotecan of symptomatic brain metastases of small-cell lung cancer also after whole-brain irradiation. a multicentre phase II study. *Eur. J. Cancer* **2002**, *38*, 1724–1729. [CrossRef] [PubMed]
19. Lorusso, V.; Galetta, D.; Giotta, F.; Rinaldi, A.; Romito, S.; Brunetti, C.; Silvestris, N.; Colucci, G. Topotecan in the treatment of brain metastases. A phase II study of GOIM (Gruppo Oncologico dell'Italia Meridionale). *Anticancer Res.* **2006**, *26*, 2259–2263. [PubMed]
20. Ardizzoni, A.; Manegold, C.; Debruyne, C.; Gaafar, R.; Buchholz, E.; Smit, E.F.; Lianes, P.; ten Velde, G.; Bosquee, L.; Legrand, C.; et al. European Organization for Research and Treatment of Cancer (EORTC) 08957 phase II study of topotecan in combination with cisplatin as second-line treatment of refractory and sensitive small cell lung cancer. *Clin. Cancer Res.* **2003**, *9*, 143.
21. Stathopoulos, G.P.; Christodoulou, C.; Stathopoulos, J.; Skarlos, D.; Rigatos, S.K.; Giannakakis, T.; Armenaki, O.; Antoniou, D.; Athanasiadis, A.; Giamboudakis, P.; et al. Second-line chemotherapy in small cell lung cancer in a modified administration of topotecan combined with paclitaxel: A phase II study. *Cancer Chemother. Pharmacol.* **2006**, *57*, 796–800. [CrossRef] [PubMed]
22. Choi, H.J.; Cho, B.C.; Shin, S.J.; Cheon, S.H.; Jung, J.Y.; Chang, J.; Kim, S.K.; Sohn, J.H.; Kim, J.H. Combination of topotecan and etoposide as a salvage treatment for patients with recurrent small cell lung cancer following irinotecan and platinum first-line chemotherapy. *Cancer Chemother. Pharmacol.* **2007**, *61*, 309–313. [CrossRef] [PubMed]
23. Kurata, T.; Kashii, T.; Takeda, K.; Seki, N.; Tsuboi, M.; Kobayashi, M.; Satoh, T.; Nakagawa, K.; Fukuoka, M. A Phase I Study of Topotecan Plus Carboplatin for Relapsed SCLC: WJTOG Trial. *J. Thorac. Oncol.* **2009**, *4*, 644–648. [CrossRef] [PubMed]
24. Evans, T.L.; Cho, B.C.; Udud, K.; Fischer, J.R.; Shepherd, F.A.; Martinez, P.; Ramlau, R.; Syrigos, K.N.; Shen, L.; Chadjaa, M.; et al. Cabazitaxel Versus Topotecan in Patients with Small-Cell Lung Cancer with Progressive Disease During or After First-Line Platinum-Based Chemotherapy. *J. Thorac. Oncol.* **2015**, *10*, 1221–1228. [CrossRef] [PubMed]
25. Christodoulou, C.; Kalofonos, H.P.; Briasoulis, E.; Bafaloukos, D.; Makatsoris, T.; Koutras, A.; Skarlos, D.V.; Samantas, E. Combination of topotecan and cisplatin in relapsed patients with small cell lung cancer: A phase II study of the hellenic cooperative oncology group (HeCOG). *Cancer Chemother. Pharmacol.* **2006**, *57*, 207–212. [CrossRef] [PubMed]

26. Suk Heist, R.; Fain, J.; Chinnasami, B.; Khan, W.; Molina, J.R.; Sequist, L.V.; Temel, J.S.; Fidias, P.; Brainerd, V.; Leopold, L.; et al. Phase I/II Study of AT-101 with Topotecan in Relapsed and Refractory Small Cell Lung Cancer. *J. Thorac. Oncol.* **2010**, *5*, 1637–1643. [CrossRef] [PubMed]
27. Paik, P.K.; Rudin, C.M.; Pietanza, M.C.; Brown, A.; Rizvi, N.A.; Takebe, N.; Travis, W.; James, L.; Ginsberg, M.S.; Juergens, R.; et al. A phase II study of obatoclax mesylate, a Bcl-2 antagonist, plus topotecan in relapsed small cell lung cancer. *Lung Cancer* **2011**, *74*, 481–485. [CrossRef] [PubMed]
28. Spigel, D.R.; Waterhouse, D.M.; Lane, S.; Legenne, P.; Bhatt, K. Efficacy and safety of oral topotecan and bevacizumab combination as second-line treatment for relapsed small-cell lung cancer: An open-label multicenter single-arm phase II study. *Clin. Lung Cancer* **2013**, *14*, 356–363. [CrossRef] [PubMed]
29. Inoue, A.; Sugawara, S.; Yamazaki, K.; Maemondo, M.; Suzuki, T.; Gomi, K.; Takanashi, S.; Inoue, C.; Inage, M.; Yokouchi, H.; et al. Randomized Phase II Trial Comparing Amrubicin With Topotecan in Patients With Previously Treated Small-Cell Lung Cancer: North Japan Lung Cancer Study Group Trial 0402. *J. Clin. Oncol.* **2008**, *26*, 5401–5406. [CrossRef]
30. Jotte, R.; Conkling, P.; Reynolds, C.; Galsky, M.D.; Klein, L.; Fitzgibbons, J.F.; McNally, R.; Renschler, M.F.; Oliver, J.W. Randomized Phase II Trial of Single-Agent Amrubicin or Topotecan as Second-Line Treatment in Patients With Small-Cell Lung Cancer Sensitive to First-Line Platinum-Based Chemotherapy. *J. Clin. Oncol.* **2011**, *29*, 287–293. [CrossRef]
31. Von Pawel, J.; Jotte, R.; Spigel, D.R.; O'Brien, M.E.R.; Socinski, M.A.; Mezger, J.; Steins, M.; Bosquée, L.; Bubis, J.; Nackaerts, K.; et al. Randomized Phase III Trial of Amrubicin Versus Topotecan As Second-Line Treatment for Patients With Small-Cell Lung Cancer. *J. Clin. Oncol.* **2014**, *32*, 4012–4019. [CrossRef] [PubMed]
32. Nogami, N.; Hotta, K.; Kuyama, S.; Kiura, K.; Takigawa, N.; Chikamori, K.; Shibayama, T.; Kishino, D.; Hosokawa, S.; Tamaoki, A.; et al. A phase II study of amrubicin and topotecan combination therapy in patients with relapsed or extensive-disease small-cell lung cancer: Okayama Lung Cancer Study Group Trial 0401. *Lung Cancer* **2011**, *74*, 80–84. [CrossRef] [PubMed]

diagnostics

MDPI

Article

Safety and Diagnostic Yield of Medical Pleuroscopy (MP) Performed under Balanced Analgosedation by a Pneumological Team Compared to Video-Assisted Thoracic Surgery (VATS): A Retrospective Controlled Real-Life Study (TORAPO)

Valentino Allocca [1], Luca Guidelli [2], Angela Galgano [3], Lucia Benedetti [2], Roberto Fabbroni [2], Andrea Bianco [1], Piero Paladini [3] and Raffaele Scala [2,*]

[1] Department of Translational Medical Sciences, "Luigi Vanvitelli" University of Campania, 80131 Naples, Italy; valentino.allocca@unicampania.it (V.A.); andrea.bianco@unicampania.it (A.B.)
[2] Pulmonology and Respiratory Intensive Care Unit, S. Donato Hospital, Cardio-Thoraco-Neuro-Vascolar Department, Usl Toscana Sudest, 52100 Arezzo, Italy; guidelliluca6@gmail.com (L.G.); lucia.benedetti@uslsudest.toscana.it (L.B.); roberto.fabbroni@uslsudest.toscana.it (R.F.)
[3] Thoracic Surgery Unit, Department of Medical, Surgical and Neuroscience Sciences, S. Maria Le Scotte Hospital, University of Siena, 53100 Siena, Italy; angela.galgano@student.unisi.it (A.G.); piero.paladini@unisi.it (P.P.)
* Correspondence: raffaele.scala@uslsudest.toscana.it

Abstract: Introduction: Medical pleuroscopy (MP) is an invasive technique that provides access to the pleural space with a rigid or semi-rigid work instrument, allowing for visualization and the obtaining of bioptic pleural samples. Using pulmonologist-based analgosedation to perform pleuroscopy is still debated for safety reasons. The aim of this real-life study is to demonstrate the safety and diagnostic yield of MP performed under balanced analgosedation by a pulmonologist team with expertise in the management of critically ill patients in the respiratory intensive care unit (RICU) and interventional pulmonology unit as compared to video-assisted thoracic surgery (VATS) performed by a thoracic surgeon team under anesthesiologist-based analgosedation. Methods: In this multicentric retrospective controlled study, the inclusion criteria were patients older than 18 years old with pleural effusion of unknown diagnosis consecutively admitted in the years 2017–2022 to the pulmonology unit and RICU of San Donato Hospital in Arezzo (Italy, Tuscany) and to the thoracic surgery unit of Santa Maria Le Scotte in Siena (Italy, Tuscany) to undergo, respectively, MP under balanced propofol-based analgosedation on spontaneous breathing with local anesthesia provided by a pulmonologist team (Group A), and VATS provided by a surgeon team under propofol-based analgosedation managed by an anesthesiologist using invasive mechanical ventilation (IMV) via endotracheal intubation (ETI) (Group B). The primary endpoints were (1) a comparison between the two groups in terms of the diagnostic yield of pleural effusion, and (2) major and minor complications of pleuroscopic procedures. The secondary endpoints were (1) the length of the pleuroscopic procedure; (2) the duration of hospitalization; (3) propofol doses; and (4) the patient's comfort after the procedure assessed using the Visual Analogue Scale (VAS). Results: We enrolled 91 patients in Group A and 116 patients in Group B. A conclusive diagnosis was obtained in 97.8% of Group A vs. 100% of Group B ($p = 0.374$). Malignant effusion was diagnosed in 59.3% of Group A and in 55.1% of Group B; $p = 0.547$. No intraoperative or postoperative mortality events or major complications were observed in Group A. The major complications observed in Group B were three major bleeding events ($p = 0.079$) and one exitus ($p = 0.315$) not related to the interventional procedure. No significant difference emerged between the two groups in terms of minor complications. The duration of the intervention was significantly lower in Group A (40.0 min $\pm$ 12.6 versus 51.5 $\pm$ 31.0; $p = 0.001$). Pain control and, therefore, patient comfort were better in Group A, with an average VAS of 0.34 $\pm$ 0.65 versus 2.58 $\pm$ 1.26, $p < 0.001$. The duration of hospitalization was lower in Group B (5.1 $\pm$ 2.6 vs. 15.5 $\pm$ 8.0, $p < 0.001$). The average overall dose of propofol administered was significantly lower in Group A (65.6 $\pm$ 35.8 mg versus 280 $\pm$ 20.0 mg; $p < 0.001$). Conclusions: This real-life study shows that the MP performed under propofol-based analgosedation by an independent pneumologist team is a safe and well-tolerated procedure with a diagnostic yield and complication rates similar to those obtained with VATS.

Keywords: medical pleuroscopy; toracoscopy; diagnostic yield; analgosedation; propofol; pulmonologist

1. Introduction

Pleural diseases are common and increasing worldwide, affecting over 3000 people per 1 million each year [1]. Malignant pleural effusion, mesothelioma, and pleural infections represent an enormous burden and a clinical challenge for respiratory physicians, with approximately 361,270 hospitalizations occurring in the United States in 2016 [2]. For this reason, it is increasingly important to improve the current techniques for the diagnosis and management of pleural diseases. Medical pleuroscopy (MP), also known as medical thoracoscopy, is an invasive technique that provides access to the pleural space with a rigid or semi-rigid work instrument, which allows for the visualization and biopsy of pleural lesions [3]. In contrast to video-assisted thoracoscopy (VATS), which is performed by a surgeon under general anesthesia, often using single-lung ventilation and via multiple ports, MP is performed by a pulmonologist using analgosedation and spontaneous breathing via a single port [4]. Currently, there is not enough evidence to establish the diagnostic test effectiveness when comparing awake thoracoscopic pleural biopsy to video-assisted thoracoscopic pleural biopsy performed under general anesthesia [5]. In the latest guidelines of the British Thoracic Society, MP finds space in numerous applications for the diagnosis of pleural effusion, for example, obtaining pleural tissue sampling, which is often necessary to achieve a definitive diagnosis in patients with pleural effusion and/or pleural thickening [5]. The major indications for MP are (1) parietal pleural biopsies for the diagnosis of mesothelioma, lung cancer, and pleural infection, such as tuberculosis, which, in a significant majority of cases, can mimic a tumor; (2) debridement of simple adhesions in early-stage empyema; and (3) talc poudrage pleurodesis for malignant pleural effusion and in selected secondary pneumothorax not suitable for surgery [5,6]. The diagnosis of tuberculosis (TB) pleurisy becomes straightforward when Mycobacterium tuberculosis is identified in the sputum, pleural fluid, or biopsy samples. In regions with a high prevalence of TB, the combination of a lymphocyte-predominant exudate and elevated adenosine–deaminase (ADA) yields a positive predictive value of 98% [7]. In areas with low TB prevalence, the absence of increased ADA and lymphocyte predominance suggests a low likelihood of TB. In such cases, pleural biopsy is recommended to confirm the diagnosis. Additionally, when there is a high prevalence of drug-resistant TB, pleural biopsy for liquid culture and susceptibility testing should be considered [7]. According to the latest BTS guidelines, MP is the preferred method for obtaining pleural tissue sampling for culture and a sensitivity test for TB pleural effusion [8,9]. In addition, pleural biopsies allow for obtaining material on which to perform polymerase chain reaction (PCR) to detect viruses, like influenza, coxsackievirus, respiratory syncytial virus (RSV), cytomegalovirus, adenovirus, human herpesvirus-8, dengue, human t-lymphotropic virus type 1 (HTLV-1), varicella, herpes simplex virus (HSV), and Epstein–Barr virus (EBV), which can be responsible for pleural effusion [10]. MP represents a valid option as a first-line approach in the diagnosis of pleural diseases, offering potential advantages in comparison to VATS. These benefits include minor invasiveness; a reduction in costs to perform the procedure, as it can be performed in an endoscopic room with a shorter duration of hospitalization; the possibility of including frail patients who are at increased risk of intubation, mechanical ventilation, and general anesthesia. Overall, this approach results in a good diagnostic yield [3,6]. MP is a relatively safe procedure with a low mortality rate and low occurrence of major complications, such as hemorrhage, persistent air leaks, port tumor dissemination, and empyema, in comparison to VATS [6,11]. Minor complications, such as transient hypoxemia and subcutaneous emphysema, are less frequently observed in patients undergoing MP instead of VATS [4,6]. According to the majority of published data and clinical field experiences, moderate to severe analgosedation during pulmonologist interventional procedures is usually performed by anesthesiologists, especially if non-midazolam-based regimens are used.

The aim of this study is to demonstrate the safety and diagnostic yield of MP performed using analgosedation by a pneumologist team in collaboration with nurses with consolidated experience in the management of airways and respiratory-critical patients in the respiratory intensive care unit (RICU) and interventional pulmonology unit in a comparative analysis, with data obtained from a series of patients undergoing pleuroscopic examination using VATS performed by a thoracic surgeon team under anesthesiologist-based analgosedation [12].

2. Materials and Methods

In this multicentric retrospective controlled study, the inclusion criteria were patients older than 18 years with pleural effusion of unknown diagnosis consecutively admitted between 2017 and 2022 to Pulmonology and RICU of "San Donato" Hospital in Arezzo (Italy, Tuscany) and to the Thoracic Surgery Unit of "Santa Maria Le Scotte" in Siena (Italy, Tuscany). The patients underwent, respectively, MP under balanced propofol-based analgosedation on spontaneous breathing with local anesthesia provided by a pulmonologist team (Group A) and VATS provided by a surgical team under propofol-based analgosedation managed by an anesthesiologist in invasive mechanical ventilation (IMV) via endotracheal intubation (ETI) (Group B). The choice between the two procedures depended solely on the hospital where the patients were admitted. Specifically, patients admitted to "San Donato" Hospital underwent MP, while those admitted to "Santa Maria Le Scotte" Hospital underwent VATS.

The exclusion criteria were as follows: (1) terminally ill patients with a life expectancy of fewer than 6 months, (2) multiorgan failure [13], (3) active bleeding from thoracic and/or extrathoracic sources, (4) American Society of Anesthesiologists (ASA) score > 3, (5) acute respiratory failure (ARF) requiring non-invasive and invasive mechanical respiratory support, (6) pregnancy, and (7) patient refusal. Informed consent was obtained from all recruited patients and the study protocol was approved by the local ethical committee (TORAPO 23776).

2.1. Medical Pleuroscopy (Group A)

Before the procedure, all patients underwent a chest X-ray followed by contrast-enhanced thoracic computerized tomography (CT) to better characterize and confirm the pleural disease. In some instances, positron emission tomography and computed tomography (PET-CT) scans were considered to aid in better identifying the target of biopsy on the pleural surface. Lung ultrasonography was performed in all patients using a convex probe (3.5–5 MHz) (Mindray Mobile Trolley UMT-160, Shenzhen, China) to define the extension and the features of the pleural effusion (i.e., septations and loculations); detect, at bedside, the point of chest access for instrument insertion; and study the motility of the diaphragm.

MP was performed in an endoscopic room under continuous assessment of heart rate (HR), respiratory rate (RR), blood pressure (BP), and three-lead ECG registered with a multiparametric monitor (Mindray Patient Monitor ePM 12, Shenzhen ,China). Full equipment for non-invasive respiratory support, including-high-flow nasal cannula (HFNC) and non-invasive ventilation (NIV), as well as for the management of airways and cardiopulmonary emergency (Guedel cannula, laryngeal mask, endotracheal tube, defibrillator) was available.

The pulmonologist team consisted of 2 respiratory physicians and 2 nurses with experience in airway management, in the management of critically ill respiratory patients, and in interventional procedures. In case of severe periprocedural complications, an anesthesiologist of the hospital was available on call.

Once the American Society of Anesthesiologists (ASA) score was established and informed consent was collected [14], the patients were placed in a lateral decubitus position with the administration of conventional oxygen by means of a Venturi mask titrated to achieve a target SpO_2 of 94–98% and 90–92%, respectively, for hypoxemic and hypercapnic patients. An intravenous line, primed with 500 mL NaCl 0.9%, was set up to increase patients' volume in case of propofol-induced arterial hypotension. Once the thoracoscope access point was identified using the ultrasound-guided mode, the local anesthetic of the intercostal cutaneous, subcutaneous, and pleural layers was performed with lidocaine 2% at a dosage of 200–300 mg.

The analgosedation procedure was set according to the protocol used in our previous study [15]. An initial intravenous dose of 1% propofol (0.5 mg/kg) was administered in a bolus followed by a continuous infusion maintenance dose of 0.5–1.0 mg/kg/h. A dose of intravenous meperidine (100 mg/2 mL) was administered at a dosage of 0.5 mg/kg. The administered drugs were then balanced to achieve a Richmond Agitation Sedation Scale (RASS) between -2 and -3 and VAS ≤ 2. The RASS is used to define the patient's level of sensorium and to titrate the sedation according to predefined target values. Structured on a 10-point scale, it defines 4 levels of increasing agitation disturbances (from +1 to +4), a neutral level (0), and 5 levels of progressively depressed sensorium (from -1 to -5) [16]. In case of insufficient sedation, discomfort, or agitation, additional boluses of propofol (10–20 mg) were administered up to a maximum of 1 mg/kg.

MP was performed with a rigid thoracoscope (Storz 4 mm, Karl Storz SE & Co. KG., Tuttlingen, Germany), and coagulation electrosurgery (Medtronic Covidien, Dublin, Ireland) was used to manage sources of bleeding. In general, 10–12 biopsies of the parietal pleura were performed. After pleuroscopy, a 20–24 Fr chest tube was placed at the thoracoscope access point (Figure 1).

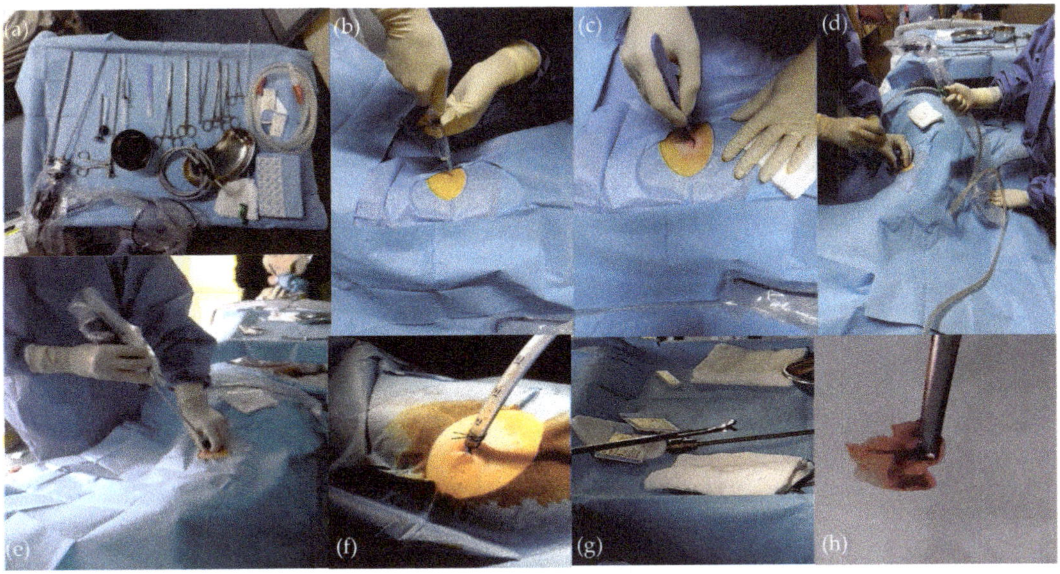

Figure 1. Medical pleuroscopy: an overview of the operation. From the top left corner: (**a**) instrumentation for MP; (**b**) local anesthetic injection; (**c**) Skin incision with a scalpel; (**d**) drainage of pleural effusion; (**e**) introduction of the optics; (**f**) chest tube drainage; (**g**) thoracoscope forceps; (**h**) pleural biopsy.

In case of hypotension, the protocol involved the rapid infusion of crystalloids (500 mL of NaCl 0.9%); in case of a lack of response to volume filling, ephedrine hydrochloride (3 mg/mL) was administered intravenously at an initial dosage ranging between 3 and 6 mg, to be increased up to a maximum of 30 mg in total depending on the hemodynamic response. The goal was to maintain the mean arterial pressure (MAP) between 60 and 100 mmHg. In the event of bradycardia complicated by hypotension, the protocol allowed the reduction of the infusion rate of propofol to 0.5 mg/kg/h. Subsequently, atropine sulfate was administered intravenously at a dosage of 0.3–0.6 mg targeted at maintaining the HR between 60 and 90 bpm. In case of acute respiratory failure (ARF), defined as $PaO_2 < 60$ mmHg with or without hypercapnia $PaCO_2 > 45$ mmHg under conventional oxygen support, the protocol considered an escalating step-by-step strategy. This included a drop in the rate of propofol infusion to 0.5 mg/kg/h, followed by a jaw dislocation maneuver and the insertion of an oropharyngeal cannula (Guedel cannula) or a laryngeal mask in case of persistent airway collapse. Subsequently, the application of HFNC or NIV was considered until the escalation to endotracheal intubation (ETI) and IMV if mandatory respiratory support was needed.

During the MP in analgosedation, we continuously monitored patients' SpO2, respiratory rate, heart rate, and blood pressure. Additionally, we frequently assessed the patients' level of sedation to maintain an RASS between −2 and −3 (indicating light to moderate sedation). While there is strong clinical evidence supporting the use of capnography in general anesthesia and moderate to deep sedation [17], there is limited evidence in the context of light–moderate sedation in non-anesthesia settings in which the available evidence does not substantiate the impact of capnography on clinical outcomes when compared to standard monitoring [18]. Unfortunately, our equipment did not include a capnograph, so if there was any suspicion of ventilation problems, we performed arterial blood gas analysis.

The recovery time of the patient was evaluated according to the achievement of an Aldrete score ≥ 9 [15]. The Aldrete score is an extensively validated scoring scale used to establish safe post-anesthetic discharge in a hospital ward or at home for patients undergoing short-term surgical procedures [19]. Once the Alderete score was ≥ 9 after the procedure, the patients were discharged from the endoscopic room and admitted to the monitored beds of Pulmonology and the RICU [19]. Two hours later, a chest X-ray was performed as well as bloodwork (i.e., chemistry, blood cell count) [13,14].

2.2. VATS (Group B)

Patients belonging to Group B underwent the same preliminary clinical, radiological, and ultrasound-based assessment as described for Group A.

In Group B, VATS was performed with a uniportal approach by a surgeon. The procedure took place after the patient was endotracheally intubated and mechanically ventilated under general anesthesia, managed by the anesthesiologist team in the operating room. The anesthetic protocol included the use of propofol 1% at 1.5–2 mg/kg intravenously administered in a bolus followed by a continuous infusion at 3–4 mg/kg/h. Fentanyl (0.1 mg/2 mL) was administered intravenously at a dosage of 1 mcg/kg as a bolus followed by an infusion of Remifentanil at 0.1 mcg/kg/h. VATS was performed with a rigid thoracoscope (Storz 10 mm) and coagulation electrosurgery (Covidien) was used to manage sources of bleeding. An endopleural drainage (24–28 Fr) was placed at the end of the procedure. Complications were managed by the anesthesiologist team according to protocols similar to those reported for Group A.

Once the Alderete score was ≥ 9 after the procedure, the patients were discharged from the endoscopic room and admitted to the monitored beds of the Thoracic Surgery Unit.

2.3. Endpoints of the Study

The primary endpoints of the study were the comparative rates between the two groups in terms of (1) the diagnostic yield of pleural effusion and (2) the major and minor complications of pleuroscopic procedures.

The major complications considered in the study were (1) ARF requiring ventilator support with HFN (high-flow nasal cannula), NIV, and/or ETI with IMV; (2) acute coronary syndrome and/or cardiogenic pulmonary edema; (3) massive bleeding; and (4) periprocedural death. Bleeding was defined as either a loss of more than one blood volume in 24 h or 50% of the total blood volume of the patient in 3 h, or blood loss > 150 mL/min [20,21].

The minor complications considered in the study were new-onset cardiac arrhythmias requiring treatment, subcutaneous emphysema, hypoxemia defined as SpO_2 < 90% for more than 1 min in conventional oxygen therapy at FiO2 of 0,50, arterial hypotension defined as systolic blood pressure (SBP) < 90 mmHg or diastolic blood pressure (DBP) < 50 mmHg in three consecutive measurements requiring volume filling and/or the use of vasoactive amines, bradycardia defined as heart rate (HR) < 50 bpm for more than two minutes and arterial hypertensive crisis defined as SBP > 170 mmHg or DBP > 100 mmHg in three consecutive measurements and requiring hypotensive therapy.

The secondary endpoints of the study were the comparative assessment of the two groups in terms of (1) the length of pleuroscopic procedure, (2) the duration of hospitalization, (3) propofol doses, and (4) the patient's comfort after the procedure assessed using the Visual Analogue Scale (VAS).

3. Statistical Analysis

The statistical analysis of the data was carried out using Microsoft Excel v. 16.78.3 and the statistical software MiniTab v. 21.4.1. The data are expressed as the mean with standard deviation (SD), or as the median with interquartile range (IQR) for continuous variables and numbers with percentages for categorical variables. The Student's *t*-test was used to compare continuous variables between the group undergoing MP and the group undergoing VATS. The chi-square test was used to compare the categorical variables between the two groups. A value of $p < 0.05$ was considered statistically significant.

4. Results

During the study time, 91 patients were enrolled in Group A and 116 patients in Group B. The baseline clinical features of the two groups are presented in Table 1. The two groups were similar for all reported parameters except for the Charlson Comorbidity Index, which was significantly greater in Group A compared to Group B (4.27 ± 2.04 versus 1.64 ± 1.52, $p < 0.001$).

Table 1. Characteristics of patients undergoing pleuroscopy; ASA: American Society of Anesthesiologists; COPD: chronic obstructive pulmonary disease; CAD: coronary artery disease; OSA: obstructive sleep apnea; CRF: chronic respiratory failure; DM: diabetes mellitus; SAH: systemic arterial hypertension; CKD: chronic kidney disease; ILD: interstitial lung disease. Data are presented as mean and standard deviation or absolute number (percentage).

	Group A	Group B	*p*
Patients, *n*	91	116	
Age, years (SD)	72 (11)	69 (11)	0.102
BMI (SD)	30 (8.72)	28 (7.45)	0.083
Male	62	79	0.997
Female	29	31	0.420
Charlson Comorbidity Index (SD)	4.27 (2.04)	1.64 (1.52)	<0.001
SpO2 % basal	92.44%	93.07%	0.138

Table 1. *Cont.*

	Group A	Group B	*p*
ASA 3, *n* (%)	48 (50%)	46 (39.66%)	0.059
ASA 2, *n* (%)	48 (50%)	70 (60.34%)	0.273
COPD, *n* (%)	21 (23.07%)	7 (6.04%)	0.001
CAD, *n* (%)	20 (21.97%)	16 (13.79%)	0.129
OSA, *n* (%)	4 (4.39%)	0	0.041
CRF *n* (%)	2 (2.19%)	5 (4.31%)	0.385
DM, *n* (%)	12 (13.18%)	29 (25%)	0.028
Arrhythmias, *n* (%)	20 (21.97%)	16 (13.79%)	0.129
SAH, *n* (%)	49 (53.84%)	0	<0.001
Heart failure, *n* (%)	6 (6.59%)	13 (11.2%)	0.239
CKD, *n* (%)	9 (9.89%)	5 (4.31%)	0.127
Neoplasm, *n* (%)	28 (30.76%)	41(35.34%)	0.486
Stroke, *n* (%)	6 (6.59%)	8 (6.89%)	0.931
ILD, *n* (%)	3 (3.29%)	1 (0.86%)	0.237

4.1. Primary Endpoints

The diagnostic yield rate did not differ between the two groups (Group A: 97.8%, Group B: 100%; $p = 0.374$). Table 2 reports the etiologic diagnosis obtained in the groups; in both groups, malignancies accounted for the greater amount of underlying pleural effusion (Group A: 59.3%, Group B: 55.1%; $p = 0.547$). Among non-malignant etiologies, the most common diagnosis was non-specific pleuritis. No major complications, either intra- or post-operative, were observed in Group A. The major complications observed in Group B were three cases of major bleeding ($p = 0.079$) and one exitus ($p = 0.315$), which occurred due to intestinal infarction after some days and was, therefore, not connected to the interventional procedure. In the comparison between the two groups, the minor complications were superimposed without any statistically significant difference. Specifically, we observed two cases of hypotension in Group A and two in Group B ($p = 0.809$); two cases of bradycardia during the procedure in Group A and none in Group B ($p = 0.153$); one hypertensive crisis in Group A and none in Group B ($p = 0.315$); and no cases of atrial fibrillation in Group A and one case in Group B ($p = 0.315$). Post-procedure subcutaneous emphysema was observed in four patients in Group A and eight patients in Group B without significant differences ($p = 0.433$). No case of hypoxemia requiring respiratory support was observed in either group (Table 3). In five cases in Group A, no biopsy could be performed due to the presence of tenacious adhesions, despite initial debridement with digitoclasia; these patients were then sent to VATS. In two cases belonging to Group A, the samples obtained were not sufficient for diagnosis (indetermined etiologies in Table 2).

Table 2. Etiologic diagnosis underlying pleural effusion obtained in the groups. NSCLC: non-small-cell lung cancer; TB: tuberculosis; SCLC: small-cell lung cancer. Data are presented as absolute values (percentages).

Diagnosis	Group A *n* (%)	Group B *n* (%)	*p*
Malignant etiologies	54 (59.3%)	64 (55.1%)	0.547
NSCLC	20 (22.0%)	17 (14.7%)	0.179
Metastases	14 (15.4%)	15 (12.9%)	0.671
Mesothelioma	17 (18.7%)	29 (25,0%)	0.270

Table 2. *Cont.*

Diagnosis	Group A *n* (%)	Group B *n* (%)	*p*
SCLC	0	1 (0.9%)	0.315
Lymphoma	3 (3.3%)	1 (0.9%)	0.237
Hemangioendothelioma	0	1 (0.9%)	0.315
Non-malignant etiologies	30 (32.9%)	51 (43.9%)	0.103
Non-specific pleuritis	16 (17.6%)	39 (33.6%)	0.007
Empyema and parapneumonic effusion	10 (10.9%%)	4 (3.4%)	0.041
TB	2 (2.2%)	4 (3.4%)	0.585
Heart failure	2 (2.2%)	0	0.153
Asbestosis	0	1 (0.9%)	0.315
Sarcoidosis	0	2 (1.72%)	0.154
Indetermined etiologies	2 (2.1%)	0 (0%)	0.153

Table 3. Complications of medical pleuroscopy in analgosedation with pneumological management versus surgical pleuroscopy complications with anesthesiologic support. Spo2: peripheral oxygen saturation. Data are presented as absolute values (percentages). N/A: not applicable.

	Group A *n* (%)	Group B *n* (%)	*p*
Major complications	0 (0%)	4 (3.44%)	0.042
Need for ventilatory support	0 (0%)	N/A	1.000
Major bleeding	0 (0%)	3 (2.58%)	0.079
Exitus	0 (0%)	1 (0.86%)	0.315
Minor complications	9 (9.9%)	11(9.5%)	0.922
Hypotension	2 (2.19%)	2 (1.72%)	0.809
Bradycardia	2 (2.19%)	0 (0%)	0.153
Subcutaneous emphysema	4 (4.39%)	8 (6.89%)	0.433
Hypoxemia (SpO2 < 90%)	0 (0%)	0 (0%)	1.000
Hypertensive crisis	1 (1.09%)	0 (0%)	0.315
Cardiac arrhythmias	0 (0%)	1 (0.86%)	0.315

4.2. Secondary Endpoints

In the comparison between the two groups, the duration of intervention was significantly lower in Group A (40.0 min $\pm$ 12.6 versus 51.5 $\pm$ 31.0; $p = 0.001$). Pain control and, therefore, patient comfort were better in Group A with an average VAS of 0.34 $\pm$ 0.65 versus 2.58 $\pm$ 1.26, $p < 0.001$ (Figures 2 and 3). The duration of hospitalization was lower in Group B than in Group A (5.1 $\pm$ 2.6 vs. 15.5 $\pm$ 8.0, $p < 0.001$). The average overall dose of propofol administered was significantly lower in Group A than in Group B (65.6 $\pm$ 35.8 mg versus 280 $\pm$ 20.0 mg; $p < 0.001$). The average dose of meperidine administered was 46.2 $\pm$ 27.0 mg, while the average dose of fentanyl administered was 70 $\pm$ 10 mcg.

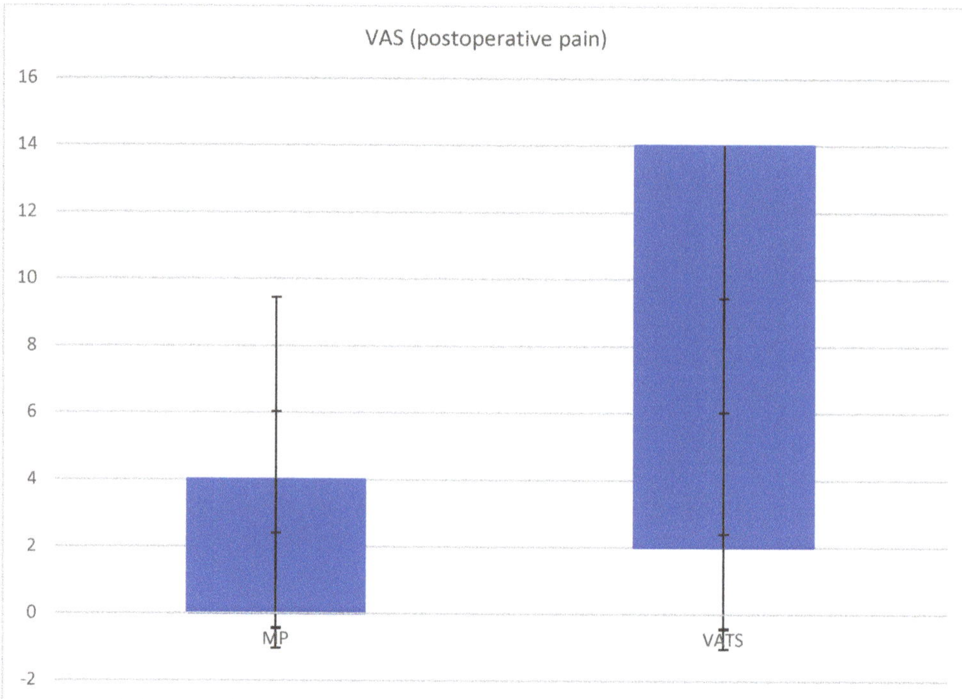

Figure 2. Postoperative pain in MP versus VATS using Visual Analogue Scale (VAS) (IQR 0–3, median 0, Group A) (IQR 0–6, median 3, Group B).

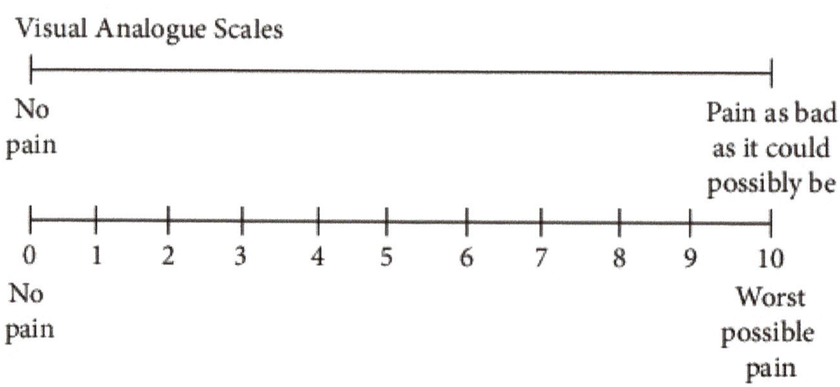

Figure 3. Visual Analogue Scale (VAS) used for evaluating postoperative pain.

5. Discussion

The aim of this controlled study was to demonstrate the safety and diagnostic yield of MP performed in analgosedation by a pulmonologist team with established experience in airway management and respiratory critically ill patients in an RICU and interventional procedures, in a comparative analysis with data obtained from a series of patients who underwent pleuroscopic examination in VATS by a surgeon team under anesthesiologist-based management. The diagnostic yield was over 97% and it was not significantly different compared to that obtained in the VATS group (100%). In our study, in both groups, malignancies accounted for the greater amount of underlying pleural effusion

(Group A: 59.3%, Group B: 55.1%; $p = 0.547$). Among the non-malignant etiologies, the most common diagnosis was that of non-specific pleuritis, described as fibrinous or inflammatory pleuritis without a specific attributable benign or malignant cause, but in subsequent follow-ups, proved to be benign in 85% of cases [22]. Another important finding to highlight is the diagnosis of TB, identified in two cases in Group A and four in Group B. After obtaining the diagnosis through PCR analysis on the biopsy sample, subsequent culture and antibiotic sensitivity testing was performed.

No deaths and no major complications, either intraoperative or postoperative, were observed in the MP group. These results are in line with literature [23–26]. In the retrospective observational study of Valsecchi et al. on 2752 analyzed MPs, the diagnostic yield was about 80%, and in more than half of the cases, the diagnosis was consistent with malignancies. The most frequent tumor was mesothelioma (about 21%), followed by metastases of breast cancer (14%). Tubercular pleurisy was the most frequent diagnosis in non-malignant pleural effusion (about 6% of cases) [23]. In Valsecchi's study, it was also estimated that MP obtained a good diagnostic yield in case of monolateral pleural effusion, but dropped to about 50% in the case of bilaterality of the effusion. These findings underline that the good selection and study of the patient to be subjected to the procedure are essential to achieve an accurate diagnosis [23]. In our study, we identified two cases of pleural effusion secondary to heart failure. The diagnosis of heart failure was subsequently made based on clinical and laboratory criteria, such as NT-proBNP. These were two cases of unilateral pleural effusion with no evidence of other pathology found in the pleural biopsies performed. Improvement in the effusion with the optimization of medical therapy confirmed the diagnostic suspicion. In another study of 19 patients who underwent MP, the diagnosis was obtained in 69% of cases (13/19 patients) and malignant etiology was observed in 100% of cases [24]. A diagnostic yield of 97.8%, very similar to that reported in our experience, was obtained in the study of Dhooria et al. conducted on 145 patients with pleural effusion undergoing MP with a rigid thoracoscope. In the same study, the use of the rigid thoracoscope was found to be superior to the use of the semi-rigid instrument [25].

Also, in a recent meta-analysis on the complications of thoracentesis and MP performed for malignant pleural effusion, the complication rate was only 0.040 (95% CI 0.029–0.052) [26].

In the British Thoracic Society (BTS) guidelines for pleural diseases, MP with a rigid thoracoscope is considered a valid procedure as well as one performed in VATS with no difference in diagnostic yield, sensitivity, or specificity [8]. Also, in the Clinical Statement of the BTS on pleural procedures, MP performed with local anesthetic is considered a safe procedure with a death rate of 0.3% [8,27]. Major complications such as massive bleeding, pleural infections, and pneumothorax occur in only 1.8% of cases, while minor complications such as subcutaneous emphysema, atrial fibrillation, mild bleeding, hypotension, and fever occur in 7.8% [8,27]. Other studies confirm the safety of MP by establishing that complications are rare, with a mortality between 0.09% and 0.24% [28–30]. In another study with the objective of evaluating MP in patients at high risk of complications, mortality was 0.28% with minor complications such as postoperative pain in 12.3% of cases and subcutaneous emphysema in 10.3%. No complications were observed in 58.9% of cases, thus establishing the safety of the procedure in high-risk patients [31]. Mild to moderate pain and minor bleeding were described in another study of 14 patients with an overall diagnostic yield of 100% [32].

In our study, in the MP group, we observed a lower, statistically insignificant rate of major bleeding compared to the VATS group and there was no statistically significant difference in terms of other complications between the two groups. The trend towards a reduction in bleeding in the MP group may be due to the use of a smaller instrument (Storz 4 mm vs. 10 mm) and, therefore, of smaller biopsy samples and smaller thoracostomy access.

The inpatient time was statically longer in patients from Group A than Group B, and this is probably due to the worse clinical conditions of patients enrolled in Group A with a

Charlson Comorbidity Index of 4.27 ± 2.04 versus 1.64 ± 1.52 in Group B ($p < 0.001$). In addition, the patients in Group B were part of a "fast-track" path for pleural effusion with hospitalization aimed at pleuroscopic intervention.

The pain was statistically better controlled in the group subjected to MP than in Group B. The reasons for the reduction in pain are likely due to the minor invasiveness of MP, the smaller size of the instrument, and, consequently, the smaller thoracostomy access [33]. The ability to perform the procedure without orotracheal intubation drastically improves patient comfort and recovery time [33]. Performing a pleuroscopy under analgosedation and spontaneous breathing also allows the enrolment of patients with multiple comorbidities who would usually be excluded from more invasive procedures.

Another advantage of our study was that MP was performed under analgosedation by a pulmonologist team without the support of the anesthesiologist. Currently, in Italy and in all European countries, the administration of propofol is authorized not only for anesthesiologists, but for all physicians with extensive experience in managing patients in intensive care units [34]. The use of a propofol-based regimen for analgesia and sedation is contemplated in the guidelines "Practice Guidelines for Sedation and Analgesia by non-Anesthesiologists", suggesting that when used for moderate sedation, the use of propofol or ketamine offers satisfactory results [35,36]. The same guidelines also suggest that when propofol is used for moderate sedation, the user should be able to intervene with rescue therapy at all levels of sedation, including general anesthesia [35,36]. Pulmonologists working in an RICU have both the experience and the knowledge in the management of possible cardio-respiratory complications induced by propofol [15]. Of course, the role of nursing staff in monitoring the patient during the procedure and in supporting pneumologists throughout the procedure must be emphasized. The pulmonologist team of the "San Donato" Hospital of Arezzo related to Pneumology and the RICU has gained significant experience in analgosedation with protocols based on propofol and meperidine after completing a specific course aimed at developing both theoretical and practical knowledge on the use of drugs for sedation and the management of potential cardio-respiratory complications. Each pulmonologist (both physicians and nurses) of the team is capable in the management of the airways and in the active support of critical patients, even with invasive mechanical ventilation; moreover, the entire team is trained in the positioning of pleural drainage of all kinds for the management of pleural pathologies [15]. All of these competences are, therefore, indispensable in preventing and promptly intervening in the event of a possible complication. It is important to note that the anesthesiologist is always available in case of any complications in an integrated path between the RICU and ICU (intensive care unit).

Analgosedation managed by a pulmonologist team is also well established in the UK. As reported in the guidelines "BTS clinical statement on pleural procedures", sedation is the responsibility of the thoracoscopist and is usually performed with the intravenous infusion of benzodiazepines (e.g., midazolam) and opioids (e.g., fentanyl). The anesthesiologist is generally involved in more complex sedations [8,27]. In real life, there is often fear regarding the use of propofol as an anesthetic for sedation and benzodiazepines are usually preferred by non-anesthesiologist teams [37]. This reluctance must be overcome, because there is already abundant literature data attesting to the safety of propofol use and it is considered, in many cases, to be superior in sedation compared to midazolam [38]. In the study conducted by Roekaerts et al. in which sedation with propofol and midazolam were compared after coronary surgery, during sedation in all patients, the hemodynamic parameters were stable except for a slight drop in systemic blood pressure for the propofol group and an increase in heart rate for the midazolam group [38], and the hemodynamic effects of both midazolam and propofol usually have no clinical significance [39]. In the work of Tschopp et al., balanced sedation with propofol was used in MP. The same authors showed that MP performed in analgosedation with propofol by non-anesthesiologist teams is a procedure that can be safely conducted without major cardio-pulmonary complications [40]. Complications such as hypoxemia and hypotension can be easily and quickly corrected in most cases [40]. In another randomized clinical trial, it appears that sedation with propofol performed by endoscopists gives the patient better satisfaction than deep sedation performed by anesthesiologists [41]. In addition, patients receiving sedation from non-anesthesiologists required significantly lower doses of propofol than those receiving sedation from anesthesiologists (94 mg versus 260 mg), with fewer side effects with statistically significant values [42,43]. Also, in our study, the doses of propofol administered were lower than those used in Group B, where sedation was the prerogative of the anesthesiologists (65 mg versus 280 mg), in the absence of major cardiorespiratory complications and with only two cases of hypotension in the 91 procedures carried out. The reduction in the dose of the sedative used obviously involves faster recovery of the patient and a lower risk of hemodynamic and respiratory side effects [44]. The reduction in the dosage of sedatives also did not affect pain control or patient comfort. In our study, in Group A, which involved a significantly lower dosage of anesthetics than Group B, the pain was better controlled with an average VAS of 0.34 ± 0.65 versus 2.58 ± 1.26.

This confirms the safety in the use of propofol as a sedative agent of choice compared to midazolam. In another non-inferiority randomized clinical trial that compared sedation with propofol in digestive endoscopic procedures performed by non-anesthetists and anesthesiologists in low-risk patient groups (ASA 1-2), the rate of complications (39%) was the same between the two groups without differences in the dose of propofol administered and without differences in amnesia and recovery time [41]. In a recent study carried out by Maffucci et al., the authors showed that sedation with propofol administered by a pneumologist team during bronchoscopic procedures was a safe practice without serious side effects. Complications occurred in only 25% of cases and were all successfully treated in the endoscopic room. Moderate sedation was achieved in 92% of subjects treated with adequate comfort and tolerance and with a recovery time of about 5–10 min on average [15]. Sedation based on propofol is now an integral part of the guidelines of gastrointestinal endoscopy with a frequency of adverse events similar to that of traditional sedation, but with a shorter recovery and duration of hospitalization [43,45]. Unlike endoscopic digestive procedures, in which the use of propofol for sedation performed by non-anesthesiologists is now consolidated [45], there is a lack of data and randomized clinical trials in the literature that clearly demonstrate the safety of sedation performed by pulmonologists in the diagnosis of pleural diseases. It is necessary, however, to highlight that the support of the anesthesiologist is indispensable in procedures of the highest risk (ASA 4) and that good collaboration between pulmonologists and anesthesiologists should be achieved.

Our study has some limitations that need to be considered. First, there was no control group in which sedation was performed by the same pulmonologist team, but with a regime based on midazolam. Second, the retrospective design of the study intrinsically has some biases like the risk of losing some relevant points such as the costs of the procedure and the real patient comfort. Finally, the study was conducted by a team with extensive experience in analgosedation with propofol and with excellent skills in airway management including invasive mechanical ventilation. The same results, therefore, may not be comparable to those of other centers with less expertise.

6. Conclusions

In summary, this study shows that MP performed under analgosedation by a pulmonologist team is a safe and well-tolerated procedure with a diagnostic yield similar to that obtained in VATS. The use of a sedation regimen with propofol and meperidine turns out to be safe with few side effects.

Unfortunately, there are few pulmonology units that routinely perform MP (estimated at approximately 30%) [46], and there are still too few pneumologists who deal first-hand with sedation. The purpose of this study was also to provide objective safety data so that this procedure becomes widely used in all pulmonology units. Too often, the management of pleural pathology is left to the thoracic surgeon, but pulmonologists must regain centrality in the management of these pathologies that increasingly present themselves in practice.

Author Contributions: Conceptualization, V.A., L.G., R.S., A.B. and P.P.; methodology, V.A., L.G. and R.S.; validation, R.S., A.B. and P.P.; writing—original draft preparation, V.A., L.G. and R.S.; writing—review and editing, V.A., R.S., A.B. and P.P.; project administration, R.S.; data curation, V.A., A.G., L.G., R.F., L.B., R.S., P.P. and A.B.; investigation, V.A., L.G., A.G., R.F., L.B., R.S., P.P. and A.B. All authors have read and agreed to the published version of the manuscript.

Funding: This research received no external funding.

Institutional Review Board Statement: The study was conducted in accordance with the Declaration of Helsinki, and approved by Ethics Committee Comitato Etico Regionale per la Sperimentaione Clinica della Regione Toscana Area Vasta Sud Est with approval code TORAPO 23776 on 20 March 2023.

Informed Consent Statement: Informed consent was obtained from all subjects involved in the study.

Data Availability Statement: The data presented in this study are available on request from the corresponding author due to privacy reasons.

Conflicts of Interest: The authors declare no conflicts of interest.

References

1. Maskell, N.; British Thoracic Society Pleural Disease Guideline Group. British Thoracic Society Pleural Disease Guidelines—2010 update. *Thorax* **2010**, *65*, 667–669. [CrossRef]
2. Mummadi, S.R.; Stoller, J.K.; Lopez, R.; Kailasam, K.; Gillespie, C.T.; Hahn, P.Y. Epidemiology of Adult Pleural Disease in the United States. *Chest* **2021**, *160*, 1534–1551. [CrossRef] [PubMed]
3. Sumalani, K.K.; Rizvi, N.A.; Asghar, A. Role of medical Thoracoscopy in the Management of Multiloculated Empyema. *BMC Pulm. Med.* **2018**, *18*, 179. [CrossRef] [PubMed]
4. Medford, A.R.L.; Bennett, J.A.; Free, C.M.; Agrawal, S. Current status of medical pleuroscopy. *Clin. Chest Med.* **2010**, *31*, 165–172. [CrossRef]
5. Roberts, M.E.; Rahman, N.M.; Maskell, N.A.; Bibby, A.C.; Blyth, K.G.; Corcoran, J.P.; Edey, A.; Evison, M.; de Fonseka, D.; Hallifax, R.; et al. British Thoracic Society Guideline for pleural disease. *Thorax* **2023**, *78* (Suppl. S3), s1–s42. [CrossRef] [PubMed]
6. Bhatnagar, R.; Maskell, N.A. Medical pleuroscopy. *Clin. Chest Med.* **2013**, *34*, 487–500. [CrossRef]
7. Shaw, J.A.; Irusen, E.M.; Diacon, A.H.; Koegelenberg, C.F. Pleural tuberculosis: A concise clinical review. *Clin. Respir. J.* **2018**, *12*, 1779–1786. [CrossRef] [PubMed]
8. Asciak, R.; Bedawi, E.O.; Bhatnagar, R.; Clive, A.O.; Hassan, M.; Lloyd, H.; Reddy, R.; Roberts, H.; Rahman, N.M. British Thoracic Society Clinical Statement on pleural procedures. *Thorax* **2023**, *78* (Suppl. S3), s43–s68. [CrossRef] [PubMed]
9. Rachelle Asciak on behalf of the BTS. BTS Clinical Statement on Pleural Procedures Draft for Public Consultation June 2022. Available online: www.brit-thoracic.org.u (accessed on 22 January 2023).
10. Nestor, J.; Huggins, T.; Kummerfeldt, C.; DiVietro, M.; Walters, K.; Sahn, S. Viral diseases affecting the pleura. *J. Clin. Virol.* **2013**, *58*, 367–373. [CrossRef]
11. Imperatori, A.; Rotolo, N.; Gatti, M.; Nardecchia, E.; De Monte, L.; Conti, V.; Dominioni, L. Peri-operative complications of video-assisted thoracoscopic surgery (VATS). *Int. J. Surg.* **2008**, *6* (Suppl. S1), S78–S81. [CrossRef]
12. Renda, T.; Scala, R.; Corrado, A.; Ambrosino, N.; Vaghi, A.; Scientific Group on Respiratory Intensive Care of the Italian Thoracic Society (ITS-AIPO). Adult Pulmonary Intensive and Intermediate Care Units: The Italian Thoracic Society (ITS-AIPO) Position Paper. *Respiration* **2021**, *100*, 1027–1037. [CrossRef] [PubMed]
13. Deitch, E.A. Multiple organ failure. Pathophysiology and potential future therapy. *Ann. Surg.* **1992**, *216*, 117–134. [CrossRef]
14. Doyle, D.J.; Hendrix, J.M.; Garmon, E.H. American Society of Anesthesiologists Classification. In *StatPearls*; StatPearls Publishing: Treasure Island, FL, USA, 2022. Available online: https://www.ncbi.nlm.nih.gov/books/NBK441940/ (accessed on 16 January 2023).
15. Maffucci, R.; Maccari, U.; Guidelli, L.; Benedetti, L.; Fabbroni, R.; Piccoli, B.; Bianco, A.; Scala, R. Pulmonologist-Administered Balanced Propofol Analgosedation during Interventional Procedures: An Italian Real-Life Study on Comfort and Safety. *Int. J. Clin. Pract.* **2022**, *2022*, 3368077. [CrossRef]
16. Sessler, C.N.; Gosnell, M.S.; Grap, M.J.; Brophy, G.M.; O'Neal, P.V.; Keane, K.A.; Tesoro, E.P.; Elswick, R.K. The Richmond Agitation-Sedation Scale: Validity and reliability in adult intensive care unit patients. *Am. J. Respir. Crit. Care Med.* **2002**, *166*, 1338–1344. [CrossRef] [PubMed]
17. Conway, A.; Douglas, C.; Sutherland, J. Capnography monitoring during procedural sedation and analgesia: A systematic review protocol. *Syst. Rev.* **2015**, *4*, 92. [CrossRef]
18. Veazie, S.; Vela, K.; Mackey, M. *Evidence Brief: Capnography for Moderate Sedation in Non-Anesthesia Settings*; Department of Veterans Affairs (US): Washington, DC, USA, 2020.
19. Aldrete, J.A.; Kroulik, D. A postanesthetic recovery score. *Anesth. Analg.* **1970**, *49*, 924–934. [CrossRef]

20. Massive Bleeding Protocols—The Transfusion Service Perspective. Available online: https://www.isbtweb.org/isbt-working-parties/clinical-transfusion/resources/patient-blood-management-resources/massive-bleeding-protocols.html (accessed on 2 January 2023).
21. Norfolk, D. (Ed.) *Handbook of Transfusion Medicine*, 5th ed.; United Kingdom Blood Services; TSO: Norwich, UK, 2023.
22. Janssen, J.; Maldonado, F.; Metintas, M. What is the significance of non-specific pleuritis? A trick question. *Clin. Respir. J.* **2018**, *12*, 2407–2410. [CrossRef]
23. Valsecchi, A.; Arondi, S.; Marchetti, G. Medical thoracoscopy: Analysis on diagnostic yield through 30 years of experience. *Ann. Thorac. Med.* **2016**, *11*, 177–182.
24. Agarwal, A.; Prasad, R.; Garg, R.; Verma, S.K.; Singh, A.; Husain, N. Medical thoracoscopy: A useful diagnostic tool for undiagnosed pleural effusion. *Indian J. Chest Dis. Allied Sci.* **2014**, *56*, 217–220. [PubMed]
25. Dhooria, S.; Singh, N.; Aggarwal, A.N.; Gupta, D.; Agarwal, R. A randomized trial comparing the diagnostic yield of rigid and semirigid thoracoscopy in undiagnosed pleural effusions. *Respir. Care* **2014**, *59*, 756–764. [CrossRef] [PubMed]
26. Martinez-Zayas, G.; Molina, S.; Ost, D.E. Sensitivity and complications of thoracentesis and thoracoscopy: A meta-analysis. *Eur. Respir. Rev.* **2022**, *31*, 220053. [CrossRef]
27. Rintoul, R.C.; Marciniak, S.J. What's new in pleural disease? *Thorax* **2023**, *78*, 1057–1058. [CrossRef]
28. Colt, H.G. Thoracoscopy. A prospective study of safety and outcome. *Chest* **1995**, *108*, 324–329. [CrossRef]
29. Viskum, K.; Enk, B. Complications of thoracoscopy. *Poumon. Coeur* **1981**, *37*, 25–28. [PubMed]
30. Lee, P.; Colt, H.G. State of the art: Pleuroscopy. *J. Thorac. Oncol.* **2007**, *2*, 663–670. [CrossRef] [PubMed]
31. Metintas, M.; Ak, G.; Yildirim, H.; Danacioglu, S.; Dundar, E.; Metintas, S. The safety of medical thoracoscopy in a group at high risk for complications. *J. Bronchol. Interv. Pulmonol.* **2013**, *20*, 224–231. [CrossRef] [PubMed]
32. Shrestha, B.K.; Adhikari, S.; Thakur, B.K.; Kadaria, D.; Tamrakar, K.K.; Devkota, M. Medical Thoracoscopy for Undiagnosed Exudative Pleural Effusion: Experience from Two Tertiary Care Hospitals of Nepal. *JNMA J. Nepal Med. Assoc.* **2020**, *58*, 158–164. [CrossRef] [PubMed]
33. Valchev, D.; Peeva, K.; Petrov, D. Are postoperative pain and patient satisfaction influenced by the number of ports in VATS for malignant pleural effusion treatment? *Postgrad. Med.* **2020**, *132*, 62–65. [CrossRef]
34. Propofol Data Sheet Agenzia Italiana del Farmaco (AIFA). 2019. Available online: https://farmaci.agenziafarmaco.gov.it/aifa/servlet/PdfDownloadServlet?pdfFileName=footer_002829_036849_RCP.pdf&sys=m0b1l3 (accessed on 9 January 2023).
35. American Society of Anesthesiologists Task Force on Sedation and Analgesia by Non-Anesthesiologists. Practice Guidelines for Sedation and Analgesia by Non-Anesthesiologists. *Anesthesiology* **2002**, *96*, 1004–1017. [CrossRef]
36. Knape, J.T.A.; Adriaensen, H.; van Aken, H.; Blunnie, W.P.; Carlsson, C.; Dupont, M.; Pasch, T.; Board of Anaesthesiology of the European Union of Medical Specialists. Guidelines for sedation and/or analgesia by non-anaesthesiology doctors. *Eur. J. Anaesthesiol.* **2007**, *24*, 563–567. [CrossRef]
37. Strohleit, D.; Galetin, T.; Kosse, N.; Lopez-Pastorini, A.; Stoelben, E. Guidelines on analgosedation, monitoring, and recovery time for flexible bronchoscopy: A systematic review. *BMC Pulm. Med.* **2021**, *21*, 198. [CrossRef]
38. Roekaerts, P.M.H.J.; Huygen, F.J.P.M.; de Lange, S. Infusion of propofol versus midazolam for sedation in the intensive care unit following coronary artery surgery. *J. Cardiothorac. Vasc. Anesth.* **1993**, *7*, 142–147. [CrossRef]
39. Magarey, J.M. Propofol or midazolam—Which is best for the sedation of adult ventilated patients in intensive care units? A systematic review. *Aust. Crit. Care* **2001**, *14*, 147–154. [CrossRef]
40. Tschopp, J.M.; Purek, L.; Frey, J.G.; Schnyder, J.M.; Diaper, J.; Cartier, V.; Licker, M. Titrated Sedation with Propofol for Medical Thoracoscopy: A Feasibility and Safety Study. *Respiration* **2011**, *82*, 451–457. [CrossRef] [PubMed]
41. Ferreira, A.; Torres, J.; Barjas, E.; Nunes, J.; Glória, L.; Ferreira, R.; Rocha, M.; Pereira, S.; Dias, S.; Santos, A.A.; et al. Non-anesthesiologist administration of propofol sedation for colonoscopy is safe in low risk patients: Results of a noninferiority randomized controlled trial. *Endoscopy* **2016**, *48*, 747–753. [CrossRef] [PubMed]
42. Poincloux, L.; Laquière, A.; Bazin, J.E.; Monzy, F.; Artigues, F.; Bonny, C.; Abergel, A.; Dapoigny, M.; Bommelaer, G. A randomized controlled trial of endoscopist vs. anaesthetist-administered sedation for colonoscopy. *Dig. Liver Dis.* **2011**, *43*, 553–558. [CrossRef] [PubMed]
43. Dumonceau, J.M. Will societies of anesthesiologists partake in the take-off of non-anesthesiologist administration of propofol? *Rev Esp Enferm. Dig.* **2017**, *109*, 87–90. [CrossRef]
44. Hahn, J.O.; Dumont, G.A.; Ansermino, J.M. A direct dynamic dose-response model of propofol for individualized anesthesia care. *IEEE Trans. Biomed. Eng.* **2012**, *59*, 571–578. [CrossRef]
45. Repici, A.; Pagano, N.; Hassan, C.; Carlino, A.; Rando, G.; Strangio, G.; Romeo, F.; Zullo, A.; Ferrara, E.; Vitetta, E.; et al. Balanced propofol sedation administered by nonanesthesiologists: The first Italian experience. *World J. Gastroenterol.* **2011**, *17*, 3818–3823. [CrossRef] [PubMed]
46. Marchetti, G.; Pinelli, V. Toracoscopia medica. *Patol. Dell'apparato Respir.* **2016**, *31*, 317–323. [CrossRef]

diagnostics

MDPI

Review

The Emerging Biomarkers in Chronic Obstructive Pulmonary Disease: A Narrative Review

Kaitlin M. Phillips, Philip F. Lavere, Nicola A. Hanania and Muhammad Adrish *

Section of Pulmonary, Critical Care, and Sleep Medicine, Baylor College of Medicine, Houston, TX 77030, USA; kaitlin.phillips@bcm.edu (K.M.P.); philip.lavere@bcm.edu (P.F.L.); hanania@bcm.edu (N.A.H.)
* Correspondence: aadrish@hotmail.com

Abstract: The burden of chronic obstructive pulmonary disease (COPD) is increasing, especially for women in low-to-middle income countries. Biomarkers provide ever-increasing diagnostic precision for COPD and show promise for primary, secondary, and tertiary disease prevention. This review describes emerging applications for biomarkers in COPD, especially as they align with the Global Initiative for Chronic Obstructive Lung Disease (GOLD) emphasis on prevention, early diagnosis, and response to therapy. These biomarkers include blood eosinophils; IgE; C-reactive protein; fibrinogen; procalcitonin; interleukins 6, 8, and 33; tumor necrosis factor alpha; and soluble receptor for advanced glycated products (sRAGE). They have been used in various ways to identify COPD endotypes, predict exacerbations, predict mortality, and monitor the response to therapy. The fraction of exhaled nitric oxide (FENO) is increasingly studied in eosinophilic COPD endotypes and can be a diagnostic and predictive non-invasive biomarker. Imaging biomarkers, especially the quantitative computerized tomography (QCT) assessment of airway remolding, functional small airway disease, air trapping, lung function, and volume surrogates, all serve as non-invasive biomarkers for screening, early detection, and disease progression. Biomarkers facilitate all the phases of COPD care from detecting early airflow obstruction to predicting exacerbation and mortality. Biomarkers will be increasingly used as precise diagnostic tools to improve the COPD outcomes. The aim of this narrative review is to summarize the recent investigations in COPD biomarkers and their clinical applications.

Keywords: biomarkers; COPD endotypes; FENO; eosinophilic COPD; quantitative computerized tomography

1. Background

Biomarkers are widely utilized tools to inform clinical decision making. In recent years, the landscape of biomarkers in chronic obstructive pulmonary disease (COPD) has evolved, but identifying clinically relevant biomarkers has been challenging due to the heterogeneity of pathophysiologic airway and alveolar injuries, respiratory symptoms, and variable COPD progressions. Many COPD biomarkers are well-described inflammatory mediators of airway and parenchymal destruction [1]. Clinical COPD phenotypes are observable traits, while COPD endotypes are the biopathological mechanistic features of an individual's disease [2]. While biomass burning smoke exposure is a significant risk factor for developing COPD, the majority of COPD biomarker research has been conducted on COPD cases related to tobacco smoking. Phenotyping COPD is straightforward, but identifying patient endotypes allows for more personalized care in the era of precision medicine. Clinical biomarkers are an important tool in describing a patient's endotype

 https://doi.org/10.3390/diagnostics15101245

in COPD. This review summarizes the recent developments in the COPD biomarkers of various media, including serum, sputum, exhaled, and radiographic features, and describes how these biomarkers can be utilized to define patient endotypes for precision medicine in clinical practice.

2. Overview of Serum Biomarkers in COPD

2.1. Blood Eosinophil Count

Eosinophils are a type of granulocyte that creates and releases pro-inflammatory mediators within the inflammation cascade. While they represent a small proportion of leukocytes in healthy patients, the amount of eosinopoiesis is increased in type 2 inflammatory states [3]. Elevated eosinophil levels (>300 cells$\cdot\mu L^{-1}$) have become a widely used indicator for type 2 inflammation, so their role as a practical and clinically useful biomarker in airway disease has exponentially grown in recent years [1]. Although elevated blood eosinophil levels are traditionally associated with asthma, patients with both asthma and COPD overlap (ACO), and some patients with COPD have elevated peripheral eosinophils [4]. Over the last few decades, an elevated blood eosinophil count has been identified as an endotype that is associated with the moderate and severe "frequent exacerbation" phenotype of COPD (requiring hospitalization) [4,5]. However, patients with elevated blood eosinophil levels have lower mortality rates both in hospitalized acute exacerbations of COPD and longitudinally over time, theoretically related to a better response to systemic corticosteroids [6,7].

An increased blood eosinophil count has also been associated with an accelerated decline in lung function. One study demonstrated that patients with elevated blood eosinophil levels demonstrated a faster decline in FEV1 by an average of 34 mL/year, regardless of their COPD status [8]. The large-scale Copenhagen General Population Study found that patients with chronic airway disease that were found to have an elevated blood eosinophil count and fractional exhaled nitric oxide (FeNO) demonstrated a more rapid decline in FEV1 [9]. However, some studies have shown that changes in serum eosinophils over time are not associated with changes in pulmonary function or health status [10] (Table 1).

The Global Initiative for Chronic Obstructive Lung Disease (GOLD) Report now recommends including blood eosinophil counts as part of the initial assessment of newly diagnosed COPD because blood eosinophil counts have proven useful pharmacodynamic biomarkers for COPD. Patients with elevated blood eosinophil levels (initially defined as at least 2% of blood eosinophils in the leukocyte differential) respond better to inhaled corticosteroid-containing regimens, specifically with reduced exacerbations, when compared to patients with normal blood eosinophil levels [11–13]. The FORWARD study also demonstrated that patients with blood eosinophil counts (defined as >279.8 in this study) who were treated with ICS in addition to LABA showed improved respiratory symptoms and pre-bronchodilator FEV1 when compared to those of similar patients using LABA alone [12]. Conversely, the FLAME study evaluated the role indacaterol–glycopyrronium versus salmeterol–fluticasone in reducing the risk of COPD exacerbations and found that the LABA-LAMA combination was more effective in preventing exacerbations in patients who had a COPD exacerbation within the last year, regardless of their blood eosinophil level (though patients with very high eosinophils as defined as >600 were excluded from analysis) [14]. This study further emphasized the role in LAMA therapy in COPD management regardless of blood eosinophil level. Later, the analysis of the ETHOS trial demonstrated that triple therapy with budesonide/glycopyrronium/formoterol fumarate dihydrate reduced the rates of moderate and severe exacerbations with blood eosinophil counts as low as ≥100 cells/mm^3 [15]. Studies like these affirm the role of the blood eosinophil count as a useful clinical biomarker for predicting response to ICS.

As such, the GOLD Report recommends that patients with COPD with blood eosinophil counts >300 cells/uL are at higher risk for exacerbation and would likely benefit from ICS in their treatment plan [16]. Some studies discuss monitoring blood eosinophil counts as a measure of the ICS treatment response, though more investigation is necessary in this area. A recent European study demonstrated that initiation of ICS demonstrated a median reduction in blood eosinophils by 197 197 cells/μL [17]. These patients notably had an average of 1.71 exacerbations per year after starting ICS treatment. A proposed algorithm involves monitoring the blood eosinophils of people on inhaled corticosteroid treatment to determine if those patients benefit from continued ICS, although this approach is theoretical based on the available data and has not been validated. In this theorized approach, if the level of blood eosinophils rises after the initiation of ICS, then the ICS should be discontinued as this predicts someone who will not benefit from ICS and would needlessly be exposed to its side effects, such as dysphonia, topical candidiasis, skeletal effects, ocular effects, and pneumonia [10]. In fact, patients with COPD and low blood eosinophil counts (those less likely to benefit from ICS) that are receiving inhaled corticosteroids are associated with a higher risk of developing pneumonia [10].

Table 1. Association of serum/blood and exhaled breath biomarkers with different COPD outcomes.

Serum/Blood Biomarkers	FEV1 Decline	Exacerbations	Emphysema on Imaging	Hospitalization	Mortality	Predicting Response to a Treatment	References
Eos	↑	↑				↑ *	[3–18]
IgE	↑	↑				↑ **	[19–22]
CRP	↔	↑		C	C	↑ ***	[23–31]
Fibrinogen	C	C		↔	↑		[4,26,27,32–34]
PCT		↑	↔			C ***	[16,25,30,35,36]
IL-6				↑	↔		[3,27,28,37–39]
IL-8		C			↑		[3,28,39]
sRAGE	↔ ****	↔	↓	↔	↔		[2,40]
CC16	↓				↔		[2,41,42]
SP-D	C	↑			↑		[2,43]
Exhaled Breath Biomarker							
FeNO	↑	↑		↑	↑	↑ *	[1,18,44–57]

↑ = positive clinical association; ↓ = inverse clinical association; ↔ = no association identified; C = conflicting evidence; Eos = blood eosinophil count, IgE = immunoglobulin E, CRP = C-reactive protein, FeNO = fractional exhaled nitric oxide, PCT = procalcitonin, IL-6 = Interleukin-6; IL-8 = Interleukin-8; sRAGE = soluble receptor for advanced glycation end products, CC16 = club cell secretory protein, SP-D = surfactant protein-D; * inhaled corticosteroids, dupilumab; ** dupilumab; *** antibiotics in acute exacerbations; **** associated with baseline lower FEV1, but not decline in FEV1.

While pharmacodynamic biomarkers have played a large role in the development of biologic targets, and thus treatments for asthma, the role of biologics in COPD is newly emerging, and currently targets molecular agents that promote eosinophil migration. IL-4 and IL-13 are inflammatory cytokines within the Th2 pathway that recruit eosinophils and contribute to goblet cell hyperplasia, mucous secretion, and airway remodeling in COPD [18]. Dupilumab, a human monoclonal antibody that blocks the IL-4 and IL-13 pathways of eosinophil recruitment, has shown a promising clinical response in patients with asthma and is now approved in the setting of COPD. The BOREAS trial, and subsequently the NOTUS trial recruited patients who were at high risk for COPD exacerbations; they had been on triple-inhaler therapy with LAMA, LABA, and ICS, demonstrated blood eosinophil counts >300, and had at least two exacerbations or one severe exacerbation within the last year [1,18]. In these studies, dupilumab was shown to decrease patients' annual COPD exacerbations and to improve FEV1 and FVC. IL-5 is another important target for biologics in asthma management given their role in promoting eosinophilic inflammation. However, biologics that target the IL-5 pathways have not shown to reduce the symptoms or exacerbations of COPD, despite a measured decline in the blood eosinophil count [18]. Hence, the utilization of the blood eosinophil count as a biomarker has limitations.

2.2. *Immunoglobulin E*

The increased production of IL-4 and IL-13 in the Th2-inflammatory state also promotes the B cell production of immunoglobulin E (IgE) after allergen exposure. Viral infections also independently promote IgE release by driving dendritic cells to recruit Th2 cells that cause IgE release [19]. The role of IgE levels in COPD exacerbations is controversial. Previously, IgE elevation was demonstrated in asthma or asthma–COPD overlap, but only limited data highlight the role of IgE in isolated COPD. Recently, the investigators who designed the COSYCONET and WISDOM cohorts reviewed specific characteristics of patients with COPD in Europe as they relate to IgE levels. Higher rates of IgE elevation are seen in patients with both COPD and asthma, especially in current smokers [20]. However, a prospective observational study in Vietnam did not show clinically significant differences in IgE levels between patients with diagnosed COPD and control patients, implying that there may be a geographic role of IgE in COPD [21]. The individuals from the Copenhagen General Population Study cohort with COPD and elevated IgE demonstrated higher rates of severe exacerbations and all-cause mortality, independent of blood eosinophil count [22]. Interestingly, higher proportions of men with COPD had elevated IgE, and increased levels of total IgE have been correlated with higher risk of exacerbations in men than women, but patient age was not correlated with the levels of IgE [20]. In the COSYCONET trial, patients with elevated IgE levels did not show significantly different baseline FEV1 values, diffusion capacities, and residual volumes at baseline, but patients with levels more than >91.5 were at risk for future decline in FEV1 [20].

There is a promising role for IgE as a predictor of the future treatment response. In the BOREAS trial, patients with baseline IgE > 100 who were treated with dupilumab had fewer exacerbations and had a greater improvement in FEV1 than those treated with placebo in the 52-week treatment period [18]. The NOTUS trial subsequently demonstrated that dupilumab decreased IgE levels over the treatment period [1]. A further investigation is needed to validate the predictive, prognostic, and pharmacodynamic roles of IgE in COPD, especially to determine if omalizumab, a monoclonal antibody that binds IgE, could have a role in COPD management in the future.

2.3. *C-Reactive Protein*

C-reactive protein (CRP) is an acute phase reactant produced by the liver in response to IL-6 activity and plays a role in activation of the complement cascade, and thus pro-inflammatory cytokines to support innate immunity and phagocytosis during acute infection [23,24]. In both research and clinical practice, CRP is the most commonly measured inflammatory biomarker and often manifests as an early sign of chronic inflammation in patients with COPD [25]. Though multiple studies have described the utilization of elevated CRP in acute exacerbations of COPD, CRP is frequently elevated in both stable and acute exacerbations of COPD [26]. However, the supportive evidence for the role of CRP as a clinical biomarker is conflicting. In the ECLIPSE study, patients with COPD showed higher rates of CRP elevation when compared to non-smokers and smokers without COPD [27]. A meta-analysis has demonstrated that higher levels of CRP in COPD are associated with increased rates of exacerbation, hospitalization, and early mortality [25,27–29]. While the

SUMMIT trial demonstrated positive association between elevated CRP and mortality, it did not demonstrate associations with FEV1 decline, exacerbations, nor hospitalizations [4].

Recent studies have demonstrated that CRP may be able to assist in managing the acute exacerbation of COPD as bacterial versus non-bacterial. In a recent meta-analysis, elevated CRP levels showed the strongest evidence of predicting bacterial COPD exacerbations (by positive sputum culture) when compared to multiple other biomarkers, such as procalcitonin, sputum IL-8, and sputum TNF-a [30]. Elevated CRP levels are more often correlated with bacterial infection in the lower airways compared to procalcitonin. A study in the United Kingdom has assessed the utility of CRP as predictive of who would benefit from antibiotics in COPD exacerbations. They demonstrated that using CRP levels to guide antibiotic therapy versus using sputum purulence showed no increase in 30-day adverse events or treatment failure [31].

2.4. Fibrinogen

Fibrinogen is a slow-reacting positive acute phase reactant influenced by increased IL-6 levels that directly determines the erythrocyte sedimentation rate (ESR). In a multitude of disease states, elevated fibrinogen levels have been traditionally associated with increased all-cause mortality. The SUMMIT trial confirmed this association with mortality in the setting of COPD specifically [4]. Fibrinogen has demonstrated both predictive and prognostic capabilities in COPD and has been approved by the United States Food & Drug Administration (FDA) to be utilized as a biomarker in clinical trials describing COPD mortality and exacerbations [32,33]. The ECLIPSE study showed that patients with COPD show higher rates of fibrinogen elevation when compared to those of non-smokers and smokers without COPD [27]. The serum fibrinogen levels are also higher in patients with both asthma and COPD [26]. The IMPACT trial demonstrates that patients with COPD and elevated fibrinogen levels demonstrated increased rates of moderate and severe exacerbations [34]. A recent meta-analysis also described that elevated levels of fibrinogen are associated with a decline in FEV1 [28]. However, the SUMMIT trial did not find significant association between elevated fibrinogen levels and FEV1 decline, exacerbations, or hospitalizations [4]. Fibrinogen is often elevated in exacerbations, but fibrinogen levels unfortunately have not demonstrated enough sensitivity nor specificity to be validated as a predictive biomarker in this setting. Fibrinogen elevation takes time, which is a limitation in its utility in the acute setting [33]. Like in the case of CRP, further research is necessary to define the association with future exacerbations.

2.5. Procalcitonin

Procalcitonin is also an acute phase reactant. Cytokines like IL-6, TNF-α, and IL-1b stimulate the synthesis of procalcitonin by thyroid C cells. Its primary utility as a biomarker has been predicting which patients with acute exacerbations of COPD would benefit from antibiotics, and the FDA has approved the use of procalcitonin for starting and stopping antibiotics for those with lower respiratory tract infections [25]. Like CRP, multiple studies have associated elevated procalcitonin levels with positive sputum cultures in bacterial acute exacerbations of COPD. A few small studies have evaluated procalcitonin-guided algorithms to reduce antibiotic exposure in COPD exacerbations. These studies demonstrated successful reduction in antibiotic exposure, but have used different cut-offs anywhere between 0.03 and 1.03 to determine who would benefit from antibiotics in the setting of acute COPD exacerbation [30,35]. One meta-analysis demonstrated that a cut-off point of 0.76 ng/mL demonstrated the best area under receiver operating characteristic curve (ROC), with sensitivity of 92.5% and specificity of 78.95% [30]. However, a French study of 300 patients admitted to the ICU with COPD exacerbations demonstrated that using a procalcitonin-guided strategy to guide antibiotic management failed to reduce antibiotic exposure [36]. Given the controversy in recent studies regarding using procalcitonin in COPD exacerbations, the GOLD Report recommends against solely using procalcitonin-based strategies to determine whether patients with COPD exacerbations would benefit from antibiotics [16].

2.6. IL-6 and IL-8

In COPD, airway exposure to noxious fumes begins inflammatory cascades that lead to an extensive environment of signaling molecules at the cellular level that eventually promotes cellular injury and apoptosis. Pro-inflammatory cytokines such as IL-6 and IL-8 are expressed at higher levels in injured airway tissue, and expression is further propagated by heat shock protein (HSP)-70 at the cellular level. IL-8 is implicated in Th1 inflammation, while IL-6 is implicated in Th2 inflammation, further describing the role of both Th1 and Th2 inflammation in COPD. Some studies on airway pathology have shown increased numbers of cells that are immunoreactive to IL-6 and IL-8 in COPD, and IL-6 and IL-8 have demonstrated potential as prognostic biomarkers in COPD [3].

Patients with COPD show higher rates of serum IL-6 elevation when compared to those of non-smokers and smokers without COPD per the ECLIPSE Study [27]. Increased serum IL-6 levels are associated with increased rates of future COPD exacerbations and hospitalizations, but not associated with mortality [28,37,38]. However, one study that utilized data from the ECLIPSE cohort showed that elevated serum IL-8 levels were associated with increased 3-year all-cause mortality in patients with COPD [28]. While one observational study detected higher levels of serum IL-8 in acute COPD exacerbations, the role of measuring serum IL-6 and IL-8 levels in acute exacerbations is unclear as previous studies have presented conflicting results [28,39]. These biomarkers are difficult to measure in clinical settings.

2.7. IL-33/ST-2

IL-33 belongs to the IL-1 superfamily of cytokines, which plays a role in innate and adaptive immunity [58]. Specifically, IL-33 is secreted by the epithelial cells and plays a role in remodeling and tissue repair, along with tissue homeostasis, which it does by repressing expression for some inflammatory genes. As an alarmin, IL-33 binds to the ST-2 (suppression of tumorigenicity 2) receptor and alerts the immune system of tissue damage and active Th2 response [59]. Neutrophils, natural killers, T cells, epithelial cells, and goblet cells also express ST-2 receptors. Increased ST-2 and IL-33 expressions have been reported in COPD and have been linked with higher risk of exacerbations [60]. The earlier data from phase 2 studies have demonstrated the safety of antibodies targeting IL-33 and ST-2, as well as improvement in lung function and exacerbation risks. A phase 2a study of anti-IL-33 agent Itepekimab showed improvement in prebronchodilator FEV1 and reduced exacerbation risk in patients with COPD [61]. A phase 2a study of anti-ST-2 Astegolimab in patients with COPD did not improve the exacerbation rates, but showed improvements in FEV1, the symptom score, and the eosinophil count [62]. Further data from several ongoing phase 3 studies are expected to shed more light on the safety and efficacy of these agents [59].

2.8. Soluble Receptor for Advanced Glycation End Products

The soluble Receptor for Advanced Glycation End products (sRAGE) is a cleaved version of a transmembrane receptor categorized as an immunoglobulin gene [40]. Mediators in the inflammatory cascade create RAGE ligands that bind to the RAGE receptor. When the RAGE receptor is bound, it promotes the further transcription of inflammatory cytokines, creating an amplification cycle for RAGE binding, and thus the generation of an inflammatory cascade [2]. Decreased levels of sRAGE have been found in multiple chronic diseases, including chronic airway disease (COPD and asthma) [4]. However, sRAGE is often elevated in acute respiratory disease, including infection-driven acute respiratory distress syndrome (ARDS) and isolated RSV infection (in experimental models at least) [40]. The levels of sRAGE have been found to be lower in smokers without COPD than non-smoking populations, which suggests a role of smoking in the modulation of the genetic expression of sRAGE [2]. Although, a study in Mexico demonstrated that the sRAGE levels are decreased in patients with COPD due to both tobacco smoking and biomass burning smoke exposure when compared to those of control groups exposed to these known substances [63]. A meta-analysis by Pratte et al. summarized the findings of sRAGE clinical associations in COPD from four large assays: COPDGene, SPIROMICS, ECLIPSE, and Pittsburgh COPD SCOOR. Reduced levels of serum sRAGE were found to be associated with severe baseline airflow obstruction by FEV1 and emphysema on imaging at the time of investigation [40]. In fact, the levels of sRAGE closely relate to the GOLD classification of COPD severity [2]. However, these cohorts demonstrated inconsistent findings when attempting to associate a reduction in sRAGE and the progression of obstruction and/or emphysema [40]. The SUMMIT trial also did not find a significant association between the levels of sRAGE and FEV1 decline, exacerbations, hospitalizations, nor mortality [4]. The relationship between the sRAGE levels and COPD suggests that COPD may be affected by genetic contributions, an important subject that warrants further investigation.

2.9. Club Cell Secretory Protein

Club cell secretory protein (CC16) is made by pulmonary tissue and secreted by club cells during a significant lung injury to provide immunosuppressant and anti-inflammatory properties [2,41]. As the number of functional club cells declines during COPD disease development, less CC16 is secreted. CC16 shows promise as a novel biomarker in COPD. Patients with COPD are found to have lower circulating levels of serum CC16 (which is been associated with the bronchoalveolar lavage levels), particularly in male patients with advanced age [42]. Lower levels of CC16 are associated with a more rapid progressive decline in FEV1 ($\geq$40 mL/yr) [2,42]. Changes in CC16 levels are not, however, associated with the progression of emphysema by imaging nor mortality in the ECLIPSE cohort. Similarly, its association with COPD exacerbations and hospitalizations has not been well studied. The role of CC16 in predicting other factors of disease progression aside from evidence of worsening airflow obstruction warrants further investigation.

2.10. Surfactant Protein-D

Surfactant protein D (SP-D) is another novel biomarker produced by type II pneumocytes that plays a role in the lungs' innate immunity [2,43]. The serum levels of SP-D increase as a result of COPD-associated lung injuries and are found at higher levels in patients with COPD when compared to those of control subjects [2,41]. The SP-D levels are elevated in acute exacerbations of COPD, and stable patients with COPD with elevated serum SP-D have shown an increased risk of exacerbation within the following year [43]. The ECLIPSE study also demonstrated associations with elevated SP-D levels and 3-year mortality. Several studies have evaluated the association between SP-D levels and changes in spirometry, but with conflicting results. Its utility as a useful and specific predictive biomarker in COPD is limited in that the levels are associated with a variety of different factors, including age, body mass index, and the male sex. THe levels are also increased in multiple different pulmonary disease states [43].

3. Overview of Lung Samples in COPD

3.1. Sputum Biomarkers

Identifying sputum biomarkers in COPD is a growing area of interest. The roles of IL-6 and IL-8 in COPD pathogenesis have been described previously, and some studies have shown increased concentrations of sputum IL-6 and IL-8 in acute COPD exacerbations and severe chronic COPD when compared to those in stable or less-severe COPD, respectively [25]. Sputum IL-8 has demonstrated a potential role of differentiating bacterial versus non-bacterial exacerbations of COPD. Several small studies have assessed the sputum IL-8 levels and their association with bacterial exacerbations of COPD, but the results of these studies have been variable [30]. Tumor necrosis factor alpha (TNF-α) is a cytokine that contributes to propagation of the Th1 inflammatory response (Figure 1). The expression of TNF-α is significantly increased by airway tissue experiencing tissue and cellular injuries in COPD. Higher numbers of TNF-α immunoreactive cells have also been found in the airway tissue of patients with COPD [3]. Elevated levels of sputum TNF-α are seen in cases of severe COPD and acute exacerbations of COPD when compared to those in stable and less-severe COPD, respectively [25]. Multiple studies have evaluated the association between increased sputum TNF-α and the evidence of bacterial infection (highest in Pseudomonas aeruginosa infections), though all these studies were completed in an outpatient setting [30]. Despite the promising role of sputum measurements, the serum measurements of TNF-α demonstrate no statistically significant associations with mortality, exacerbations, or hospitalizations [28].

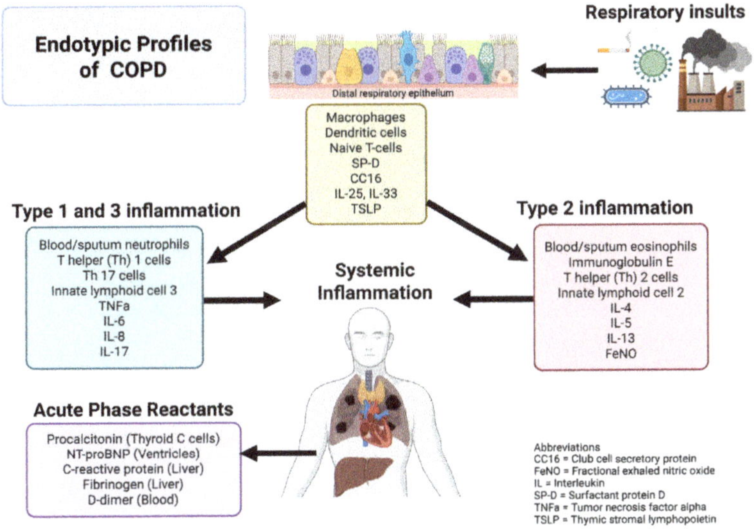

Figure 1. Endotypic profile of COPD (created with www.biorender.com).

Many other sputum biomarkers have been associated with COPD, but the data regarding their diagnostic, predictive, prognostic, and therapeutic utility are limited. Sputum studies have shown that a subset of patients with COPD have eosinophil counts of >3%. Similarly, elevated eosinophil counts have also been noted in airway biopsy studies on patients with COPD exacerbation. Therefore, eosinophilic inflammation has emerged as a treatable trait in COPD due to the large number of targeted therapies being studied. Some studies have indicated that blood eosinophils correlate with sputum eosinophils to a lesser degree in COPD compared to patients with asthma. Elevated sputum eosinophils were associated with increased risks of exacerbation and lung function decline in the SPIROMICS cohort [64]. A small Colombian study reviewed the sputum biomarkers of women with COPD secondary to tobacco or wood burning smoke and demonstrated that the women with COPD demonstrate higher sputum levels of IL-8, metalloproteinase 9 (MMP-9), and chemokine ligand 5 (CCL5) than the control groups [65]. The investigators also reported higher levels of sputum CCL5 in COPD from tobacco smoking than disease secondary to wood burning, implicating a potential role for this biomarker to delineate between these two disease drivers, though more investigations are warranted. This study assessed the variation in sputum vascular endothelial growth factor (VEGF) levels in COPD and control patients, but did not detect a significant difference. However, a small Japanese study compared the VEGF levels in control patients and those with emphysema and chronic bronchitis and demonstrated that lower VEGF levels were associated with a decline in airflow obstruction in emphysema, and higher VEGF levels were associated with worsening obstruction in chronic bronchitis [66]. Further research is needed to evaluate the utility of these biomarkers in clinical practice.

3.2. *Invasive Biomarkers*

Many studies have also begun to explore the utility of invasive pulmonary samples, such as bronchoalveolar lavage (BAL) and lung biopsy, in identifying the clinically relevant biomarkers in COPD. Most studies that have analyzed BAL samples from patients with COPD have found higher levels of various inflammatory proteins when compared to those of healthy patients. One study evaluated the extracellular vesicle-associated protein and microRNA profiles in BAL samples from patients with COPD and found higher expression levels of 284 proteins, many of which were found to be involved in inflammatory mechanisms [67]. Few studies have assessed the clinical utility of this biomarker. BAL glutathione levels have been found to be reduced in acute exacerbations of COPD when compared to those of patients with COPD [25]. In a bronchoscopy sub-study of the SPIROMICS cohort, the investigators aimed to assess a BAL protein signature associated with FEV1 decline in COPD. They did not find a particular protein signature in the BAL protein analysis of this 25-patient cohort that predicted FEV1 decline given that their chosen assay had limited utility in studying BAL samples [68]. Unfortunately, many of the assays used to detect serum biomarkers have not been validated for BAL and lung biopsy samples, which has created barriers to research progress in this area. The role of lung biopsy in identifying the clinically relevant biomarkers in COPD has had limited investigation, and further research in this area is warranted.

3.3. *FENO in COPD*

Fractional exhaled nitric oxide (FENO) is a biomarker of Th-2 airway inflammation and has been extensively studied in asthma [44]. The emerging data show promise using FENO as a biomarker to endotype COPD. Two large clinical trials recently evaluated dupilumab for patients with eosinophilic-type COPD (defined by blood eosinophil counts >300 cells/uL). Interestingly, although >50% of the patients in both the trials had FENO < 20 ppb, the averaged baseline FENO for all the patients was 24 ppb, suggesting heterogenous endotypes of eosinophilic COPD with a range of FENO values. When the patients were stratified by FENO > 20, the NOTUS trial showed FEV1 improvements on dupilumab at 12 weeks, but not 52 weeks. Throughout the trial, FENO was reduced by roughly 10 ppb during therapy. These results are contrasted in the BOREAS trial, which when stratified by FENO > 20, the patients on dupilumab had sustained improvements in FEV1 and reduced moderate-to-severe COPD exacerbations at 52 weeks. Similar to NOTUS, FENO was roughly reduced by 10 ppb during treatment [1,18].

In addition to pharmacodynamic and monitoring applications, FENO can be used as a diagnostic, prognostic, and predictive biomarker in COPD. FENO may distinguish asthma from COPD or characterize asthma–COPD overlap (ACO) [45–48]. Non-smokers or former smokers with COPD have a lower FENO than those with asthma [49]. A 1-year prospective biomarker study found that measuring FENO combined with blood eosinophil counts improved the diagnostic accuracy for severe COPD exacerbations compared to that of each biomarker alone [50]. Other research suggests FENO does not improve the diagnostic utility of exacerbation, but it does distinguish COPD from ACO [51]. Persistent FENO $\geq$ 20 ppb in patients with stable COPD is associated with an increased risk of future exacerbations [52]. One study shows patients with COPD and elevated FENO levels experienced greater symptom control on ICS + LABA; however, even so, an older study suggests ICS has no effect on COPD lung function [53,54]. Patients with severe COPD who no longer smoke displayed concordant reductions in FENO alongside systemic markers of inflammation (CRP, IL-6, and IL-8), while receiving ICS [55]. A study investigated the effects of bronchodilation on the measurement of FENO and showed an increase in measured value after inhaled bronchodilator therapy [56]. In a small cohort of patients with stable COPD caused by either tobacco smoke or biomass burning smoke exposure, the FENO level was equally elevated compared to that of healthy controls [57].

3.4. Biomarkers of Exhaled Breath Condensate

Exhaled breath condensate (EBC) is a non-invasive method of assessing airway inflammation and oxidative stress in the airways, with potential implications in COPD diagnosis, phenotyping, and disease monitoring [69]. Several EBC biomarkers have been studied, including, but not limited to, leukotriene B4 (LTB4); prostaglandin E2 (PGE 2); 8-isoprostanein; nitric oxide (NO) metabolite; hydrogen peroxide (H_2O_2); lactate; formate; acetate; butyrate; malondialdehyde (MDA); matrix metalloproteinase-12 (MMP-12); neutrophil elastase (NE); and the tissue inhibitor of metalloproteinase-4 (TIMP-4); and cytokines such as IL-6, IL-10, and TNF-α [69–71].

The EBC profiles have shown the potential to detect subclinical COPD. In a study of 300 community participants, EBC, MDA, and lactate were able to discriminate even early-stage COPD in previously undiagnosed patients [71]. Additionally, the analysis of EBC profiles has been shown to differentiate between asthma and COPD. A study showed increases in ethanol and methanol levels and a significantly lower level of formate in the EBC of patients with COPD compared to patients with asthma [72]. Similarly, a decrease in methanol level along with improvement in walk distance and dyspnea was seen in the EBC of patients with COPD undergoing pulmonary rehabilitation [73].

Exhaled breath biomarkers have also been linked with COPD disease severity and exacerbation. A study comparing 161 patients with stable COPD and 112 controls showed that the EBC pH was lower in the patients with COPD [74]. Hydrogen peroxide, which is a direct marker of air-space oxidative burden, was noted to be elevated in the EBC of smokers and COPD subjects compared to that of non-smokers [75]. Another study evaluating neutrophil chemotactic activity and a neutrophil chemoattractant in EBC demonstrated higher levels in outpatients with COPD exacerbation compared to those of stable patients with COPD [76]. EBC can also identify bacterial nucleic acid in patients with COPD with exacerbations [77]. A prospective study of 68 patients evaluated their exhaled breath profiles before, during, and after COPD exacerbation and demonstrated classification accuracies of 71% for baseline vs. exacerbation and of 78% for exacerbation vs. recovery [78]. Elevated levels of H_2O_2 and 8-isoprostanein have also been shown to be linked with dyspnea sensation in COPD [79].

The study of EBC biomarkers is an evolving field, but challenges remain with regard to identifying reliable and clinically useful exhaled breath biomarkers that are accurate and effective for their intended use. Additional limitations include the lack of standardized interpretation for these biomarkers, as well as the need for a well-equipped laboratory to ensure an accurate performance [80].

4. Overview of Imaging Biomarkers for COPD

Numerous thoracic computerized tomography (CT) features are emerging biomarkers for COPD. Quantitative CT (QCT) imaging reliably characterizes the changes in obstructive lung disease, including air trapping, hyperinflation, mucus plugging, airway remodeling, and emphysema severity, among others. Advanced QCT techniques elucidate more nuanced and dynamic disease classifications within COPD. For example, parametric response mapping (PRM) identifies the heterogenous air trapping regions between emphysema and functional small airway disease (fSAD) and strongly correlates with all the stages of COPD severity [81–83]. Research on QCT for COPD has proliferated due to the availability of imaging datasets from several large cohort studies and provides new tools for COPD screening, early diagnosis, monitoring disease progression, predicting lung function decline, and prognosticating exacerbations and mortality (Table 2).

Table 2. Association of imaging biomarkers with different COPD outcomes.

Imaging Biomarker	Lung Function Decline	Risk for Exacerbation	Functional Status Decline	Mortality	References
Mucus Plugs	↑	↑	↑	↑	[84–88]
Airway Wall Thickness	↑	↑	↑	↑	[89–97]
Total Airway Counts	↑				[94]
Emphysema	↑	↑		↑	[98–107]
fSAD	↑				[108]
Pulmonary Artery Pruning	↑		↑	↑	[109,110]
Enlarging Pulmonary Arteries			↑	↑	[111]

↑ = positive clinical association; fSAD = functional small airway disease.

4.1. Changes in Airway Anatomy

In patients with COPD, mucus plugs obstructing medium-to-large airways are associated with higher mortality [84]. Silent mucus plugs are very common in patients who smoke tobacco [85]. In patients without cough or phlegm, silent mucus plugs are associated with poor disease control and a worse functional status [86]. Mucus plugs are linked to airflow limitation [87,88]. Airway wall thickness was associated with impaired quality of life, but not arterial oxygen tension [89,90]. In addition, airway wall thickness was associated with increased mortality as emphysema progressed [91]. Pathological airway remodeling with diminished airway branching complexity was prognostic of respiratory morbidity and lung function [92]. Ex-smokers were found to have thinner airway walls, which was linked to higher total (presumably recovered) airway counts [93]. Decreasing total airway counts are related to lung function decline, serving as an early imaging biomarker for COPD progression [94]. Changes in the airway-surface-area-to-volume ratio are associated with respiratory morbidity, COPD progression, and mortality [95]. A cohort of 400 patients with COPD with increasing Pi10 values (normalized index of airway thickness) could predict exacerbation risk, lung function decline, and mortality [96]. Current or former smokers with normal spirometry, but airway wall thickening on QTC, have exacerbations and functional limitations [97].

4.2. Developing and Progressing COPD: Emphysema and Lung Function

In one cohort of patients with clinical symptoms of COPD and significant emphysema, spirometry was normal 10% of the time [98]. Visual emphysema in smokers at GOLD stage 0 predicted structural physiological disease progression [99]. QCT determines the percentage of low-level attenuation areas on expiration, indicating regions of air trapping, which are strongly associated with airflow obstruction on spirometry [100]. QCT on emphysema is correlated with the diffusion capacity of the lungs for carbon dioxide [101]. In a 5-year cohort of patients actively smoking tobacco, most emphysema progression was not accounted for by the FEV1 trends, but rather the QCT measures of air trapping [102]. Pan lobular emphysema is associated with greater airflow obstruction, increased respiratory symptoms, and systemic inflammation [103]. Severe centrilobular emphysema was linked to worsening diffusion capacity and higher mortality in severe GOLD stage [104]. Higher grades of emphysema on visual grading and artificial intelligence deep learning were associated with disease progression across all the GOLD stages [105,106]. Emphysema progression on QCT is associated with increased all-cause respiratory mortality [107].

Functional small airway disease (fSMAD) is associated with FEV1 decline in mild-to-moderate COPD [112]. The spatial analysis of coalescing fSMAD pockets predicted lung function decline and identified regions of emphysema onset [108]. The regions of fSAD to emphysema transition may serve as an imaging biomarker that identifies otherwise healthy patients at risk for developing COPD [113]. Decline in lung density was associated with the serum biomarkers SP-D and sRAGE [114]. Centrilobular emphysema is associated with sRAGE [115]. Declines in quantified total lung capacity are associated with an FEV1 increase in PRISm, but FEV1 decreases through GOLD stages 3–4 [116]. Over 10 years, patients actively smoking with emphysema had the greatest decrease in lung density [117]. The QCT markers linked with spirometry and the development of COPD include FEV1 and airway wall thickness, FEV1/FVC and emphysema, air trapping and residual volume, functional small airway disease and FEV1, and lung volumes with COPD progression [112,118,119].

Disease progression modeling characterizes developing COPD subtypes. One study found a "tissue to airway" pattern of fSAD and emphysema before large airway changes (airway wall thickness and airway wall area), whereas the "airway to tissue" pattern was reversed. Such QTC information applied across multiple biomarker domains, and indeed 30% of healthy smokers, showed features of both COPD pathologic patterns [120]. Two indices of CT-measured lung volume in patients smoking tobacco were successfully descriptive of emphysematous pathology (high TLC) versus airway disease/airflow obstruction (high FRC/TLC) [121].

4.3. Predicting Exacerbations by Imaging

The QCT markers of lung tissue texture and airway structure were predictive of exacerbations and performed better than BODE or exacerbation history [122,123]. Airway wall area and wall thickness were associated with the frequent COPD exacerbation phenotype in patients with stable disease [92]. Even when QCT was obtained for non-pulmonary reasons, the incidental finding of emphysema and airway wall thickness predicted future severe COPD exacerbations [124].

4.4. Pulmonary Vasculature and Other Imaging Markers

Some patients with mild COPD were found to have reduced pulmonary microvascular blood flow in non-emphysematous lung regions, a potentially distinct process, or in the "airway to tissue" subtype [125]. Pulmonary artery strain is reduced in COPD [126]. An increased residual volume in patients with COPD was linked to a larger pulmonary artery area [127]. Pulmonary artery pruning was associated with progressing emphysema, rapid lung function decline, increased residual volume, exercise capacity, and COPD mortality [109,110]. The total cross-sectional area of sub-segmental pulmonary artery vessels was correlated with the degree of emphysema [128]. Enlarging pulmonary arteries in patients with COPD were associated with lower exercise tolerance and increased mortality [111]. Hyperinflation is associated with coronary artery disease in patients who smoke [129].

4.5. Identifying Undiagnosed COPD by Imaging

COPD is highly prevalent in lung cancer screening populations [130]. Almost 2/3 of patients in a lung cancer screening cohort were found to have undiagnosed COPD [131]. In another large cohort of patients undergoing general lung health screening, almost 20% were found to have undiagnosed COPD [132]. One study found low-dose expiratory CT scans for male smokers undergoing lung cancer screening had 63% sensitivity and 88% specificity for diagnosing COPD [133].

5. Clinical Utility of COPD Biomarkers

Physiological biomarkers have varying degrees of utility in the clinical evaluation and management of COPD. Blood eosinophil count is a crucial biomarker for identifying patients with type 2 inflammation who are more likely to benefit from inhaled corticosteroids, as well as biologics like dupilumab, and it also carries prognostic information concerning exacerbation frequency and lung function decline [1,18] (Figure 2). As such, the GOLD Report guidelines now recommend the measurement of blood eosinophil as a component of routine COPD evaluation [16]. IgE is emerging as an important prognostic biomarker and one that may also predict the response to dupilumab [18]. The systemic inflammatory markers CRP and fibrinogen offer predictive and prognostic values for exacerbations and mortality, with CRP also showing promise in guiding antibiotic use in the management of exacerbations. The role of procalcitonin in guiding antibiotic therapy in COPD exacerbations remains debated, but it remains a sound strategy to de-escalate antibiotics based on decreasing procalcitonin levels. The cytokines IL-6 and IL-8 have demonstrated prognostic associations with exacerbations, hospitalizations, and mortality, though their use in acute settings is less clear. Novel biomarkers, such as IL-33/ST-2, sRAGE, CC16, and SP-D, provide potential prognostic insights into exacerbation risk, disease severity, lung function decline, and mortality, but require further investigation of how this insight tailors management. Sputum biomarkers are being investigated for their ability to reflect airway inflammation and predict exacerbations, while invasive biomarkers from bronchoalveolar lavage and lung biopsies require more exploration. FeNO is proving useful in distinguishing COPD from asthma/ACO, predicting the exacerbation risk, and potentially monitoring and predicting the response to therapies targeting type 2 inflammation [44–48]. Exhaled breath condensate analysis offers a non-invasive approach to assess airway inflammation and oxidative stress, though the identification of reliable clinical biomarkers in this domain remains a challenge.

Currently, there are only a few clinical applications for imaging biomarkers in COPD. The guidelines highlight QCT for planning endobronchial valve placement or lung reduction surgery [16]. Specifically for bronchial valves, QCT identifies the emphysema lobe level, emphysema distribution, adjacent lobar volume, and fissure integrity, which are critical for proper patient and valve site selection [134–136]. Another clinical application for imaging biomarkers uses annual low-dose CT (LDCT) scans for lung cancer screening. One study reported 63% sensitivity and 88% specificity of LDCT for detecting COPD [133]. A very high prevalence of COPD has been identified in other patient cohorts undergoing annual LDCT, highlighting the clinical utility of LDCT for COPD diagnosis and population-level case findings [131,132,137].

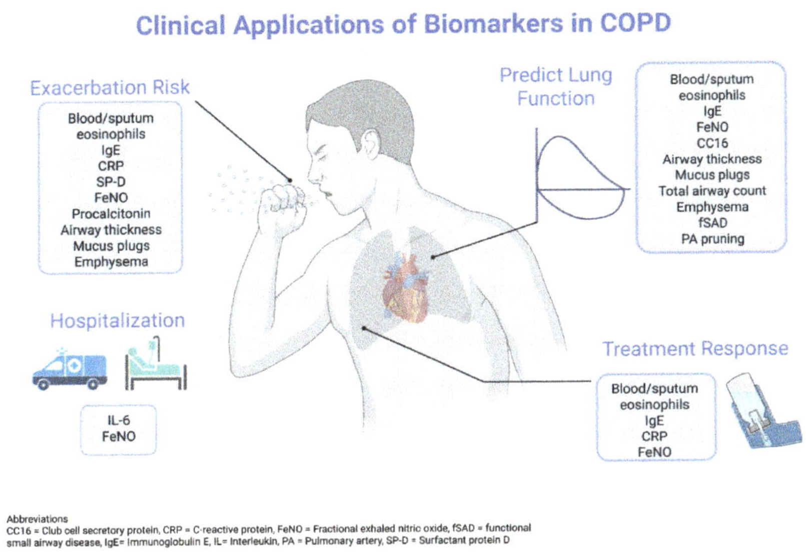

Figure 2. Clinical applications of biomarkers in COPD (created with www.biorender.com).

In a few short years, we anticipate other emerging QCT techniques to enter clinical medicine. Emphysema percent quantification (somewhat analogous to the left ventricular ejection fraction) identifies the percent of low-level attenuation areas in pulmonary parenchyma, which are associated with COPD symptoms, airflow obstruction, disease progression, and mortality [100]. The systematic grading of airway mucus plug burden or small airway counts should become easier with artificially intelligent automated radiology software, and these findings have been associated with airflow obstruction, symptoms, and exacerbations [138]. Future challenges remain, such as defining optimal imaging protocols considering age, sex, and race, where meaningful clinical differences exist; proper artificial intelligence training on large imaging datasets; and further technological advancements in scanning resolution and software processing power.

Despite the growing interest in COPD biomarkers, several challenges hinder their widespread clinical application and incorporation into practice guidelines. Conflicting evidence across studies for these biomarkers creates uncertainty regarding their reliability and utility. A significant limitation is the lack of validation in large, prospective studies to confirm the clinical significance of these biomarkers and to establish reliable cut-off values for clinical decision making. Furthermore, the heterogeneity of COPD contributes to inconsistencies in the research findings. COPD is a complex disease that involves multiple inflammatory pathways and clinical phenotypes, so relying on individual biomarkers may not appropriately capture the multifaceted nature of the disease. Many of these biomarkers are altered in a multitude of other disease states, and there are a lack of validated data to support the specificity in utilizing these biomarkers in the clinical setting of COPD management. Lastly, while the emerging targeted therapies have highlighted the need for effective biomarkers, the translation of the research findings into routine clinical practice often faces hurdles related to standardization and cost-effectiveness. Overall, further research will be essential to validate the clinical utility of COPD biomarkers and to develop biomarker-driven strategies for precision COPD care.

6. Differential Biomarkers Between Tobacco Smoke and Biomass Burning Smoke Exposure

It is estimated that almost half of COPD worldwide is due to non-tobacco exposure [139]. Although the majority of COPD biomarker work has been conducted on patients exposed to tobacco smoke, (TS-COPD) there are emerging biomarker data for COPD related to biomass burning smoke exposure (BS-COPD). A cohort of over 700 patients from Morrocco who self-reported cumulative biomass burning smoke exposure (wood heating for men and cooking biomass for women) revealed increased odds of COPD with higher biomass burning smoke exposure indices [140]. The natural history of BS-COPD is different from that of tobacco smoking, with less parenchymal destruction, more chronic bronchitis, and the slower decline of lung function [141,142].

One small study found greater bronchial hyperresponsiveness in BS-COPD compared to TS-COPD [143]. Another study directly compared the cytokine profiles between 29 women with TS-COPD and 31 women with BS-COPD and found the level of chemokine ligand 5 was higher in the TS-COPD group [65]. Additional research has uncovered distinct features of BS-COPD compared to those of healthy controls; one cross sectional study of 180 women in rural Peru found that kitchen concentrations of black carbon were associated with increased blood levels of the pro-inflammatory cytokine TNF-alpha [144]. Another cohort of 140 women from East India who cook exclusively with biomass burning smoke exposure had elevated serum levels of TNF-alpha and neutrophils compared to those of women who cook with cleaner petroleum gases, and this finding was replicated in a larger cohort study involving over 1000 participants [145]. Conversely, a smaller case–control study from South India found that serum TNF-alpha levels were much higher in 40 patients with tobacco smoke COPD versus those of 40 patients with BS-COPD [146]. What is clear, however, is that more research is needed to understand the major BS-COPD phenotype.

7. Conclusions

Biomarkers have significant potential for improving the diagnosis, prognosis, and management of COPD. Serum, sputum, imaging, and exhaled biomarkers are becoming increasingly important tools in not only predicting COPD outcomes, but predicting the treatment response to evolving therapies in COPD. The integration of these biomarkers into clinical assessments promotes personalized treatment strategies, enabling more targeted interventions and better patient outcomes. However, important unmet needs remain, including the stability of biomarkers over time, the predictive ability of composite biomarkers, and the lack of well-defined type 1 biomarkers. Importantly, additional research that would clinically validate these various biomarkers longitudinally is warranted. An ongoing investigation of novel biomarkers, including other known inflammatory mediators and genetic markers, may also uncover new therapeutic targets, improve prognostic guidance, and further enhance the understanding of COPD pathogenesis. There has been immense development in investigating the role of biomarkers in COPD, but further research is needed to continue to progress COPD care and improve the quality of life for affected patients.

Author Contributions: All authors K.M.P., P.F.L., N.A.H. and M.A. participated in conceptualization; methodology; validation; resources; data curation; writing—original draft preparation; writing—review and editing; visualization; and project administration. All authors have read and agreed to the published version of the manuscript.

Funding: No funding was received for this study. M.A. is a consultant for AstraZeneca. N.A.H. received honoraria for serving as an advisor or consultant for GSK, AstraZeneca, Sanofi, Regeneron, Amgen, Genentech, Novartis, and Teva. His institution received research grant support on his behalf from GSK, Genentech, Sanofi, Teva, Novartis, and AstraZeneca. M.A. serves as a consultant for

AstraZeneca. No financial compensation has been received for this role. This research received no external funding.

Institutional Review Board Statement: This study did not require IRB approval.

Informed Consent Statement: Not applicable.

Data Availability Statement: No new data were created for this study.

Conflicts of Interest: The authors declare no conflicts of interest.

References

1. Bhatt, S.P.; Rabe, K.F.; Hanania, N.A.; Vogelmeier, C.F.; Bafadhel, M.; Christenson, S.A.; Papi, A.; Singh, D.; Laws, E.; Patel, N.; et al. NOTUS Study Investigators. Dupilumab for COPD with Blood Eosinophil Evidence of Type 2 Inflammation. *N. Engl. J. Med.* **2024**, *390*, 2274–2283. [CrossRef] [PubMed]
2. Stockley, R.A.; Halpin, D.M.G.; Celli, B.R.; Singh, D. Chronic Obstructive Pulmonary Disease Biomarkers and Their Interpretation. *Am. J. Respir. Crit. Care Med.* **2019**, *199*, 1195–1204. [CrossRef] [PubMed]
3. Vitenberga, Z.; Pilmane, M.; Babjoniševa, A. An Insight into COPD Morphopathogenesis: Chronic Inflammation, Remodeling, and Antimicrobial Defense. *Medicina* **2019**, *55*, 496. [CrossRef] [PubMed] [PubMed Central]
4. Celli, B.R.; Anderson, J.A.; Brook, R.; Calverley, P.; Cowans, N.J.; Crim, C.; Dixon, I.; Kim, V.; Martinez, F.J.; Morris, A.; et al. Serum biomarkers and outcomes in patients with moderate COPD: A substudy of the randomised SUMMIT trial. *BMJ Open Respir. Res.* **2019**, *6*, e000431. [CrossRef] [PubMed] [PubMed Central]
5. Vedel-Krogh, S.; Nielsen, S.F.; Lange, P.; Vestbo, J.; Nordestgaard, B.G. Blood Eosinophils and Exacerbations in Chronic Obstructive Pulmonary Disease. The Copenhagen General Population Study. *Am. J. Respir. Crit. Care Med.* **2016**, *193*, 965–974. [CrossRef] [PubMed]
6. Zhang, Y.; Liang, L.R.; Zhang, S.; Lu, Y.; Chen, Y.Y.; Shi, H.Z.; Lin, Y.X. Blood Eosinophilia and Its Stability in Hospitalized COPD Exacerbations are Associated with Lower Risk of All-Cause Mortality. *Int. J. Chronic Obstr. Pulm. Dis.* **2020**, *15*, 1123–1134. [CrossRef] [PubMed] [PubMed Central]
7. Prudente, R.; Ferrari, R.; Mesquita, C.B.; Machado, L.H.S.; Franco, E.A.T.; Godoy, I.; Tanni, S.E. Peripheral Blood Eosinophils and Nine Years Mortality in COPD Patients. *Int. J. Chronic Obstr. Pulm. Dis.* **2021**, *16*, 979–985. [CrossRef] [PubMed] [PubMed Central]
8. Tan, W.C.; Bourbeau, J.; Nadeau, G.; Wang, W.; Barnes, N.; Landis, S.H.; Kirby, M.; Hogg, J.C.; Sin, D.D.; CanCOLDCollaborative Research Group. Authors would also like to thank the men women who participated in the study individuals in the CanCOLDCollaborative Research Group not listed as authors: High eosinophil counts predict decline in FEV1: Results from the CanCOLDstudy. *Eur. Respir. J.* **2021**, *57*, 2000838. [CrossRef] [PubMed]
9. Çolak, Y.; Afzal, S.; Marott, J.L.; Vestbo, J.; Nordestgaard, B.G.; Lange, P. Type-2 inflammation and lung function decline in chronic airway disease in the general population. *Thorax* **2024**, *79*, 349–358. [CrossRef] [PubMed] [PubMed Central]
10. Mathioudakis, A.G.; Bate, S.; Sivapalan, P.; Jensen, J.S.; Singh, D.; Vestbo, J. Rethinking Blood Eosinophils for Assessing Inhaled Corticosteroids Response in COPD: A Post Hoc Analysis from the FLAME Trial. *Chest* **2024**, *166*, 987–997. [CrossRef] [PubMed] [PubMed Central]
11. Pascoe, S.; Locantore, N.; Dransfield, M.T.; Barnes, N.C.; Pavord, I.D. Blood eosinophil counts, exacerbations, and response to the addition of inhaled fluticasone furoate to vilanterol in patients with chronic obstructive pulmonary disease: A secondary analysis of data from two parallel randomised controlled trials. *Lancet Respir. Med.* **2015**, *3*, 435–442, Erratum in *Lancet Respir. Med.* **2015**, *3*, e19. [CrossRef] [PubMed]
12. Siddiqui, S.H.; Guasconi, A.; Vestbo, J.; Jones, P.; Agusti, A.; Paggiaro, P.; Wedzicha, J.A.; Singh, D. Blood Eosinophils: A Biomarker of Response to Extrafine Beclomethasone/Formoterol in Chronic Obstructive Pulmonary Disease. *Am. J. Respir. Crit. Care Med.* **2015**, *192*, 523–525. [CrossRef] [PubMed] [PubMed Central]
13. Pavord, I.D.; Lettis, S.; Locantore, N.; Pascoe, S.; Jones, P.W.; Wedzicha, J.A.; Barnes, N.C. Blood eosinophils and inhaled corticosteroid/long-acting β-2 agonist efficacy in COPD. *Thorax* **2016**, *71*, 118–125. [CrossRef] [PubMed] [PubMed Central]
14. Wedzicha, J.A.; Banerji, D.; Chapman, K.R.; Vestbo, J.; Roche, N.; Ayers, R.T.; Thach, C.; Fogel, R.; Patalano, F.; Vogelmeier, C.F.; et al. Indacaterol-Glycopyrronium versus Salmeterol-Fluticasone for COPD. *N. Engl. J. Med.* **2016**, *374*, 2222–2234. [CrossRef] [PubMed]
15. Bafadhel, M.; Rabe, K.F.; Martinez, F.J.; Singh, D.; Darken, P.; Jenkins, M.; Aurivillius, M.; Patel, M.; Dorinsky, P. Benefits of Budesonide/Glycopyrronium/Formoterol Fumarate Dihydrate on COPD Exacerbations, Lung Function, Symptoms, and Quality of Life Across Blood Eosinophil Ranges: A Post-Hoc Analysis of Data from ETHOS. *Int. J. Chronic Obstr. Pulm. Dis.* **2022**, *17*, 3061–3073. [CrossRef]
16. Global Strategy for the Diagnosis, Management and Prevention of Chronic Obstructive Pulmonary Disease: 2024 Report. Available online: www.goldcopd.org (accessed on 1 March 2025).

17. Baraldi, F.; Bartlett-Pestle, S.; Allinson, J.P.; Macleod, M.; Mah, J.; Bloom, C.; Brady-Green, A.; Marion, A.; Papi, A.; Wedzicha, J.A.; et al. Blood Eosinophil Count Stability in COPD and the Eosinophilic Exacerbator Phenotype. *Am. J. Respir. Crit. Care Med.* 2025; *epub ahead of print*. [CrossRef] [PubMed]
18. Bhatt, S.P.; Rabe, K.F.; Hanania, N.A.; Vogelmeier, C.F.; Cole, J.; Bafadhel, M.; Christenson, S.A.; Papi, A.; Singh, D.; Laws, E.; et al. Dupilumab for COPD with Type 2 Inflammation Indicated by Eosinophil Counts. *N. Engl. J. Med.* **2023**, *389*, 205–214. [CrossRef] [PubMed]
19. Karakioulaki, M.; Papakonstantinou, E.; Goulas, A.; Stolz, D. The Role of Atopy in COPD and Asthma. *Front. Med.* **2021**, *8*, 674742. [CrossRef] [PubMed] [PubMed Central]
20. Lommatzsch, M.; Speer, T.; Herr, C.; Jörres, R.A.; Watz, H.; Müller, A.; Welte, T.; Vogelmeier, C.F.; Bals, R.; COSYCONET Study Group. IgE is associated with exacerbations and lung function decline in COPD. *Respir. Res.* **2022**, *23*, 1. [CrossRef]
21. Hai, C.N.; Ba, T.T.; Duc, T.B.; Xuan, C.H.; Manh, T.V. Serum immunoglobulin levels in group E of chronic obstructive pulmonary disease: Insights for clinical management and immunoglobulin therapy strategies. *BMC Pulm. Med.* **2024**, *24*, 381. [CrossRef] [PubMed] [PubMed Central]
22. Çolak, Y.; Ingebrigtsen, T.S.; Nordestgaard, B.G.; Marott, J.L.; Lange, P.; Vestbo, J.; Afzal, S. Plasma immunoglobulin E and risk of exacerbation and mortality in chronic obstructive pulmonary disease: A contemporary population-based cohort. *Ann. Allergy Asthma Immunol.* **2022**, *129*, 490–496. [CrossRef] [PubMed]
23. Nehring, S.M.; Goyal, A.; Patel, B.C. C Reactive Protein. In *StatPearls*; StatPearls Publishing: Treasure Island, FL, USA, 2025. Available online: https://www.ncbi.nlm.nih.gov/books/NBK441843/ (accessed on 25 April 2025).
24. Cleland, D.A.; Eranki, A.P. Procalcitonin. In *StatPearls*; StatPearls Publishing: Treasure Island, FL, USA, 2025. Available online: https://www.ncbi.nlm.nih.gov/books/NBK539794/ (accessed on 25 April 2025).
25. Munuswamy, R.; De Brandt, J.; Burtin, C.; Derave, W.; Aumann, J.; Spruit, M.A.; Michiels, L. Monomeric CRP is Elevated in Patients with COPD Compared to Non-COPD Control Persons. *J. Inflamm. Res.* **2021**, *14*, 4503–4507. [CrossRef] [PubMed] [PubMed Central]
26. Pantazopoulos, I.; Magounaki, K.; Kotsiou, O.; Rouka, E.; Perlikos, F.; Kakavas, S.; Gourgoulianis, K. Incorporating Biomarkers in COPD Management: The Research Keeps Going. *J. Pers. Med.* **2022**, *12*, 379. [CrossRef] [PubMed] [PubMed Central]
27. Papi, A.; Magnoni, M.S.; Muzzio, C.C.; Benso, G.; Rizzi, A. Phenomenology of COPD: Interpreting phenotypes with the ECLIPSE study. *Monaldi Arch. Chest Dis.* **2016**, *83*, 721. [CrossRef] [PubMed]
28. Fermont, J.M.; Masconi, K.L.; Jensen, M.T.; Ferrari, R.; Di Lorenzo, V.A.P.; Marott, J.M.; Schuetz, P.; Watz, H.; Waschki, B.; Müllerova, H.; et al. Biomarkers and clinical outcomes in COPD: A systematic review and meta-analysis. *Thorax* **2019**, *74*, 439–446. [CrossRef] [PubMed] [PubMed Central]
29. Zemans, R.L.; Jacobson, S.; Keene, J.; Kechris, K.; Miller, B.E.; Tal-Singer, R.; Bowler, R.P. Multiple biomarkers predict disease severity, progression and mortality in COPD. *Respir. Res.* **2017**, *18*, 117. [CrossRef] [PubMed] [PubMed Central]
30. Hoult, G.; Gillespie, D.; Wilkinson, T.M.A.; Thomas, M.; Francis, N.A. Biomarkers to guide the use of antibiotics for acute exacerbations of COPD (AECOPD): A systematic review and meta-analysis. *BMC Pulm. Med.* **2022**, *22*, 194. [CrossRef] [PubMed] [PubMed Central]
31. Prins, H.J.; Duijkers, R.; van der Valk, P.; Schoorl, M.; Daniels, J.M.A.; van der Werf, T.S.; Boersma, W.G. CRP-guided antibiotic treatment in acute exacerbations of COPD in hospital admissions. *Eur. Respir. J.* **2019**, *53*, 1802014. [CrossRef] [PubMed]
32. Miller, B.E.; Tal-Singer, R.; Rennard, S.I.; Furtwaengler, A.; Leidy, N.; Lowings, M.; Martin, U.J.; Martin, T.R.; Merrill, D.D.; Snyder, J.; et al. Plasma Fibrinogen Qualification as a Drug Development Tool in Chronic Obstructive Pulmonary Disease. Perspective of the Chronic Obstructive Pulmonary Disease Biomarker Qualification Consortium. *Am. J. Respir. Crit. Care Med.* **2016**, *193*, 607–613. [CrossRef] [PubMed]
33. Gruys, E.; Toussaint, M.J.; Niewold, T.A.; Koopmans, S.J. Acute phase reaction and acute phase proteins. *J. Zhejiang Univ. Sci. B* **2005**, *6*, 1045–1056. [CrossRef] [PubMed] [PubMed Central]
34. Singh, D.; Criner, G.J.; Dransfield, M.T.; Halpin, D.M.G.; Han, M.K.; Lange, P.; Lettis, S.; Lipson, D.A.; Mannino, D.; Martin, N.; et al. InforMing the PAthway of COPD Treatment (IMPACT) trial: Fibrinogen levels predict risk of moderate or severe exacerbations. *Respir. Res.* **2021**, *22*, 130. [CrossRef] [PubMed] [PubMed Central]
35. Sivapalan, P.; Jensen, J.U. Biomarkers in Chronic Obstructive Pulmonary Disease: Emerging Roles of Eosinophils and Procalcitonin. *J. Innate Immun.* **2022**, *14*, 89–97. [CrossRef] [PubMed] [PubMed Central]
36. Daubin, C.; Valette, X.; Thiollière, F.; Mira, J.P.; Hazera, P.; Annane, D.; Labbe, V.; Floccard, B.; Fournel, F.; Terzi, N.; et al. Procalcitonin algorithm to guide initial antibiotic therapy in acute exacerbations of COPD admitted to the ICU: A randomized multicenter study. *Intensive Care Med.* **2018**, *44*, 428–437. [CrossRef] [PubMed] [PubMed Central]
37. Huang, H.; Huang, X.; Zeng, K.; Deng, F.; Lin, C.; Huang, W. Interleukin-6 is a Strong Predictor of the Frequency of COPD Exacerbation Within 1 Year. *Int. J. Chronic Obstr. Pulm. Dis.* **2021**, *16*, 2945–2951. [CrossRef] [PubMed] [PubMed Central]

38. Wedzicha, J.A.; Seemungal, T.A.; MacCallum, P.K.; Paul, E.A.; Donaldson, G.C.; Bhowmik, A.; Jeffries, D.J.; Meade, T.W. Acute exacerbations of chronic obstructive pulmonary disease are accompanied by elevations of plasma fibrinogen and serum IL-6 levels. *Thromb. Haemost.* **2000**, *84*, 210–215. [PubMed]
39. Zhang, J.; Bai, C. The Significance of Serum Interleukin-8 in Acute Exacerbations of Chronic Obstructive Pulmonary Disease. *Tanaffos* **2018**, *17*, 13–21. [PubMed] [PubMed Central]
40. Pratte, K.A.; Curtis, J.L.; Kechris, K.; Couper, D.; Cho, M.H.; Silverman, E.K.; DeMeo, D.L.; Sciurba, F.C.; Zhang, Y.; Ortega, V.E.; et al. Soluble receptor for advanced glycation end products (sRAGE) as a biomarker of COPD. *Respir. Res.* **2021**, *22*, 127. [CrossRef] [PubMed] [PubMed Central]
41. Serban, K.A.; Pratte, K.A.; Bowler, R.P. Protein Biomarkers for COPD Outcomes. *Chest* **2021**, *159*, 2244–2253. [CrossRef] [PubMed] [PubMed Central]
42. Park, H.Y.; Churg, A.; Wright, J.L.; Li, Y.; Tam, S.; Man, S.P.; Tashkin, D.; Wise, R.A.; Connett, J.E.; Sin, D.D. Club cell protein 16 and disease progression in chronic obstructive pulmonary disease. *Am. J. Respir. Crit. Care Med.* **2013**, *188*, 1413–1419. [CrossRef]
43. Lock-Johansson, S.; Vestbo, J.; Sorensen, G.L. Surfactant protein D, club cell protein 16, pulmonary and activation-regulated chemokine, C-reactive protein, and fibrinogen biomarker variation in chronic obstructive lung disease. *Respir. Res.* **2014**, *15*, 147. [CrossRef]
44. Loewenthal, L.; Menzies-Gow, A. FeNO in Asthma. *Semin. Respir. Crit. Care Med.* **2022**, *43*, 635–645. [CrossRef] [PubMed]
45. Li, M.; Yang, T.; He, R.; Li, A.; Dang, W.; Liu, X.; Chen, M. The Value of Inflammatory Biomarkers in Differentiating Asthma-COPD Overlap from COPD. *Int. J. Chronic Obstr. Pulm. Dis.* **2020**, *15*, 3025–3037. [CrossRef] [PubMed] [PubMed Central]
46. Donohue, J.F.; Herje, N.; Crater, G.; Rickard, K. Characterization of airway inflammation in patients with COPD using fractional exhaled nitric oxide levels: A pilot study. *Int. J. Chronic Obstr. Pulm. Dis.* **2014**, *9*, 745–751. [CrossRef] [PubMed] [PubMed Central]
47. Alcázar-Navarrete, B.; Romero-Palacios, P.J.; Ruiz-Sancho, A.; Ruiz-Rodriguez, O. Diagnostic performance of the measurement of nitric oxide in exhaled air in the diagnosis of COPD phenotypes. *Nitric Oxide* **2016**, *54*, 67–72. [CrossRef] [PubMed]
48. Guo, Y.; Hong, C.; Liu, Y.; Chen, H.; Huang, X.; Hong, M. Diagnostic value of fractional exhaled nitric oxide for asthma-chronic obstructive pulmonary disease overlap syndrome. *Medicine* **2018**, *97*, e10857. [CrossRef] [PubMed] [PubMed Central]
49. Reddel, H.K.; Vestbo, J.; Agustí, A.; Anderson, G.P.; Bansal, A.T.; Beasley, R.; Bel, E.H.; Janson, C.; Make, B.; Pavord, I.D.; et al. NOVELTY study investigators. Heterogeneity within and between physician-diagnosed asthma and/or COPD: NOVELTY cohort. *Eur. Respir. J.* **2021**, *58*, 2003927. [CrossRef] [PubMed] [PubMed Central]
50. Alcázar-Navarrete, B.; Díaz-Lopez, J.M.; García-Flores, P.; Ortega-Antelo, M.; Aguilar-Cruz, I.; Ruiz-Rodríguez, O.; Santiago-Diaz, P.; Romero Palacios, P.J. T2 Biomarkers as Predictors of Exacerbations of Chronic Obstructive Pulmonary Disease. *Arch. Bronconeumol.* **2022**, *58*, 595–600. [CrossRef] [PubMed]
51. Río Ramírez, M.T.; Juretschke Moragues, M.A.; Fernández González, R.; Álvarez Rodríguez, V.; Aznar Andrés, E.; Zabaleta Camino, J.P.; Romero Pareja, R.; Esteban de la Torre, A. Value of Exhaled Nitric Oxide (FeNO) And Eosinophilia During the Exacerbations of Chronic Obstructive Pulmonary Disease Requiring Hospital Admission. *COPD* **2018**, *15*, 369–376. [CrossRef] [PubMed]
52. Alcázar-Navarrete, B.; Ruiz Rodríguez, O.; Conde Baena, P.; Romero Palacios, P.J.; Agusti, A. Persistently elevated exhaled nitric oxide fraction is associated with increased risk of exacerbation in COPD. *Eur. Respir. J.* **2018**, *51*, 1701457. [CrossRef] [PubMed]
53. Wu, Y.K.; Su, W.L.; Huang, C.Y.; Yang, M.C.; Chen, S.Y.; Lan, C.C. Treatment of chronic obstructive pulmonary disease in patients with different fractional exhaled nitric oxide levels. *Medicine* **2018**, *97*, e11922. [CrossRef] [PubMed] [PubMed Central]
54. Zietkowski, Z.; Kucharewicz, I.; Bodzenta-Lukaszyk, A. The influence of inhaled corticosteroids on exhaled nitric oxide in stable chronic obstructive pulmonary disease. *Respir. Med.* **2005**, *99*, 816–824. [CrossRef] [PubMed]
55. Kunisaki, K.M.; Rice, K.L.; Janoff, E.N.; Rector, T.S.; Niewoehner, D.E. Exhaled nitric oxide, systemic inflammation, and the spirometric response to inhaled fluticasone propionate in severe chronic obstructive pulmonary disease: A prospective study. *Ther. Adv. Respir. Dis.* **2008**, *2*, 55–64. [CrossRef] [PubMed]
56. Amer, M.; Cowan, J.; Gray, A.; Brockway, B.; Dummer, J. Effect of Inhaled β2-Agonist on Exhaled Nitric Oxide in Chronic Obstructive Pulmonary Disease. *PLoS ONE* **2016**, *11*, e0157019. [CrossRef] [PubMed] [PubMed Central]
57. Silva-Gallardo, J.; Sansores, R.H.; Ramírez-Venegas, A.; Robles Hernández, R.E.; Centeno-Saenz, G.I.; Hernández-Zenteno, R.J. Fractional Exhaled Nitric Oxide (FeNO) in Biomass Smoke-Associated Chronic Obstructive Pulmonary Disease. *Med. Sci.* **2024**, *12*, 52. [CrossRef] [PubMed] [PubMed Central]
58. Dinarello, C.A. Immunological and inflammatory functions of the interleukin-1 family. *Annu. Rev. Immunol.* **2009**, *27*, 519–550. [CrossRef] [PubMed]
59. Riera-Martínez, L.; Cànaves-Gómez, L.; Iglesias, A.; Martin-Medina, A.; Cosío, B.G. The Role of IL-33/ST2 in COPD and Its Future as an Antibody Therapy. *Int. J. Mol. Sci.* **2023**, *24*, 8702. [CrossRef]
60. Xia, J.; Zhao, J.; Shang, J.; Li, M.; Zeng, Z.; Zhao, J.; Wang, J.; Xu, Y.; Xie, J. Increased IL-33 Expression in Chronic Obstructive Pulmonary Disease. *Am. J. Physiol. Lung Cell. Mol. Physiol.* **2015**, *308*, L619–L627. [CrossRef]

61. Rabe, K.F.; Celli, B.R.; Wechsler, M.E.; Abdulai, R.M.; Luo, X.; Boomsma, M.M.; Staudinger, H.; Horowitz, J.E.; Baras, A.; Ferreira, M.A.; et al. Safety and efficacy of itepekimab in patients with moderate-to-severe COPD: A genetic association study and randomised, double-blind, phase 2a trial. *Lancet Respir. Med.* **2021**, *9*, 1288–1298. [CrossRef]
62. Yousuf, A.J.; Mohammed, S.; Carr, L.; Ramsheh, M.Y.; Micieli, C.; Mistry, V.; Haldar, K.; Wright, A.; Novotny, P.; Parker, S.; et al. Astegolimab, an anti-ST2, in chronic obstructive pulmonary disease (COPD-ST2OP): A phase 2a, placebo-controlled trial. *Lancet Respir. Med.* **2022**, *10*, 469–477. [CrossRef]
63. Fricke-Galindo, I.; García-Carmona, S.; Alanis-Ponce, J.; Pérez-Rubio, G.; Ramírez-Venegas, A.; Montiel-Lopez, F.; Robles-Hernández, R.; de Jesús Hernández-Zenteno, R.; Rea, D.V.P.; Bautista-Becerril, B.; et al. sRAGE levels are decreased in plasma and sputum of COPD secondary to biomass-burning smoke and tobacco smoking: Differences according to the rs3134940 AGER variant. *Heliyon* **2024**, *10*, e28675. [CrossRef]
64. Hastie, A.T.; Martinez, F.J.; Curtis, J.L.; Doerschuk, C.M.; Hansel, N.N.; Christenson, S.; Putcha, N.; Ortega, V.E.; Li, X.; Barr, R.G.; et al. SPIROMICS investigators. Association of sputum and blood eosinophil concentrations with clinical measures of COPD severity: An analysis of the SPIROMICS cohort. *Lancet Respir. Med.* **2017**, *5*, 956–967. [CrossRef] [PubMed] [PubMed Central]
65. Giraldo-Montoya, Á.M.; Torres-Duque, C.A.; Giraldo-Cadavid, L.F.; Laucho-Contreras, M.E.; González-Flórez, A.; Santos, A.M.; Tuta-Quintero, E.A.; Celli, B.R.; González-García, M. Sputum Biomarkers in Wood and Tobacco Smoke Etiotypes of Chronic Obstructive Pulmonary Disease. *Int. J. Chronic Obstr. Pulm. Dis.* **2023**, *19*, 1–10. [CrossRef] [PubMed]
66. Kanazawa, H.; Asai, K.; Hirata, K.; Yoshikawa, J. Possible effects of vascular endothelial growth factor in the pathogenesis of chronic obstructive pulmonary disease. *Am. J. Med.* **2003**, *114*, 354–358. [CrossRef] [PubMed]
67. Bartel, S.; Wolters, J.C.; Noor, H.; Rafie, K.; Fang, J.; Kirchner, B.; Nolte-THoen, E.; Pfaffl, M.W.; Rutgers, S.; Timens, W.; et al. Altered Extracellular Vesicle-Derived Protein and microRNA Signatures in Bronchoalveolar Lavage Fluid from Patients with Chronic Obstructive Pulmonary Disease. *Cells* **2024**, *13*, 945. [CrossRef] [PubMed] [PubMed Central]
68. DiLillo, K.M.; Norman, K.C.; Freeman, C.M.; Christenson, S.A.; Alexis, N.E.; Anderson, W.H.; Barjaktarevic, I.Z.; Barr, R.G.; Comellas, A.P.; Bleecker, E.R.; et al. A blood and bronchoalveolar lavage protein signature of rapid FEV_1 decline in smoking-associated COPD. *Sci. Rep.* **2023**, *13*, 8228. [CrossRef] [PubMed] [PubMed Central]
69. Borrill, Z.L.; Roy, K.; Singh, D. Exhaled breath condensate biomarkers in COPD. *Eur. Respir. J.* **2008**, *32*, 472–486. [CrossRef]
70. Ratiu, I.A.; Ligor, T.; Bocos-Bintintan, V.; Mayhew, C.A.; Buszewski, B. Volatile Organic Compounds in Exhaled Breath as Fingerprints of Lung Cancer, Asthma and COPD. *J. Clin. Med.* **2020**, *10*, 32. [CrossRef] [PubMed]
71. Freund, R.; Sauvain, J.J.; Suarez, G.; Wild, P.; Charreau, T.; Debatisse, A.; Sakthithasan, K.; Jouannique, V.; Pralong, J.A.; Guseva Canu, I. Discriminative potential of exhaled breath condensate biomarkers with respect to chronic obstructive pulmonary disease. *J. Occup. Med. Toxicol.* **2024**, *19*, 10. [CrossRef]
72. Maniscalco, M.; Paris, D.; Melck, D.J.; Molino, A.; Carone, M.; Ruggeri, P.; Caramori, G.; Motta, A. Differential diagnosis between newly diagnosed asthma and COPD using exhaled breath condensate metabolomics: A pilot study. *Eur. Respir. J.* **2018**, *51*, 1701825. [CrossRef]
73. Maniscalco, M.; Paris, D.; Cuomo, P.; Fuschillo, S.; Ambrosino, P.; Tramice, A.; Palomba, L.; Motta, A. Metabolomics of COPD Pulmonary Rehabilitation Outcomes via Exhaled Breath Condensate. *Cells* **2022**, *11*, 344. [CrossRef]
74. Papaioannou, A.I.; Loukides, S.; Minas, M.; Kontogianni, K.; Bakakos, P.; Gourgoulianis, K.I.; Alchanatis, M.; Papiris, S.; Kostikas, K. Exhaled breath condensate pH as a biomarker of COPD severity in ex-smokers. *Respir. Res.* **2011**, *12*, 67. [CrossRef]
75. Inonu, H.; Doruk, S.; Sahin, S.; Erkorkmaz, U.; Celik, D.; Celikel, S.; Seyfikli, Z. Oxidative stress levels in exhaled breath condensate associated with COPD and smoking. *Respir. Care* **2012**, *57*, 413–419. [CrossRef] [PubMed]
76. Corhay, J.L.; Moermans, C.; Henket, M.; Nguyen Dang, D.; Duysinx, B.; Louis, R. Increased of exhaled breath condensate neutrophil chemotaxis in acute exacerbation of COPD. *Respir. Res.* **2014**, *15*, 115. [CrossRef]
77. Zakharkina, T.; Koczulla, A.R.; Mardanova, O.; Hattesohl, A.; Bals, R. Detection of microorganisms in exhaled breath condensate during acute exacerbations of COPD. *Respirology* **2011**, *16*, 932–938. [CrossRef] [PubMed]
78. Van Velzen, P.; Brinkman, P.; Knobel, H.H.; Van den Berg, J.W.K.; Jonkers, R.E.; Loijmans, R.J.; Prins, J.M.; Sterk, P.J. Exhaled Breath Profiles Before, During and After Exacerbation of COPD: A Prospective Follow-Up Study. *COPD* **2019**, *16*, 330–337. [CrossRef] [PubMed]
79. Koutsokera, A.; Kostikas, K.; Nicod, L.P.; Fitting, J.W. Pulmonary biomarkers in COPD exacerbations: A systematic review. *Respir. Res.* **2013**, *14*, 111. [CrossRef]
80. Brinkman, P.; Wilde, M.; Ahmed, W.; Wang, R.; van der Schee, M.; Abuhelal, S.; Schaber, C.; Cunoosamy, D.; Clarke, G.W.; Maitland-van der Zee, A.H.; et al. Fulfilling the Promise of Breathomics: Considerations for the Discovery and Validation of Exhaled Volatile Biomarkers. *Am. J. Respir. Crit. Care Med.* **2024**, *210*, 1079–1090. [CrossRef]
81. Galbán, C.J.; Han, M.K.; Boes, J.L.; Chughtai, K.A.; Meyer, C.R.; Johnson, T.D.; Galbán, S.; Rehemtulla, A.; Kazerooni, E.A.; Martinez, F.J.; et al. Computed tomography-based biomarker provides unique signature for diagnosis of COPD phenotypes and disease progression. *Nat. Med.* **2012**, *18*, 1711–1715. [CrossRef] [PubMed] [PubMed Central]

82. Pompe, E.; van Rikxoort, E.M.; Schmidt, M.; Rühaak, J.; Estrella, L.G.; Vliegenthart, R.; Oudkerk, M.; de Koning, H.J.; van Ginneken, B.; de Jong, P.A.; et al. Parametric response mapping adds value to current computed tomography biomarkers in diagnosing chronic obstructive pulmonary disease. *Am. J. Respir. Crit. Care Med.* **2015**, *191*, 1084–1086. [CrossRef] [PubMed]
83. Pompe, E.; Galbán, C.J.; Ross, B.D.; Koenderman, L.; Ten Hacken, N.H.; Postma, D.S.; van den Berge, M.; de Jong, P.A.; Lammers, J.J.; Mohamed Hoesein, F.A. Parametric response mapping on chest computed tomography associates with clinical and functional parameters in chronic obstructive pulmonary disease. *Respir. Med.* **2017**, *123*, 48–55. [CrossRef] [PubMed] [PubMed Central]
84. Diaz, A.A.; Orejas, J.L.; Grumley, S.; Nath, H.P.; Wang, W.; Dolliver, W.R.; Yen, A.; Kligerman, S.J.; Jacobs, K.; Manapragada, P.P.; et al. Airway-Occluding Mucus Plugs and Mortality in Patients with Chronic Obstructive Pulmonary Disease. *JAMA* **2023**, *329*, 1832–1839. [CrossRef] [PubMed] [PubMed Central]
85. Dunican, E.M.; Elicker, B.M.; Henry, T.; Gierada, D.S.; Schiebler, M.L.; Anderson, W.; Barjaktarevic, I.; Barr, R.G.; Bleecker, E.R.; Boucher, R.C.; et al. Mucus Plugs and Emphysema in the Pathophysiology of Airflow Obstruction and Hypoxemia in Smokers. *Am. J. Respir. Crit. Care Med.* **2021**, *203*, 957–968. [CrossRef] [PubMed] [PubMed Central]
86. Mettler, S.K.; Nath, H.P.; Grumley, S.; Orejas, J.L.; Dolliver, W.R.; Nardelli, P.; Yen, A.C.; Kligerman, S.J.; Jacobs, K.; Manapragada, P.P.; et al. Silent Airway Mucus Plugs in COPD and Clinical Implications. *Chest* **2024**, *166*, 1010–1019. [CrossRef] [PubMed] [PubMed Central]
87. Tanabe, N.; Shimizu, K.; Shima, H.; Wakazono, N.; Shiraishi, Y.; Terada, K.; Terada, S.; Oguma, T.; Sakamoto, R.; Suzuki, M.; et al. Computed tomography mucus plugs and airway tree structure in patients with chronic obstructive pulmonary disease: Associations with airflow limitation, health-related independence and mortality. *Respirology* **2024**, *29*, 951–961. [CrossRef] [PubMed]
88. Okajima, Y.; Come, C.E.; Nardelli, P.; Sonavane, S.K.; Yen, A.; Nath, H.P.; Terry, N.; Grumley, S.A.; Ahmed, A.; Kligerman, S.; et al. Luminal Plugging on Chest CT Scan: Association with Lung Function, Quality of Life, and COPD Clinical Phenotypes. *Chest* **2020**, *158*, 121–130. [CrossRef] [PubMed] [PubMed Central]
89. Gietema, H.A.; Edwards, L.D.; Coxson, H.O.; Bakke, P.S.; ECLIPS Einvestigators. Impact of emphysema airway wall thickness on quality of life in smoking-related COPD. *Respir. Med.* **2013**, *107*, 1201–1209. [CrossRef] [PubMed]
90. Saure, E.W.; Bakke, P.S.; Lind Eagan, T.M.; Aanerud, M.; Jensen, R.L.; Grydeland, T.B.; Johannessen, A.; Nilsen, R.M.; Thorsen, E.; Hardie, J.A. Diffusion capacity and CT measures of emphysema and airway wall thickness—Relation to arterial oxygen tension in COPD patients. *Eur. Clin. Respir. J.* **2016**, *3*, 29141. [CrossRef] [PubMed] [PubMed Central]
91. Johannessen, A.; Skorge, T.D.; Bottai, M.; Grydeland, T.B.; Nilsen, R.M.; Coxson, H.; Dirksen, A.; Omenaas, E.; Gulsvik, A.; Bakke, P. Mortality by level of emphysema and airway wall thickness. *Am. J. Respir. Crit. Care Med.* **2013**, *187*, 602–608. [CrossRef] [PubMed]
92. Bodduluri, S.; Puliyakote, A.S.K.; Gerard, S.E.; Reinhardt, J.M.; Hoffman, E.A.; Newell, J.D., Jr.; Nath, H.P.; Han, M.K.; Washko, G.R.; San José Estépar, R.; et al. Airway fractal dimension predicts respiratory morbidity mortality in COPD. *J. Clin. Investig.* **2018**, *128*, 5374–5382, Erratum in *J. Clin. Investig.* **2018**, *128*, 5676. [CrossRef] [PubMed] [PubMed Central]
93. Wyszkiewicz, P.V.; Sharma, M.; Desaigoudar, V.; Cunningham, I.A.; McCormack, D.G.; Abdelrazek, M.A.; Kirby, M.; Parraga, G. Reduced Total Airway Count and Airway Wall Tapering after Three-Years in Ex-Smokers. *COPD* **2023**, *20*, 186–196. [CrossRef] [PubMed]
94. Kirby, M.; Tanabe, N.; Tan, W.C.; Zhou, G.; Obeidat, M.; Hague, C.J.; Leipsic, J.; Bourbeau, J.; Sin, D.D.; Hogg, J.C.; et al. Total Airway Count on Computed Tomography the Risk of Chronic Obstructive Pulmonary Disease Progression Findings from a Population-based Study. *Am. J. Respir. Crit. Care Med.* **2018**, *197*, 56–65. [CrossRef] [PubMed]
95. Bodduluri, S.; Kizhakke Puliyakote, A.; Nakhmani, A.; Charbonnier, J.P.; Reinhardt, J.M.; Bhatt, S.P. Computed Tomography-based Airway Surface Area-to-Volume Ratio for Phenotyping Airway Remodeling in Chronic Obstructive Pulmonary Disease. *Am. J. Respir. Crit. Care Med.* **2021**, *203*, 185–191. [CrossRef] [PubMed] [PubMed Central]
96. Kahnert, K.; Jörres, R.A.; Kauczor, H.U.; Alter, P.; Trudzinski, F.C.; Herth, F.; Jobst, B.; Weinheimer, O.; Nauck, S.; Mertsch, P.; et al. Standardized airway wall thickness Pi10 from routine CT scans of COPD patients as imaging biomarker for disease severity, lung function decline, and mortality. *Ther. Adv. Respir. Dis.* **2023**, *17*, 17534666221148663. [CrossRef] [PubMed] [PubMed Central]
97. Woodruff, P.G.; Barr, R.G.; Bleecker, E.; Christenson, S.A.; Couper, D.; Curtis, J.L.; Gouskova, N.A.; Hansel, N.N.; Hoffman, E.A.; Kanner, R.E.; et al. Clinical Significance of Symptoms in Smokers with Preserved Pulmonary Function. *N. Engl. J. Med.* **2016**, *374*, 1811–1821. [CrossRef] [PubMed] [PubMed Central]
98. Lutchmedial, S.M.; Creed, W.G.; Moore, A.J.; Walsh, R.R.; Gentchos, G.E.; Kaminsky, D.A. How Common Is Airflow Limitation in Patients with Emphysema on CT Scan of the Chest? *Chest* **2015**, *148*, 176–184. [CrossRef] [PubMed] [PubMed Central]
99. Oh, A.S.; Baraghoshi, D.; Lynch, D.A.; Ash, S.Y.; Crapo, J.D.; Humphries, S.M. COPDGene Investigators. Emphysema Progression at CT by Deep Learning Predicts Functional Impairment and Mortality: Results from the COPDGene Study. *Radiology* **2022**, *304*, 672–679. [CrossRef] [PubMed] [PubMed Central]

100. Schroeder, J.D.; McKenzie, A.S.; Zach, J.A.; Wilson, C.G.; Curran-Everett, D.; Stinson, D.S.; Newell JDJr Lynch, D.A. Relationships between airflow obstruction and quantitative CT measurements of emphysema, air trapping, and airways in subjects with and without chronic obstructive pulmonary disease. *AJR Am. J. Roentgenol.* **2013**, *201*, W460–W470. [CrossRef] [PubMed] [PubMed Central]
101. Grydeland, T.B.; Thorsen, E.; Dirksen, A.; Jensen, R.; Coxson, H.O.; Pillai, S.G.; Sharma, S.; Eide, G.E.; Gulsvik, A.; Bakke, P.S. Quantitative CT measures of emphysema and airway wall thickness are related to D(L)CO. *Respir. Med.* **2011**, *105*, 343–351. [CrossRef] [PubMed]
102. Pompe, E.; Strand, M.; van Rikxoort, E.M.; Hoffman, E.A.; Barr, R.G.; Charbonnier, J.P.; Humphries, S.; Han, M.K.; Hokanson, J.E.; Make, B.J.; et al. COPDGene Investigators. Five-year Progression of Emphysema and Air Trapping at CT in Smokers with and Those without Chronic Obstructive Pulmonary Disease: Results from the COPDGene Study. *Radiology* **2020**, *295*, 218–226. [CrossRef] [PubMed] [PubMed Central]
103. Wilgus, M.L.; Abtin, F.; Markovic, D.; Tashkin, D.P.; Phillips, J.E.; Buhr, R.G.; Flynn, M.J.; Dembek, M.; Cooper, C.B.; Barjaktarevic, I. Panlobular emphysema is associated with COPD disease severity: A study of emphysema subtype by computed tomography. *Respir. Med.* **2022**, *192*, 106717. [CrossRef] [PubMed]
104. Shiraishi, Y.; Tanabe, N.; Shimizu, K.; Oguma, A.; Shima, H.; Sakamoto, R.; Yamazaki, H.; Oguma, T.; Sato, A.; Suzuki, M.; et al. Stronger Associations of Centrilobular Than Paraseptal Emphysema with Longitudinal Changes in Diffusing Capacity and Mortality in COPD. *Chest* **2023**, *164*, 327–338. [CrossRef] [PubMed]
105. El Kaddouri, B.; Strand, M.J.; Baraghoshi, D.; Humphries, S.M.; Charbonnier, J.P.; van Rikxoort, E.M.; Lynch, D.A. Fleischner Society Visual Emphysema CT Patterns Help Predict Progression of Emphysema in Current and Former Smokers: Results from the COPDGene Study. *Radiology* **2021**, *298*, 441–449. [CrossRef] [PubMed] [PubMed Central]
106. Oh, A.S.; Strand, M.; Pratte, K.; Regan, E.A.; Humphries, S.; Crapo, J.D.; Lynch, D.A.; Genetic Epidemiology of COPDGene Investigators. Visual Emphysema at Chest CT in GOLD Stage 0 Cigarette Smokers Predicts Disease Progression: Results from the COPDGene Study. *Radiology* **2020**, *296*, 641–649. [CrossRef] [PubMed] [PubMed Central]
107. Ash, S.Y.; San José Estépar, R.; Fain, S.B.; Tal-Singer, R.; Stockley, R.A.; Nordenmark, L.H.; Rennard, S.; Han, M.K.; Merrill, D.; Humphries, S.M.; et al. Relationship between Emphysema Progression at CT and Mortality in Ever-Smokers: Results from the COPDGene and ECLIPSE Cohorts. *Radiology* **2021**, *299*, 222–231. [CrossRef] [PubMed] [PubMed Central]
108. Bell, A.J.; Pal, R.; Labaki, W.W.; Hoff, B.A.; Wang, J.M.; Murray, S.; Kazerooni, E.A.; Galban, S.; Lynch, D.A.; Humphries, S.M.; et al. Local heterogeneity of normal lung parenchyma and small airways disease are associated with COPD severity and progression. *Respir. Res.* **2024**, *25*, 106. [CrossRef] [PubMed] [PubMed Central]
109. Pistenmaa, C.L.; Nardelli, P.; Ash, S.Y.; Come, C.E.; Diaz, A.A.; Rahaghi, F.N.; Barr, R.G.; Young, K.A.; Kinney, G.L.; Simmons, J.P.; et al. Pulmonary Arterial Pruning and Longitudinal Change in Percent Emphysema and Lung Function: The Genetic Epidemiology of COPD Study. *Chest* **2021**, *160*, 470–480. [CrossRef] [PubMed] [PubMed Central]
110. Washko, G.R.; Nardelli, P.; Ash, S.Y.; Vegas Sanchez-Ferrero, G.; Rahaghi, F.N.; Come, C.E.; Dransfield, M.T.; Kalhan, R.; Han, M.K.; Bhatt, S.P.; et al. Arterial Vascular Pruning, Right Ventricular Size, and Clinical Outcomes in Chronic Obstructive Pulmonary Disease. A Longitudinal Observational Study. *Am. J. Respir. Crit. Care Med.* **2019**, *200*, 454–461. [CrossRef] [PubMed] [PubMed Central]
111. Chen, H.; Shu, T.; Wang, L.; Yang, L.; Hu, C.; Du, S.; Wei, H. Pulmonary artery enlargement predicts poor survival in patients with COPD: A meta-analysis. *Pulm. Circ.* **2022**, *12*, e12099. [CrossRef] [PubMed] [PubMed Central]
112. Bhatt, S.P.; Soler, X.; Wang, X.; Murray, S.; Anzueto, A.R.; Beaty, T.H.; Boriek, A.M.; Casaburi, R.; Criner, G.J.; Diaz, A.A.; et al. Association between Functional Small Airway Disease and FEV1 Decline in Chronic Obstructive Pulmonary Disease. *Am. J. Respir. Crit. Care Med.* **2016**, *194*, 178–184. [CrossRef] [PubMed] [PubMed Central]
113. Wang, J.M.; Bell, A.J.; Ram, S.; Labaki, W.W.; Hoff, B.A.; Murray, S.; Kazerooni, E.A.; Galban, S.; Hatt, C.R.; Han, M.K.; et al. Topologic Parametric Response Mapping Identifies Tissue Subtypes Associated with Emphysema Progression. *Acad. Radiol.* **2024**, *31*, 1148–1159. [CrossRef] [PubMed] [PubMed Central]
114. Coxson, H.O.; Dirksen, A.; Edwards, L.D.; Yates, J.C.; Agusti, A.; Bakke, P.; Calverley, P.M.; Celli, B.; Crim, C.; Duvoix, A.; et al. The presence and progression of emphysema in COPD as determined by CT scanning and biomarker expression: A prospective analysis from the ECLIPSE study. *Lancet Respir. Med.* **2013**, *1*, 129–136. [CrossRef] [PubMed]
115. Klont, F.; Horvatovich, P.; Bowler, R.P.; van Rikxoort, E.; Charbonnier, J.P.; Kwiatkowski, M.; Lynch, D.A.; Humphries, S.; Bischoff, R.; Ten Hacken, N.H.T.; et al. Plasma sRAGE levels strongly associate with centrilobular emphysema assessed by HRCT scans. *Respir. Res.* **2022**, *23*, 15. [CrossRef] [PubMed] [PubMed Central]
116. Diaz, A.A.; Strand, M.; Coxson, H.O.; Ross, J.C.; San Jose Estepar, R.; Lynch, D.; van Rikxoort, E.M.; Rosas, I.O.; Hunninghake, G.M.; Putman, R.K.; et al. Disease Severity Dependence of the Longitudinal Association Between CT Lung Density and Lung Function in Smokers. *Chest* **2018**, *153*, 638–645. [CrossRef] [PubMed] [PubMed Central]

117. Baraghoshi, D.; Strand, M.; Humphries, S.M.; San José Estépar, R.; Vegas Sanchez-Ferrero, G.; Charbonnier, J.P.; Latisenko, R.; Silverman, E.K.; Crapo, J.D.; Lynch, D.A. Quantitative CT Evaluation of Emphysema Progression over 10 Years in the COPDGene Study. *Radiology* **2023**, *307*, e222786, Erratum in *Radiology* **2023**, *309*, e239028. [CrossRef] [PubMed] [PubMed Central]
118. Mohamed Hoesein, F.A.; de Jong, P.A.; Lammers, J.W.; Mali, W.P.; Mets, O.M.; Schmidt, M.; de Koning, H.J.; Aalst Cv Oudkerk, M.; Vliegenthart, R.; Ginneken Bv van Rikxoort, E.M.; et al. Contribution of CT quantified emphysema, air trapping and airway wall thickness on pulmonary function in male smokers with and without COPD. *COPD* **2014**, *11*, 503–509. [CrossRef] [PubMed]
119. Arjomandi, M.; Zeng, S.; Barjaktarevic, I.; Barr, R.G.; Bleecker, E.R.; Bowler, R.P.; Buhr, R.G.; Criner, G.J.; Comellas, A.P.; Cooper, C.B.; et al. Radiographic lung volumes predict progression to COPD in smokers with preserved spirometry in SPIROMICS. *Eur. Respir. J.* **2019**, *54*, 1802214. [CrossRef] [PubMed] [PubMed Central]
120. Young, A.L.; Bragman, F.J.S.; Rangelov, B.; Han, M.K.; Galbán, C.J.; Lynch, D.A.; Hawkes, D.J.; Alexander, D.C.; Hurst, J.R.; COPDGene Investigators. Disease Progression Modeling in Chronic Obstructive Pulmonary Disease. *Am. J. Respir. Crit. Care Med.* **2020**, *201*, 294–302. [CrossRef] [PubMed] [PubMed Central]
121. Zeng, S.; Luo, G.; Lynch, D.A.; Bowler, R.P.; Arjomandi, M. Lung volumes differentiate the predominance of emphysema versus airway disease phenotype in early COPD: An observational study of the COPDGene cohort. *ERJ Open Res.* **2023**, *9*, 00289–02023, Erratum in *ERJ Open Res.* **2023**, *9*, 50289–52023. [CrossRef] [PubMed] [PubMed Central]
122. Chaudhary, M.F.A.; Hoffman, E.A.; Guo, J.; Comellas, A.P.; Newell, J.D., Jr.; Nagpal, P.; Fortis, S.; Christensen, G.E.; Gerard, S.E.; Pan, Y.; et al. Predicting severe chronic obstructive pulmonary disease exacerbations using quantitative CT: A retrospective model development and external validation study. *Lancet Digit. Health* **2023**, *5*, e83–e92. [CrossRef] [PubMed] [PubMed Central]
123. Lin, Y.; Sang, L.; Wang, J.; Chen, Y.; Lai, J.; Zhu, X.; Yang, Y.; Zhang, Z.; Liu, Y.; Wen, S.; et al. Analysis of Airway Thickening and Serum Cytokines in COPD Patients with Frequent Exacerbations: A Heart of the Matter. *Int. J. Chronic Obstr. Pulm. Dis.* **2023**, *18*, 2353–2364. [CrossRef] [PubMed] [PubMed Central]
124. Jairam, P.M.; van der Graaf, Y.; Lammers, J.W.; Mali, W.P.; de Jong, P.A.; PROVIDI Study Group. Incidental findings on chest CT imaging are associated with increased COPD exacerbations and mortality. *Thorax* **2015**, *70*, 725–731. [CrossRef] [PubMed]
125. Hueper, K.; Vogel-Claussen, J.; Parikh, M.A.; Austin, J.H.; Bluemke, D.A.; Carr, J.; Choi, J.; Goldstein, T.A.; Gomes, A.S.; Hoffman, E.A.; et al. Pulmonary Microvascular Blood Flow in Mild Chronic Obstructive Pulmonary Disease and Emphysema. The MESA COPD Study. *Am. J. Respir. Crit. Care Med.* **2015**, *192*, 570–580. [CrossRef] [PubMed] [PubMed Central]
126. Liu, C.Y.; Parikh, M.; Bluemke, D.A.; Balte, P.; Carr, J.; Dashnaw, S.; Poor, H.D.; Gomes, A.S.; Hoffman, E.A.; Kawut, S.M.; et al. Pulmonary artery stiffness in chronic obstructive pulmonary disease (COPD) and emphysema: The Multi-Ethnic Study of Atherosclerosis (MESA) COPD Study. *J. Magn. Reson. Imaging* **2018**, *47*, 262–271. [CrossRef] [PubMed] [PubMed Central]
127. Poor, H.D.; Kawut, S.M.; Liu, C.Y.; Smith, B.M.; Hoffman, E.A.; Lima, J.A.; Ambale-Venkatesh, B.; Michos, E.D.; Prince, M.R.; Barr, R.G. Pulmonary hyperinflation due to gas trapping and pulmonary artery size: The MESA COPD Study. *PLoS ONE* **2017**, *12*, e0176812. [CrossRef] [PubMed] [PubMed Central]
128. Matsuoka, S.; Washko, G.R.; Dransfield, M.T.; Yamashiro, T.; San Jose Estepar, R.; Diaz, A.; Silverman, E.K.; Patz, S.; Hatabu, H. Quantitative CT measurement of cross-sectional area of small pulmonary vessel in COPD: Correlations with emphysema and airflow limitation. *Acad. Radiol.* **2010**, *17*, 93–99. [CrossRef] [PubMed] [PubMed Central]
129. Chandra, D.; Gupta, A.; Kinney, G.L.; Fuhrman, C.R.; Leader, J.K.; Diaz, A.A.; Bon, J.; Barr, R.G.; Washko, G.; Budoff, M.; et al. The Association Between Lung Hyperinflation and Coronary Artery Disease in Smokers. *Chest* **2021**, *160*, 858–871. [CrossRef] [PubMed] [PubMed Central]
130. Goffin, J.R.; Pond, G.R.; Puksa, S.; Tremblay, A.; Johnston, M.; Goss, G.; Nicholas, G.; Martel, S.; Bhatia, R.; Liu, G.; et al. Chronic obstructive pulmonary disease prevalence and prediction in a high-risk lung cancer screening population. *BMC Pulm. Med.* **2020**, *20*, 300. [CrossRef] [PubMed] [PubMed Central]
131. Ruparel, M.; Quaife, S.L.; Dickson, J.L.; Horst, C.; Tisi, S.; Hall, H.; Taylor, M.N.; Ahmed, A.; Shaw, P.J.; Burke, S.; et al. Prevalence, Symptom Burden, and Underdiagnosis of Chronic Obstructive Pulmonary Disease in a Lung Cancer Screening Cohort. *Ann. Am. Thorac. Soc.* **2020**, *17*, 869–878. [CrossRef] [PubMed] [PubMed Central]
132. Tisi, S.; Dickson, J.L.; Horst, C.; Quaife, S.L.; Hall, H.; Verghese, P.; Gyertson, K.; Bowyer, V.; Levermore, C.; Mullin, A.M.; et al. Detection of COPD in the SUMMIT Study lung cancer screening cohort using symptoms and spirometry. *Eur. Respir. J.* **2022**, *60*, 2200795. [CrossRef] [PubMed] [PubMed Central]
133. Mets, O.M.; Buckens, C.F.; Zanen, P.; Isgum, I.; van Ginneken, B.; Prokop, M.; Gietema, H.A.; Lammers, J.W.; Vliegenthart, R.; Oudkerk, M.; et al. Identification of chronic obstructive pulmonary disease in lung cancer screening computed tomographic scans. *JAMA* **2011**, *306*, 1775–1781. [CrossRef] [PubMed]
134. Criner, G.J.; Delage, A.; Voelker, K.; Hogarth, D.K.; Majid, A.; Zgoda, M.; Lazarus, D.R.; Casal, R.; Benzaquen, S.B.; Holladay, R.C.; et al. Improving Lung Function in Severe Heterogenous Emphysema with the Spiration Valve System (EMPROVE). A Multicenter, Open-Label Randomized Controlled Clinical Trial. *Am. J. Respir. Crit. Care Med.* **2019**, *200*, 1354–1362. [CrossRef]
135. Klooster, K.; ten Hacken, N.H.; Hartman, J.E.; Kerstjens, H.A.; van Rikxoort, E.M.; Slebos, D.J. Endobronchial Valves for Emphysema without Interlobar Collateral Ventilation. *N. Engl. J. Med.* **2015**, *373*, 2325–2335. [CrossRef]

136. Caviezel, C.; Froehlich, T.; Schneiter, D.; Muehlematter, U.; Frauenfelder, T.; Guglielmetti, L.C.; Opitz, I.; Weder, W. Identification of target zones for lung volume reduction surgery using three-dimensional computed tomography rendering. *ERJ Open Res.* **2020**, *6*, 00305–02020. [CrossRef] [PubMed]
137. Horsfall, L.J.; Clarke, C.S.; Nazareth, I.; Ambler, G. The value of blood-based measures of liver function and urate in lung cancer risk prediction: A cohort study and health economic analysis. *Cancer Epidemiol.* **2023**, *84*, 102354. [CrossRef]
138. Raoof, S.; Shah, M.; Braman, S.; Agrawal, A.; Allaqaband, H.; Bowler, R.; Castaldi, P.; DeMeo, D.; Fernando, S.; Hall, C.S.; et al. Lung Imaging in COPD Part 2: Emerging Concepts. *Chest* **2023**, *164*, 339–354. [CrossRef] [PubMed]
139. Yang, I.A.; Jenkins, C.R.; Salvi, S.S. Chronic obstructive pulmonary disease in never-smokers: Risk factors, pathogenesis, and implications for prevention and treatment. *Lancet Respir. Med.* **2022**, *10*, 497–511. [CrossRef] [PubMed]
140. Harch, I.E.; Garcia-Larsen, V.; Benmaamar, S.; Nejjari, C.; Biaze, M.E.; Benjelloun, M.C.; Rhazi, K.E. Association between biomass exposure and COPD occurrence in Fez, Morocco: Results from the BOLD study. *BMJ Open Respir. Res.* **2024**, *11*, e002409. [CrossRef]
141. Ortiz-Quintero, B.; Martínez-Espinosa, I.; Pérez-Padilla, R. Mechanisms of Lung Damage and Development of COPD Due to Household Biomass-Smoke Exposure: Inflammation, Oxidative Stress, MicroRNAs, and Gene Polymorphisms. *Cells* **2022**, *12*, 67. [CrossRef]
142. Moran-Mendoza, O.; Pérez-Padilla, J.R.; Salazar-Flores, M.; Vazquez-Alfaro, F. Wood smoke-associated lung disease: A clinical, functional, radiological and pathological description. *Int. J. Tuberc. Lung Dis.* **2008**, *12*, 1092–1098.
143. González-García, M.; Torres-Duque, C.A.; Bustos, A.; Jaramillo, C.; Maldonado, D. Bronchial hyperresponsiveness in women with chronic obstructive pulmonary disease related to wood smoke. *Int. J. Chronic Obstr. Pulm. Dis.* **2012**, *7*, 367–373. [CrossRef]
144. Fandiño-Del-Rio, M.; Kephart, J.L.; Williams, K.N.; Malpartida, G.; Boyd Barr, D.; Steenland, K.; Koehler, K.; Checkley, W. Household air pollution and blood markers of inflammation: A cross-sectional analysis. *Indoor Air.* **2021**, *31*, 1509–1521. [CrossRef]
145. Banerjee, A.; Mondal, N.K.; Das, D.; Ray, M.R. Neutrophilic Inflammatory Response and Oxidative Stress in Premenopausal Women Chronically Exposed to Indoor Air Pollution from Biomass Burning. *Inflammation* **2011**, *35*, 671–683. [CrossRef] [PubMed]
146. Dutta, A.; Ray, M.R.; Banerjee, A. Systemic inflammatory changes and increased oxidative stress in rural Indian women cooking with biomass fuels. *Toxicol. Appl. Pharmacol.* **2012**, *261*, 255–262. [CrossRef] [PubMed]

diagnostics MDPI

Review

Recognition of Idiopathic Inflammatory Myopathies Underlying Interstitial Lung Diseases

Giulia Morina [1], Domenico Sambataro [2], Alessandro Libra [1], Stefano Palmucci [3], Michele Colaci [4], Gaetano La Rocca [5], Francesco Ferro [5], Linda Carli [5], Chiara Baldini [5], Santa Valentina Liuzzo [1], Carlo Vancheri [1] and Gianluca Sambataro [6,*]

1 Department of Clinical and Experimental Medicine, Regional Referral Center for Rare Lung Diseases, Policlinico "G.Rodolico-San Marco", University of Catania, 95123 Catania, Italy; giulia.morina@tiscali.it (G.M.); alessandrolibra@outlook.it (A.L.); valeliu92@gmail.com (S.V.L.); vancheri@unict.it (C.V.)
2 Artroreuma S.R.L., Outpatient Clinic Associated with the Regional Health System, Mascalucia, 95030 Catania, Italy; d.sambataro@hotmail.it
3 Unità Operativa Semplice Dipartimentale di Imaging Polmonare e Tecniche Radiologiche Avanzate (UOSD IPTRA), Department of Medical Surgical Sciences and Advanced Technologies "GF Ingrassia", University Hospital Policlinico "G.Rodolico-San Marco", University of Catania, 95123 Catania, Italy; spalmucci@unict.it
4 Internal Medicine Unit, Division of Rheumatology, Department of Clinical and Experimental Medicine, Cannizzaro Hospital, University of Catania, 95123 Catania, Italy; michele.colaci@unict.it
5 Rheumatology Unit, Department of Clinical and Experimental Medicine, University of Pisa, 56126 Pisa, Italy; gaelarocca94@gmail.com (G.L.R.); francescoferrodoc@gmail.com (F.F.); 81clinda@gmail.com (L.C.); chiara.baldini74@gmail.com (C.B.)
6 Department of Medicine and Surgery, University of Enna "Kore", 94100 Enna, Italy
* Correspondence: dottorsambataro@gmail.com

Abstract: Interstitial Lung Disease (ILD) is one of the most common causes of mortality in idiopathic Inflammatory Myopathies (IIM). Despite these conditions being commonly associated with proximal weakness, skin rashes and arthritis, ILD can be the first or the sole clinical feature in up to 60% of patients, potentially leading to incorrect diagnosis. The early recognition of an underlying IIM in ILD patients can allow for prompt treatment, which could potentially stabilize or even improve the lung disease, also avoiding the development of other clinical features associated with the condition. The objective of this review is to describe the clinical, serological and radiological features associated with IIM-ILD, mainly focusing on dermatomyositis and antisynthetase syndrome.

Keywords: interstitial lung disease; idiopathic inflammatory myopathies; dermatomyositis; polymyositis; antisynthetase syndrome; autoantibodies

1. Introduction

The term "Interstitial Lung Disease" (ILD) defines the deposition of cells and/or extracellular matrix in the lung interstitium. While in some cases, the disease may regress or stabilize, in other cases, especially in the presence of a fibrotic phenotype, it can progress, leading to decreased quality of life or even death due to respiratory failure [1].

Over 200 different conditions are associated with ILD, such as environmental exposure (smoke, dust, drugs), genetic factors, infections and autoimmune conditions. Among the latter group, Connective Tissue Diseases (CTDs) are responsible for about 25% of ILDs [2]. In particular, about 6% of the total ILD is caused by a subgroup of CTDs called Idiopathic Inflammatory Myopathies (IIM) [3].

The diagnosis of an IIM underlying ILD is not straightforward because it requires tight collaboration between rheumatologists and pulmonologists for the recognition of clinical and serological signs, which are often nuanced. Moreover, as seen in primary Sjögren's Syndrome (pSS), ILD may be the first or even the sole clinical manifestation of IIM reported by patients in 20–60% of cases [4–7]. Those patients are commonly classified as idiopathic, especially in the presence of fibrotic features identified by High-Resolution Computed Tomography (HRCT) or biopsy [8,9].

Despite ILD being a common manifestation of IIM, it was not included in the classification criteria for IIM or DM [10–12], which are currently the most widely used criteria for the classification of these patients, contributing further to the diagnostic delay.

Clearly, an incorrect diagnosis may delay an appropriate therapeutic approach, which may have a potential impact on survival.

The objective of this review is to describe the diagnostic approach to ILD patients, with a focus on the identification of potentially underlying IIM.

2. Classification of Idiopathic Inflammatory Myopathies

The term IIM encompasses a group of CTDs characterized by proximal weakness and prevalent inflammation of muscles. The group initially included only Polymyositis (PM) and Dermatomyositis (DM), according to the criteria proposed by Bohan and Peter in 1975 [10,11]. In the following years, Inclusion Body Myositis (IBM), Immune-Mediated necrotizing Myopathy (IMNM) and Antisynthetase Syndrome (ASyS) were added [12–14], and some authors suggested including Overlap myositis [15], conditions in which patients have myositis associated with another CTD. Increased knowledge of myositis histology, along with the discovery of a number of autoantibodies serving as disease markers, has given rise to suggestions of removing PM from this classification, as most patients proved to be more correctly classified as IBM, IMNM, or ASyS [14].

At present, the sole classification criteria available for the recognition of these conditions are the historical Bohan & Peter Criteria [10,11], while ASyS is classified according to expert opinion-based criteria [16], pending the forthcoming criteria endorsed by the American College of Rheumatology (ACR) and the European Alliance for Associations for Rheumatology (EULAR) [17].

As DM and ASyS are responsible for the vast majority of IIM-ILD, this review mainly focuses on these two entities.

3. Clinical Features of Dermatomyositis and Antisynthetase Syndrome

The clinical manifestations of DM and ASyS are mainly associated with skin and muscle involvement. These features can be very useful for diagnosis. However, it should be noted that up to 40% of non-Jo1 + ASyS patients will only show ILD during their entire clinical course [18]. A comprehensive list of possible clinical features is reported below.

3.1. Unexplained Fever

Fever is quite common in IIM, with a prevalence ranging from 19–41% [17,19]. This symptom seems to be more common in DM than in ASyS, probably due to its underlying pathogenic mechanism driven by interferon 1 [20]. Generally, it is described as an increased oral body temperature of 37.5 °C for 3 days [19]. Significantly higher values can help differentiate an autoimmune origin from an infectious one, although this is not always straightforward. Some infections, such as COVID-19 pneumonia, can induce an inflammatory ILD very similar to that expected in IIM-ILD and can even be associated with the production of anti-MDA5 autoantibodies [21,22]. Of note is the fact that the presence of fever is mainly associated with the presence of consolidations on HRCT and an increased risk of acute exacerbation [23].

3.2. Raynaud's Phenomenon

Raynaud's Phenomenon (RP) is a relatively common sign in both healthy individuals and CTD patients. RP results from the vasoconstriction of terminal circulation (mainly in the hands, but also feet, ears and nose), with a color change from white to blue and red (at least the first two phases must be present to define the condition) [24]. The prevalence of RP is 3–5% in healthy subjects; it is almost always present in Systemic Sclerosis (SSc), while it affects about 20–30% of IIM patients [25–27]. Nailfold Videocapillaroscopy (NVC) is a very useful tool to determine whether RP is isolated ("primary RP") or associated with an underlying autoimmune disease: in the presence of SSc, NVC findings include giant capillaries and/or avascular areas [28]. NVC can also be pathologic in IIM, with the possible presence of bushy capillaries [29]. However, differently from SSc, NVC positivity is not associated with the presence of RP in IIM patients [30,31]. Therefore, it could be useful to perform NVC on all ILD patients for whom IIM is suspected.

3.3. Inflammatory Arthritis

Inflammatory arthritis, together with myositis and ILD, is part of the classic triad used to define the typical clinical pattern of ASyS [32]. Prevalence ranges from about 15% in DM patients to 45% in ASyS patients [17]. Different patterns of arthritis have been described: if present at the onset of ASyS (about 30% of cases), arthritis shows a distribution and clinical behavior similar to Rheumatoid Arthritis (RA), with symmetrical involvement of the small joints and the potential presence of erosions. If inflammatory arthritis develops during the disease course (about 15% of cases), it resembles a more common arthritis seen in CTDs (asymmetrical, without erosions) [33].

3.4. Myositis

Myositis is present in up to 70% of ASyS cases during the clinical course, and it is virtually always present in DM, although some DM cases, mainly those associated with anti-MDA5 positivity, can present a clinically amyopathic phenotype [17,34]. The presence of myositis may be suggested by increased levels of muscle enzymes: Lactic Dehydrogenase (LDH), Aspartate and Alanine Transferase (AST and ALT), and other more specific markers such as Creatine Phosphokinase (CPK), myoglobin and aldolase [35]. The most important enzyme for diagnosing myositis is CPK, although aldolase may be increased with normal CPK levels, suggesting early muscle cell regeneration [36]. Clinically, myositis typically involves the proximal muscles of the arms and legs, as well as neck flexors, pharynx and the proximal esophagus, leading to fatigue, proximal weakness, and, more rarely, dysphagia [37]. These two latter symptoms are considered very important for a correct diagnosis and were included in the classification criteria for IIM [12]. The clinical picture often also includes myalgia, reported by about 40% of patients [17]. While myalgia is common, it is deemed nonspecific and was not included in the classification criteria for any IIM. However, it could be diagnostically valuable, especially in patients presenting with ILD. ILD-IIM patients often show little or no increase in muscle enzymes, making diagnosis difficult [38]. However, ILD patients with myalgia have been found to test positive for myositis autoantibodies in up to 68% of cases [39].

Myositis can be confirmed by electromyography (EMG), commonly showing a myopathic pattern, or through imaging: ultrasound can suggest the presence of edema, while Magnetic Resonance Imaging (MRI) can provide more specific details on muscle inflammation and damage. Finally, Positron Emission Tomography (PET) can be useful both in evaluating muscle inflammation and as a potential screening tool for associated cancer [40,41]. Indeed, IIM patients present an increased risk of various cancers within 3 years of diagnosis [42].

Rarely, confirmation of myositis may require histological examination. DM is histologically characterized by CD4+ cells in the endomysium, B lymphocytes in perivascular areas, endothelial hyperplasia with thrombi, reduced capillary density, and perifascicular atrophy. In AsyS, macrophages predominate, with a similar distribution of CD4+ cells, B cells in the perimysium, and CD8 cells in the peri and endomysium. Perifascicular regions typically show necrosis [43,44].

3.5. Skin Involvement

Skin involvement is very important for diagnosing ILD-IIM. ASyS is typically associated with hyperkeratotic rashes, while DM is associated with erythematous lesions. Among the hyperkeratotic rashes, Mechanic's Hands and Hiker's Feet are notable. They are characterized by cracking and hyperkeratosis of the fingers (generally the first three) and the palms and plantar soles, respectively [45]. Regarding erythematous rashes, the most important are Gottron Papules. They are red-to-violaceous papules on the extensor surfaces of joints, primarily the hands. The lesion is considered pathognomonic for DM [46]. Another nearly pathognomonic feature is the Heliotrope Rash, a violaceous, dusky rash around the eyes, with or without associated edema. Other erythematous-violaceus rashes and macules may appear on the elbows and knees (Gottron Sign), the posterior neck and shoulders (Shawl Sign), the anterior neck and chest (V sign) and the thighs (Holster Sign) [46]. DM can also be associated with facial erythema involving the nasolabial folds and calcinosis [46].

It is important to highlight that, when facing a patient with ILD, the presence of florid erythematous skin involvement, especially if not associated with significant myositis, should raise a suspicion of Amyopathic DM, a condition generally associated with positivity for anti-MDA5 antibodies and a severe risk of rapid progression [47]. A collection of typical skin rashes associated with IIM is reported in Figures 1 and 2.

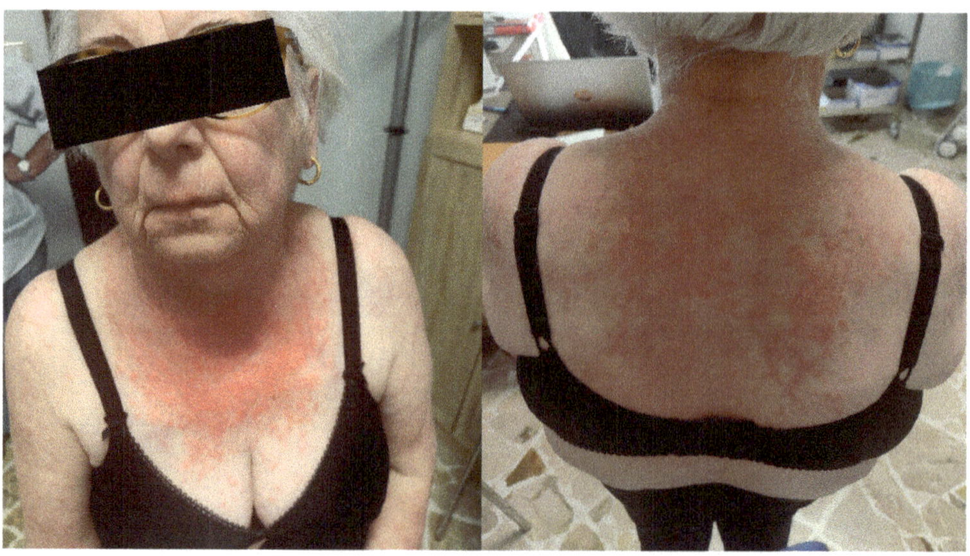

Figure 1. Shawl Sign present in the last year in a 68-year-old woman with DM.

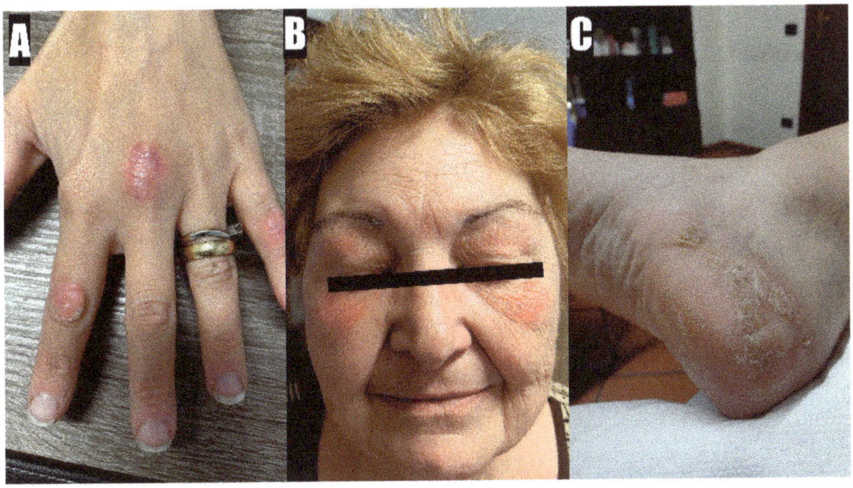

Figure 2. (**A**): Gottron Papules that first appeared a week before in a 43-year-old woman with DM; (**B**): Heliotropic Rash present in the last 6 months in a 64-year-old woman with DM; (**C**): Hiker's Feet, undated, in a 53-year-old woman with ASyS.

4. Autoantibodies in Dermatomyositis and Antisynthetase Syndrome

Autoantibodies play a crucial role in diagnosing IIM in the context of ILD potentially associated with IIM. In recent years, many new myositis-related autoantibodies have been described. The majority of these are associated with the presence of ILD, and each autoantibody typically correlates with a specific clinical presentation [48].

4.1. Types and Classification of Myositis Autoantibodies

Myositis autoantibodies are commonly divided into two categories, namely Myositis Associated Autoantibodies (MAA) and Myositis Specific Autoantibodies (MSA). The former includes anti-PM/Scl, anti-Ku, anti-U1RNP and anti-Ro52kD. These autoantibodies are often present in overlap conditions or may also be found in CTDs other than IIM. In contrast, MSA is regarded as highly specific for different IIM subtypes [49]. Table 1 summarizes the main IIM-related autoantibodies, categorizing them according to the specific disease subtype with which they are associated [50].

Table 1. Autoantibodies in Idiopathic inflammatory Myopathies.

Class	Subclass	Autoantibody	Target	ANA Staining	Condition	Clinical Picture
MAA	MAA	Anti-Pm/Scl [51,52]	Anti-Pm/scl exosome complex (100kD or 75 kD)	Nucleolar	Scleromyositis	ILD, DM skin rashes, myositis (100 kD), Scleroderma features (75 kD)
		Anti-Ku [53]	Ku (DNA binding protein)	Fine speckled	Scleromyositis, but also other CTD	SSc features, myositis
		Anti-Ro52kD [54]	TRIM21	Negative, fine speckled or cytoplasmic	ASyS, pSS but also other CTDs	Severe ILD, commonly associated with ASA
		Anti-U1RNP [55]	Subunit 1 of Ribonucleoprotein	Coarse speckled	MCTD	Mixed Connective Tissue Disease (ILD, arthritis, myositis)

Table 1. *Cont.*

Class	Subclass	Autoantibody	Target	ANA Staining	Condition	Clinical Picture
MSA	IBMA	Anti-CN1A [56]	Cytosolic 5′-nucleotidase 1A	none	IBM	Weakness with mild increased CPK
	IMNMA	Anti-SRP [57]	Signal Recognition Particle	Cytoplasmic	IMNM	Myositis with little or no histologic inflammation
		Anti-HMGCR [57]	3Hydroxy-2-methilglutaryl-Coenzyme A reductase	Negative or cytoplasmic speckled	IMNM	Myositis with little or no histologic inflammation
	DMA	Anti-MJ/NXP2 [58]	NXP2	Multiple nuclear dots or speckled	DM	Classic DM, severe myositis at onset, possible association with cancer
		Anti-Tif1γ [59]	Transcription intermediary factor 1 γ	Fine speckled	DM (cancer-associated)	DM, mainly cancer-related. Possible hypo-amyopathic
		Anti-Mi2 [60]	Nucleosome remodeling deacetylase complex	Fine speckled	DM	Classic DM, mild
		Anti-MDA5 [61]	Melanoma Differentiation associated protein 5	Negative or faint cytoplasmic	DM (CADM)	Clinically amyopathic, potentially severe ILD and rashes
		Anti-SAE1 [62]	Small ubiquitin-like modifier activating enzyme	Speckled	DM	Classic DM
	ASA	Anti-Jo1 [34]	Hystidil-tRNA synthetase	Cytoplasmic fine speckled	ASyS	ASyS, classic triad ILD-inflammatory arthritis and myositis
		PL-7 [63]	Threonyl-tRNA synthetase	Cytoplasmic dense fine speckled	ASyS	ASyS, prevalence of ILD (potentially severe)
		PL-12 [63]	Alanyl-tRNA synthetase	Cytoplasmic dense fine speckled	ASyS	ASyS, prevalence of ILD (potentially severe)
		Anti-EJ [63]	Glycyl-tRNA synthetase	Cytoplasmic speckled	ASyS	ASyS, prevalence of ILD (potentially severe)
		Anti-OJ [64]	soleucyl-tRNA synthetase	Cytoplasmic speckled	ASyS	ASyS with prevalent ILD or even ILD alone
		Anti-KS [64]	Asparaginyl-tRNA synthetase	Cytoplasmic speckled	ASyS	ASyS with prevalent ILD or even ILD alone
		Anti-Zo [65]	Phenylalanyl-tRNA synthetase	Cytoplasmic speckled	ASyS	ILD commonly alone
		Anti-Ha [65]	Tyrosyl-tRNA synthetase	Cytoplasmic speckled	ASyS	ILD commonly alone

Legend: ANA: Antinuclear Antibody; ASA: Antisynthetase Antibodies; ASYS: Antisynthetase Syndrome; CADM: Clinically Amyopathic Dermatomyositis; CPK: Creatine Phosphokinase; CTD: Connective Tissue Disease; DM: dermatomyositis; DMA: Dermatomyositis Antibodies; IBM: Inclusion Body Myositis; IBMA: Inclusion Body Myositis Antibodies; ILD: Interstitial Lung Disease; IMNM: Immune Mediated Necrotizing myositis; IMNMA: Immune Mediated Necrotizing myositis Antibodies; MCTD: Mixed Connective Tissue Disease; MSA: Myositis Specific Antibodies; SSc: Systemic Sclerosis.

Considering MAA from a pneumological perspective, the most important autoantibodies are probably anti-Ro52kD. They may be detected in virtually all CTD subtypes, but they are associated with Antisynthetase Antibodies (ASA) in up to 50% of cases of ASyS, portending a worse prognosis in ILD patients [54]. In clinical practice, it is not uncommon to find anti-Ro52kD autoantibodies in the sera of ILD patients in the absence of any extra-pulmonary sign of CTDs. These patients are usually classified as "Interstitial Pneumonia with Autoimmune Features" (IPAF) [66]. In these cases, a thorough assessment of the autoimmune profile, including IIM-related autoantibody testing, is warranted to detect potentially associated ASA. Moreover, IPAF patients should undergo a prompt follow-up to promote early recognition of potential progression toward specific CTDs [67,68].

The remaining MAA are anti-Pm/Scl, anti-Ku and anti-U1RNP. The first two are associated with an overlap condition between SSc and myositis called "scleromyositis", while anti-U1RNP is the typical marker of Mixed Connective Tissue Disease (MCTD), a rare condition characterized by an overlap between myositis, Systemic Lupus Erythematosus and SSc. ILD is generally present in scleromyositis and MCTD in an established clinical picture; therefore, the diagnosis is generally quite straightforward.

ILD secondary to IIM (ILD-IIM) is closely associated with MSA, mainly in the subclass of Dermatomyositis Antibodies (DMA) and ASA. DMA include anti-MJ/NXP2, anti- Tif1γ, anti-Mi2, anti-MDA5 and anti-SAE. All of these autoantibodies are associated with significant erythematous skin involvement. Myositis is generally more severe in anti-MJ/NXP2, whereas anti-MDA5 is associated with a severe interferon-1-mediated interstitial lung disease, potentially rapidly progressive with a poor prognosis [34]. The frequent associations of DM with cancer should also be highlighted. Of note, cancer-related DM is generally associated with the presence of anti- Tif1γ autoantibodies [59].

ASAs are a group of MSAs representing the serological hallmark of ASyS. The classic triad of inflammatory arthritis, ILD, and myositis is more common with anti-Jo1 positivity (which is also the most common ASA) [32]. Other autoantibodies are rarer but are associated with severe ILD, which can be the sole ASyS manifestation during the entire course of the disease [18]. Since anti-Jo1 is the only ASA detected by commonly available commercial kits for Extractable Nuclear Antigen, specific testing for non-jo1-ASA should be performed whenever the suspicion of secondary ILD is raised [68].

4.2. Interpretation and Reliability of Myositis Autoantibodies

Correct interpretation of MSA/MAA is crucial in the management of ILD patients. Therefore, it is important to highlight current limitations in autoantibody testing. The current gold standard for the assessment of MSA/MAA is Immunoprecipitation (IP), which is an expensive, time-consuming technique and, therefore, not available on a large scale. Moreover, incorrect results are possible even with IP testing due to the potential co-migration of other proteins affecting the recognition of some autoantibodies (e.g., anti-MDA5, anti-NXP2, anti-SRP, anti-SAE) [48]. A certain variability has been shown in IP performed in different centers [69]. Other techniques, such as Line Blot Immunoassay (LIA), can recognize different autoantibodies at the same time in a semi-quantitative manner. They are easy and cheap to perform, but they are also burdened by a significant number of incorrect results. It is relatively common to see false positivity for MSAs such as anti-TIF1γ, anti-MDA5, anti-Mi2, and false negativity for rare ASA [70–72]. The optimal approach could be a periodic test in selected patients, followed by confirmation through IP in patients who test positive on LIA for MSA/MAA. However, other measures should also be employed to improve the reliability of LIA. First of all, MSA/MAA testing shows low reliability in populations with a low pre-test probability of IIM diagnosis [72]. Correlation with the expected clinical picture is useful in the general population [73] but probably less so with ILD patients since isolated ILD is very common in MSA-positive patients.

Correlation with the expected Antinuclear Antibody (ANA) pattern has also been shown to improve the reliability of LIA results [74]. The expected ANA pattern is reported for each autoantibody in Table 1. Moreover, false positivity is more common for low autoantibody titers and rare in the presence of high titers [75]. Inconclusive results, such as the "grey zone" in LIA, should be treated as negative. Finally, MSAs are considered mutually exclusive [49]: the association of multiple MSAs and the association of anti-Pm/scl with MSA is extremely uncommon [76]. However, some authors reported the coexistence of multiple autoantibodies, repeatedly confirmed and correctly associated with the clinical picture in about 10% of patients [77,78]. This is in line with clinical practice, where it is relatively common to find patients with seropositive ASyS showing clinical features associated with other conditions (mainly SSc). Therefore, positivity for multiple autoantibodies should be treated with caution by carefully considering the overall clinical picture, autoantibody titers and the expected ANA pattern. Importantly, the high risk of false positivity and the possibility of a potential overlap syndrome should be considered.

5. ILD Associated with Dermatomyositis and Antisynthetase Syndrome

ILD is a major cause of mortality in IIM, with an increased mortality rate of around 40% [79]. Respiratory failure is pathologically linked with the infiltration of inflammatory cells and/or the deposition of extracellular matrix in the lung interstitium. However, weakness of the respiratory muscles can also contribute [80]. As pointed out before, ILD can be the main or even the sole clinical feature in the context of IIM. Therefore, recognition of the underlying autoimmune disease can be difficult. Of note, patients with isolated lung involvement tend to be referred to lung units for respiratory symptoms prior to any rheumatologic assessment [17,81].

It should be noted that from the clinical point of view, respiratory symptoms of IIM-ILD patients are the same as those reported for other ILD patients (dyspnea, dry cough), but Velcro crackles on chest auscultation are uncommon, given the rarity of a UIP pattern in these conditions [3]. Attention should be given to the possibility that a significant proportion could be asymptomatic, as well as the severity of ILD in IIM and the myositis disease activity [3]. In addition, Pulmonary Function Tests (PFTs) can be sensitive but not specific. They can be normal or show a restrictive pattern characterized by a reduction in forced vital capacity and diffuse lung capacity for carbon monoxide. It is of great interest that a restrictive pattern can also be due to the inflammatory involvement of respiratory muscles rather than ILD. In these cases, the evaluation of maximal inspiratory and respiratory pressure (respectively, Total Lung Capacity and Residual Volume) has been proposed [82]. In any case, clinics, objective exams, and PFTs can be used as a first diagnostic assessment, but appropriate imaging is required. Their role had greater value during follow-up.

Chest X-ray can recognize parenchymal abnormalities, but usually when present at advanced stages [83]. HRCT is, therefore, the gold standard for detecting ILD, monitoring the response to treatment and identifying disease progression and/or complications. Importantly, HRCT allows early diagnosis of ILD and is useful to localize pulmonary lesions, estimating their extent and assess the fibrotic burden. HRCT is also crucial for prognostic purposes, as it can highlight the presence of ventilation defects and aspiration pneumonia. However, radiological findings alone are not sufficient for the proper diagnosis and classification of ILD patients; therefore, a multidisciplinary approach is warranted.

5.1. Radiological Pattern of ILD in IIM

The most common ILD pattern associated with the presence of IIM is Nonspecific Interstitial Pneumonia (NSIP), with or without superimposed Organizing Pneumonia (OP) (Figure 3). The NSIP pattern can be divided into cellular and fibrotic. Both are characterized by the presence of bibasilar and symmetric ground-glass opacities, with the fibrotic form adding fine reticulations. OP is characterized by ground-glass consolidations and rarely by reticulation, bronchiectasis, interstitial nodules, interlobular thickening of septa, halo sign, inverted halo sign or airspace nodules. OP lesions have a subpleural/peribronchovascular and perilobular distribution, with bilateral involvement, especially in the middle-basal lung regions [84]. These two patterns are present in about 66% and 32% of DM patients, respectively. However, when considering only Rapidly Progressive ILD-DM cases, the proportion of NSIP and OP patterns shifts to 30% and 70%, respectively, clearly suggesting that consolidations should be considered a marker of potential rapid progression, commonly associated with the presence of anti-MDA5 antibodies [85]. These autoantibodies are also associated with a possible acute Diffuse Alveolar Damage (DAD) pattern [85,86]. This pattern, although rare, is very similar to that observed in COVID-19 pneumonia, potentially associated with pneumothorax and/or subcutaneous emphysema [87]. The Usual Interstitial Pneumonia (UIP) pattern is also uncommon in DM-ILD, being present in about 2% of patients [86]. It is characterized by traction bronchiectasis, honeycombing and thickening of bronchovascular bundles [86].

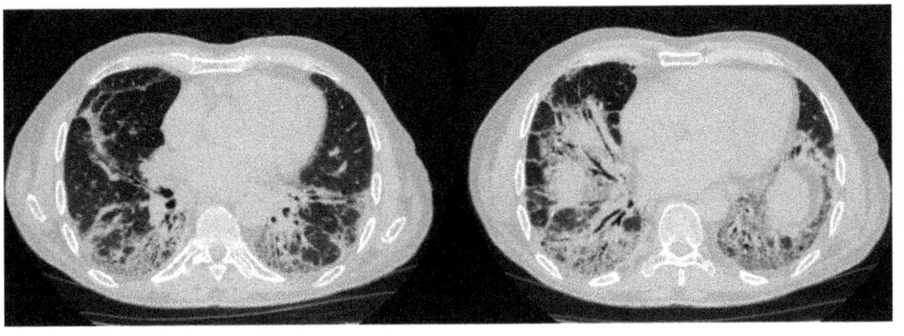

Figure 3. NSIP + OP pattern in a 57-year-old male at the time of diagnosis of DM.

ILD is also the main clinical sign of ASyS, affecting about 80% of patients in the largest cohort of ASYS available in the literature [17]. The most commonly encountered pattern, characterizing about 60% of cases, is also NSIP with or without associated OP. However, a UIP pattern is also possible, with a prevalence of about 12% of patients [7]. Mainly in the presence of non-Jo1 ASyS antibodies, ILD can be the first, or even the only, clinical feature associated with the condition. Notably, the prognosis seems to be more severe in the presence of anti-PL7 antibodies due to the higher risk of ILD progression [17].

5.2. Progressive Fibrosing Phenotype in IIM-ILD

The concept of progressive fibrosing phenotype has recently gained significant value in the management of ILD patients, especially since the antifibrotic agent nintedanib was shown to slow the progression of fibrotic ILD other than Idiopathic Pulmonary Fibrosis (IPF) and its use was subsequently approved for the treatment of these patients [88]. Unfortunately, it is difficult to directly extrapolate information for the treatment of IIM patients from the INBUILD trial, as the proportion of IIM-ILD patients recruited is unclear. However, the concept is of great interest, as it has been demonstrated that about 30% of non-IPF patients with a fibrotic phenotype on HRCT share a similar prognosis to classic IPF patients despite the underlying condition [89]. Progressive fibrosing ILD is not limited to UIP patients but is potentially also connected to fibrotic NSIP and OP, being defined by specific criteria including a relative decline of Forced Vital Capacity (FVC) of 10% or a decline of FVC between 5–10% combined with worsening of the fibrotic extension in HRCT or respiratory symptoms within 24 months [88].

It is reported that about 25% of ASyS-ILD present a progressive fibrosing phenotype, mainly associated with the seropositivity for non-anti-Jo1 antibodies and the presence of reticular opacities on HRCT. This group could benefit from antifibrotic treatment in addition to the classic immunosuppressive drugs [90].

There are no available data on PPF in the context of ILD-DM. However, some authors reported on the administration of the antifibrotic agent pirfenidone to clinically amyopathic dermatomyositis patients, showing no effects in the acute phase but a trend for the reduction of mortality in the sub-acute phase [91].

Overall, these data highlight the importance of the correct recognition of ILD-IIM patients in order to avoid their misclassification as indeterminate or idiopathic ILD cases with the consequent risk of receiving incomplete treatment.

6. The Problem of Interstitial Pneumonia with Autoimmune Features

IPAF was proposed as a research classification to include ILD patients with some clinical and/or serological features of autoimmune disease that do not meet classification criteria for specific CTDs [66]. The concept resembles the idea of Undifferentiated CTD, used in rheumatology, and it could potentially include patients where the association between the ILD and autoimmune characteristics is entirely stochastic, but also patients where ILD is the first or even the sole clinical feature of a well-defined autoimmune disease [92]. Since its publication in 2015, the IPAF classification has also become common in clinical practice, with several concerns mainly regarding IIM. In fact, the criteria are divided into clinical, serological, and morphological domains, and classification requires at least one criterion from two different domains. In the clinical domain, IPAF criteria include Gottron Papules, which are considered pathognomonic for DM. Also, in the serological domain, the IPAF classification includes autoantibodies highly specific for IIM, such as anti-MDA and all of the ASA. The morphological domain includes all possible HRCT patterns of ILD, excluding the UIP pattern, which is deemed to be insufficiently associated with autoimmune conditions. The main difficulty is that ILD is not considered in the historical Bohan and Peter criteria for DM and is not mentioned even in the 2017 criteria for IIM. ASyS still lacks validated classification criteria (although they are expected to be available in early 2025). Therefore, ILD patients with positivity for ASA might be classified as ASyS or IPAF depending on the clinical setting (the first being a reasonable diagnosis in a rheumatology unit and the second in a respiratory setting), with possible implications on treatment. This is especially true for UIP-like patients with ASA positivity. This condition is likely sufficient for a rheumatologist to classify as ASyS (and therefore suggest immunosuppressive treatment) but not sufficient for a classification as IPAF. Consequently, in some respiratory settings, these patients may be classified as IPF and treated with antifibrotic drugs. Retrospective studies on IPAF have reported a surprisingly high prevalence of ASA positivity, whereas this number is almost absent in prospective studies [67,93]. As expected, IPAF patients with MSA positivity show a prognosis similar to established ILD-IIM, further suggesting that a true positivity for MSA in ILD patients could be enough to make a specific diagnosis [94]. Diagnosis is particularly important in patients with fibrotic ILD, as those treated with antifibrotic medications have an almost identical prognosis to IPF, while those treated with immunosuppressive therapy show benefits even in a fibrotic phenotype [95,96].

In light of this, ILD patients with pathognomonic clinical features and/or a true positivity for highly specific MSA should be considered to be affected by IIM and treated accordingly. The IPAF classification has the merit of highlighting the potential autoimmune pathway underlying ILD; however, it requires revision of its criteria in light of new knowledge, mainly regarding MSA.

7. Conclusions

The recognition of an underlying systemic autoimmune disease in ILD patients is crucial, opening the way to significant therapeutic opportunities. Immunosuppressive drugs can stabilize or even improve lung involvement, also potentially preventing the development of other typical signs of the disease. This is particularly important for IIM because ILD is potentially severe and commonly even the sole clinical manifestation associated with the autoimmune condition.

Screening for a possible underlying IIM should be proposed to ILD patients showing an NSIP and/or OP pattern, but it should not be limited to these because about 30% of ILD-IIM patients show a UIP or indeterminate pattern.

All patients with ILD should be clinically evaluated for the presence of proximal weakness, myalgia and typical IIM skin rashes, which are of great diagnostic value. Recognizing them may not be easy for physicians who are not properly trained; therefore, we suggest close collaboration between pulmonologists and rheumatologists working together in lung units or at least on a multidisciplinary team. Indeed, this collaboration could be of mutual benefit.

In addition, all patients with ILD should be evaluated for the presence of serological clues of possible underlying IIM. A significant increase in muscular enzymes such as transaminases, CPK, aldolase and myoglobin (1.5–2× the upper limit) could be meaningful, even if not associated with proximal weakness.

Finally, the autoimmune profile should be thoroughly assessed in all patients with particular attention to the testing for MSA/MAAs in the presence of ANA positivity with a nucleolar or cytoplasmic pattern, positivity for anti-Ro52kD and/or clinical features suggestive of IIM, even if classic autoimmunity tests (such as ANA and ENA) are negative. However, positivity for MSA/MAA should be carefully evaluated for possible false positivity, particularly in the presence of borderline positivity and lack of compelling clinical features. A possible approach is summarized in Figure 4.

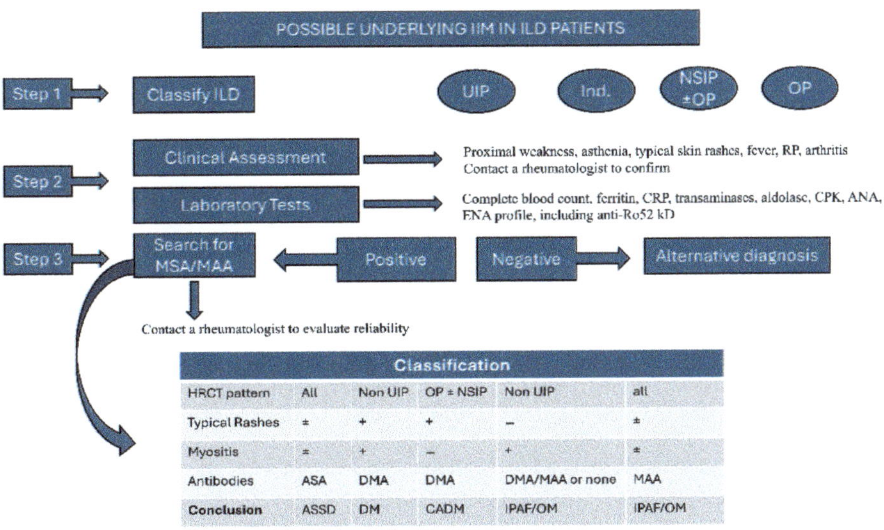

Figure 4. ANA: Antinuclear Antibody; ASA: Antisynthetase Antibodies; ASyS: Antisynthetase Syndrome; CADM: Clinically Amyopathic DM; CPK: Creatine Phosphokinase; DM: Dermatomyositis;

DMA: DM Antibodies; ENA: Extractable Nuclear Antigen; IIM: Idiopathic Inflammatory Myopathy; IPAF: Interstitial Pneumonia with Autoimmune Features; HRCT: High-Resolution Computed Tomography; ILD: Interstitial Lung Disease; Ind: Indeterminate; MAA: Myositis Associated Antibodies; MSA: Myositis Associated Antibodies; NSIP: Nonspecific Interstitial Pneumonia; OM: Overlap Myositis; OP: Organizing Pneumonia; RP: Raynaud's Phenomenon; UIP: Usual Interstitial Pneumonia. +: present; −: absent; ±: possible.

Author Contributions: Conceptualization, G.M. and G.S.; methodology, G.S. and D.S.; software, G.M.; validation, G.M., G.S. and C.V.; formal analysis, C.V. and G.S.; investigation, D.S., G.M., A.L., S.V.L., D.S., S.P., M.C., G.L.R., F.F., C.B., G.S. and L.C.; resources G.S.; data curation, G.M.; writing—original draft preparation, D.S., G.M., A.L., F.F., L.C., G.L.R. and S.V.L.; writing—review and editing, S.P., M.C., C.B., C.V. and G.S.; visualization, G.S.; supervision, G.S. and C.V.; project administration, G.S.; funding acquisition, none. All authors have read and agreed to the published version of the manuscript.

Funding: This research received no external funding.

Institutional Review Board Statement: Not applicable.

Informed Consent Statement: Written informed consent has been obtained from the patients to publish this paper.

Data Availability Statement: Not applicable.

Conflicts of Interest: G.S. reports personal fees from Boehringer Ingelheim and Gentili outside the submitted work; S.P. reports personal fees from Boehringer Ingelheim, Delphi International srl, and F. Hoffmann-La Roche Ltd. outside the submitted work; C.V. is on the F. Hoffmann-La Roche Ltd. and Boehringer Ingelheim Scientific boards. He received consulting fees and/or speaker fees from AstraZeneca, Boehringer Ingelheim, C.F. Hoffmann-La Roche Ltd., and Menarini. G.M., D.S., S.V.L., A.L., G.L.R., F.F., C.B., M.C. declare no conflicts of interest.

References

1. Wijsenbeek, M.; Suzuki, A.; Maher, T.M. Interstitial Lung Diseases. *Lancet* **2022**, *400*, 769–786. [CrossRef] [PubMed]
2. Maher, T.M. Interstitial Lung Diseases: A review. *JAMA* **2024**, *331*, 1655–1665. [CrossRef] [PubMed]
3. Saketoo, L.A.; Ascherman, D.P.; Cottin, V.; Christopher-Stine, L.; Danoff, S.K.; Oddis, C.V. Interstitial Lung Disease in Idiopathic Inflammatory Myopathy. *Curr. Rheumatol.* **2020**, *6*, 108–119. [CrossRef]
4. La Rocca, G.; Ferro, F.; Sambataro, G.; Elefante, E.; Fulvio, G.; Navarro, I.C.; Moretti, M.; Romei, C.; Mosca, M.; Baldini, C. Interstitial Lung Disease Phenotypes and Predictive Risk Factors in Primary Sjögren's Syndrome. *J. Clin. Med.* **2024**, *13*, 4963. [CrossRef]
5. La Rocca, G.; Ferro, F.; Sambataro, G.; Elefante, E.; Fonzetti, S.; Fulvio, G.; Navarro, I.C.; Mosca, M.; Baldini, C. Primary-Sjögren's-Syndrome-Related Interstitial Lung Disease: A Clinical Review Discussing Current Controversies. *J. Clin. Med.* **2023**, *12*, 3428. [CrossRef]
6. Sambataro, G.; Ferro, F.; Orlandi, M.; Sambataro, D.; Torrisi, S.E.; Quartuccio, L.; Vancheri, C.; Baldini, C.; Matucci Cerinic, M. Clinical, morphological features and prognostic factors associated with interstitial lung disease in primary Sjögren's syndrome: A systematic review from the Italian Society of Rheumatology. *Autoimmun. Rev.* **2020**, *19*, 102447. [CrossRef]
7. Cavagna, L.; Trallero-Araguas, E.; Meloni, F.; Cavazzana, I.; Rojas-Serrano, J.; Feist, E.; Zanframundo, G.; Morandi, V.; Meyer, A.; Pereira da Silva, J.A.; et al. Influence of Antisynthetase Antibodies Specificities on Antisynthetase Syndrome clinical spectrum time course. *J. Clin. Med.* **2019**, *8*, 2013. [CrossRef] [PubMed]
8. Manfredi, A.; Sebastiani, M.; Cerri, S.; Cassone, G.; Bellini, P.; Della Casa, G.; Luppi, F.; Ferri, C. Prevalence and characterization of non-sicca onset primary Sjögren's syndrome with interstitial lung involvement. *Clin. Rheumatol.* **2017**, *36*, 1261–1268. [CrossRef] [PubMed]
9. La Rocca, G.; Ferro, F.; Bulleri, A.; Fulvio, G.; Fonzetti, S.; Donati, V.; Romei, C.; Mosca, M.; Baldini, C. Glandular involvement in primary Sjögren's syndrome patients with interstitial lung disease-onset and sicca-onset, a single centre cross-sectional study. *Clin. Exp. Rheumatol.* **2022**, *40*, 2344–2349. [CrossRef] [PubMed]
10. Bohan, A.; Peter, J.B. Polymyositis and dermatomyositis (first of two parts). *N. Eng. J. Med.* **1975**, *292*, 344–347. [CrossRef]
11. Bohan, A.; Peter, J.B. Polymyositis and dermatomyositis (second of two parts). *N. Eng. J. Med.* **1975**, *292*, 403–407. [CrossRef]

12. Lundberg, I.E.; Tjarnlund, A.; Bottai, M.; Werth, V.P.; Pilkington, C.; de Visser, M.; Alfredsson, L.; Amato, A.A.; Barohn, R.J.; Liang, M.H.; et al. 2017 European League Against Rheumatism/American College of Rheumatology classification criteria for adult and juvenile idiopathic inflammatory myopathies and their major subgroups. *Ann. Rheum. Dis.* **2017**, *76*, 1955–1964. [CrossRef] [PubMed]
13. Griggs, R.C.; Askanas, V.; DiMauro, S.; Engel, A.; Karpati, G.; Mendell, J.R.; Rowland, L.P. Inclusion body myositis and myopathies. *Ann. Neurol.* **1995**, *38*, 705–713. [CrossRef]
14. Mariampillai, K.; Granger, B.; Amelin, D.; Guiguet, M.; Hachulla, E.; Maurier, F.; Meyer, A.; Tohmé, A.; Charuel, J.L.; Musset, L.; et al. Development of a new classification system for idiopathic inflammatory myopathies based on clinical manifestations and myositis-specific autoantibodies. *JAMA Neurol.* **2019**, *75*, 1528–1537. [CrossRef] [PubMed]
15. Nuno-Nuno, L.; Joven, B.E.; Carreira, P.E.; Maldonado-Romero, V.; Larena Grijalba, C.; Llorente Cubas, I.; Tomero, E.; Barbadillo-Mateos, M.C.; Garcia de la Pena Lefebvre, P.; López-Longo, F.J.; et al. Overlap myosityis, a distinct entity beyond primary inflammatory myositis: A retrospective analysis of lange cohort from the REMICAM registry. *Int. J. Rheum. Dis.* **2019**, *22*, 1393–1401. [CrossRef] [PubMed]
16. Witt, L.J.; Curran, J.J.; Strek, M.E. The diagnosis and treatment of Antisynthetase syndrome. *Clin. Pulm. Med.* **2016**, *23*, 218–226. [CrossRef] [PubMed]
17. Faghihi-Kashani, S.; Yoshida, A.; Bozan, F.; Zanframundo, G.; Rozza, D.; Loganathan, A.; Dourado, E.; Sambataro, G.; Bauer-Ventura, I.; Bae, S.S.; et al. Clinical Characteristics of Anti-Synthetase Syndrome: Analysis from the CLASS project. *Arthritis Rheumatol.* **2024**, *in press*. [CrossRef]
18. Aggarwal, E.; Cassidy, E.; Fertingh, N.; Koontz, D.C.; Lucas, M.; Ascherman, D.P.; Oddis, C.V. Patients with non-Jo1 anti-tRNA-synthetase autoantibodies have worse survival than Jo-1 positive patients. *Ann. Rheum. Dis.* **2014**, *73*, 227–232. [CrossRef] [PubMed]
19. Zuo, Y.; Ye, L.; Liu, M.; Li, S.; Liu, W.; Chen, F.; Lu, X.; Gordon, P.; Wang, G.; Shu, X. Clinical significance of radiological patterns of HRCT and their association with macrophage activation in dermatomyositis. *Rheumatology* **2020**, *59*, 2829–2837. [CrossRef] [PubMed]
20. La Rocca, G.; Ferro, F.; Baldini, C.; Libra, A.; Sambataro, D.; Colaci, M.; Malatino, L.; Palmucci, S.; Vancheri, C.; Sambataro, G. Targeting intracellular pathways in idiopathic inflammatory myopathies: A narrative review. *Front. Med.* **2023**, *10*, 1158768. [CrossRef]
21. Wang, G.; Wang, W.; Wang, Y.; Liu, C.; Wang, L.; Chen, H.; Jiao, T.; Hu, C.; Lei, X.; Guo, L.; et al. Presence of anti-MDA5 antibody and its value for the clinical assessment in patients with COVID-19: A retrospective cohort study. *Front. Immunol.* **2021**, *12*, 791348. [CrossRef] [PubMed]
22. Cavagna, L.; Ferro, F.; Zanframundo, G.; La Rocca, G.; Puxeddu, I. Idiopathic inflammatory myopathies and COVID-19: An intriguing liaison? *Clin. Exp. Rheumatol.* **2023**, *41*, 217–220. [CrossRef] [PubMed]
23. Liu, T.; Chen, H.; Shi, Y.; Xu, W.; Yuan, F. Fever in the initial stage of IIM patients: An early clinical warning sign for AE-ILD. *Adv. Rheumatol.* **2023**, *63*, 31. [CrossRef]
24. Maverakis, E.; Patel, F.; Kronenberg, D.G.; Chung, L.; Fiorentino, D.; Allanore, Y.; Guiducci, S.; Hesselstrand, R.; Hummers, L.K.; Duong, C.; et al. International consensus criteria for the diagnosis of Raynaud's Phenomenon. *J. Autoimmun.* **2014**, *48–49*, 60–65. [CrossRef] [PubMed]
25. Wigley, F.M.; Flavahan, N.A. Raynaud's Phenomenon. *N. Eng. J. Med.* **2016**, *375*, 556–565. [CrossRef] [PubMed]
26. Hughes, M.; Allanore, Y.; Chung, L.; Pauling, J.D.; Denton, C.P.; Matucci-Cerinic, M. Raynaud phenomenon and digital ulcers in systemic sclerosis. *Nat. Rev. Reumatol.* **2020**, *16*, 208–211. [CrossRef]
27. Park, E.H.; Hwang, W.C.; Lee, Y.; Lee, E.Y.; Lee, E.B.; Song, Y.W.; Park, J.K. Raynaud's phenomenon and anti-nuclear antibody are associated with pulmonary function decline in patients with dermatomyositis and polymyositis. *Int. J. Rheum. Dis.* **2019**, *22*, 507–515. [CrossRef] [PubMed]
28. Piette, Y.; Reynaert, V.; Vanhaecke, A.; Bonroy, C.; Gutermuth, J.; Sulli, A.; Cutolo, M.; Smith, V. Standardised interpretation of capillaroscopy in autoimmune idiopathic inflammatory myopathies: A structured review on behalf of the EULAR study group on microcirculation in Rheumatic Diseases. *Autoimmun. Rev.* **2022**, *21*, 103087. [CrossRef] [PubMed]
29. Bertolazzi, C.; Cutolo, M.; Smith, V.; Gutierrez, M. State of the art on nailfold capillaroscopy in dermatomyositis and polymyositis. *Semin. Arthritis Rheum.* **2017**, *47*, 432–444. [CrossRef]
30. Sebastiani, M.; Triantafyllias, K.; Manfredi, A.; Gonzalez-Gay, M.A.; Palmou-Fontana, N.; Cassone, G.; Drott, U.; Delbruck, C.; Rojas-Serrano, J.; Bertolazzi, C.; et al. Nailfold Capillaroscopy Characteristics of Antisynthetase Syndrome and Possible clinical associations: Results of a multicenter international study. *J. Rheumatol.* **2019**, *46*, 279–284. [CrossRef] [PubMed]
31. Sambataro, D.; Sambataro, G.; Libra, A.; Vignigni, G.; Pino, F.; Fagone, E.; Fruciano, M.; Gili, E.; Pignataro, F.; Del Papa, N.; et al. Nailfold videocapillaroscopy is a useful tool to recognize definite forms of systemic sclerosis and idiopathic inflammatory myositis in Interstitial Lung Disease Patients. *Diagnostics* **2020**, *10*, 253. [CrossRef]

32. Cavagna, L.; Nuno, L.; Scirè, C.A.; Govoni, M.; Lopez Longo, F.J.; Franceschini, F.; Neri, R.; Castaneda, S.; Sifuentes Giraldo, W.A.; Caporali, R.; et al. Clinical spectrum time course in anti-Jo1 positive antisynthetase syndrome: Results from an international retrospective multicenter study. *Medicine* **2015**, *94*, e1144. [CrossRef] [PubMed]
33. Gonzalez-Gay, M.A.; Montecucco, C.; Selva-O'Callaghan, A.; Trallero-Araguas, E.; Molberg, O.; Andersson, H.; Rojas-Serrano, J.; Perez-Roman, D.I.; Bauhammer, J.; Fiehn, C.; et al. Timing of onset affects arthritis presentation pattern in antisynthetase syndrome. *Clin. Exp. Rheumatol.* **2018**, *36*, 44–49.
34. Cavagna, L.; Meloni, F.; Meyer, A.; Sambataro, G.; Belliato, M.; De Langhe, E.; Cavazzana, I.; Pipitone, N.; Triantafyllias, K.; Mosca, M.; et al. Clinical spectrum time course in non-Asian patients positive for anti-MDA5 antibodies. *Clin. Exp. Rheumatol.* **2022**, *40*, 274–283. [CrossRef] [PubMed]
35. Volochayev, R.; Csako, G.; Wesley, R.; Rider, L.G.; Miller, F.W. Laboratory test abnormalities are common in polymyositis and dermatomyositis and differ among clinical and demographic groups. *Open Rheumatol. J.* **2012**, *6*, 54–63. [CrossRef] [PubMed]
36. Cavazzana, I.; Fredi, M.; Selmi, C.; Tincani, A. The clinical and histological spectrum of Idiopathic Inflammatory Myopathies. *Clin. Rev. Allerg. Immunol.* **2017**, *52*, 88–98. [CrossRef] [PubMed]
37. Casciola-Rosen, L.; Hall, J.C.; Mammen, A.L.; Christopher-Stine, L.; Rosen, A. Isolated elevation of aldolase in the serum of myosityis patients: A potential biomarker of damaged early regenerating muscle cells. *Clin. Exp. Rheumatol.* **2012**, *30*, 548–553.
38. Hengstman, G.J.D.; Brouwer, R.; Vree Egberts, W.T.M.; Seelig, H.P.; Jongen, P.J.H.; Van Venrooij, W.J.; Van Engelen, B.G.M. Clinical and serological characteristics of 125 Dutch myositis patients. *J. Neurol.* **2002**, *249*, 65–75. [CrossRef] [PubMed]
39. Sambataro, G.; Ferrara, C.A.; Spadaro, C.; Torrisi, S.E.; Vignigni, G.; Vancheri, A.; Muscato, G.; Del Papa, N.; Colaci, M.; Malatino, L.; et al. A new method for the assessment of myalgia in interstitial lung disease: Association with positivity for myositis-specific and mytositis-associated antibodies. *Diagnostics* **2022**, *12*, 1139. [CrossRef]
40. Blijham, P.J.; Hengstman, G.J.D.; Hama-Amin, A.D.; van Engelen, B.G.M.; Zwarts, M.J. Needle electromyographic findings in 98 patients with myositis. *Eur. Neurol.* **2006**, *55*, 183–188. [CrossRef]
41. Tan, A.L.; Di Matteo, A.; Wakefield, R.J.; Biglands, J. Update on muscle imaging in myositis. *Curr. Opin. Rheumatol.* **2023**, *35*, 395–403. [CrossRef] [PubMed]
42. Dobloug, G.C.; Garen, T.; Brunborg, C.; Gran, J.T.; Molberg, O. Survival and cancer risk in an unselected and complete Norwegian Idiopathic inflammatory myopathy cohort. *Semin. Arthritis Rheum.* **2015**, *45*, 301–308. [CrossRef] [PubMed]
43. Dalakas, M.C. Muscle biopsy findings in inflammatory myopathies. *Rheum. Dis. Clin. N. Am.* **2002**, *28*, 779–798. [CrossRef]
44. Mescam-Mancini, L.; Allenbach, Y.; Hervier, B.; Devilliers, H.; Mariampillay, K.; Dubourg, O.; Maisonobe, T.; Gherardi, R.; Mezin, P.; Preusse, C.; et al. Anti-Jo1 antibody-positive patients show a characteristic necrotizing perifascicular myositis. *Brain* **2015**, *138*, 2485–2492. [CrossRef] [PubMed]
45. Cox, J.T.; Gullotti, D.M.; Mecoli, C.A.; Lahouti, A.H.; Albayda, J.; Paik, J.; Danoff, S.K.; Mammen, A.L.; Christopher-Stine, L. "Hiker's feet": A novel cutaneuous finding in the inflammatory myopathies. *Clin. Rheumatol.* **2017**, *36*, 1683–1686. [CrossRef]
46. Santmyire-Rosenberger, B.; Dugan, E.M. Skin involvement in dermatomyositis. *Curr. Opin. Rheumatol.* **2003**, *15*, 714–722. [CrossRef]
47. Bailey, E.E.; Fiorentino, D.F. Amyopathic dermatomyositis: Definitions, diagnosis, and management. *Curr. Rheumatol. Rep.* **2014**, *16*, 465. [CrossRef]
48. Satoh, M.; Tanaka, S.; Ceribelli, A.; Calise, S.J.; Chan, E.K.L. A Comprehensive overview on myositis-specific antibodies: New and old biomarkers in Idiopathic Inflammatory Myopathy. *Clin. Rev. Allergy Immunol.* **2017**, *52*, 1–19. [CrossRef] [PubMed]
49. McHugh, N.J.; Tansley, S.L. Autoantibodies in myositis. *Nat. Rev. Rheumatol.* **2018**, *14*, 290–302. [CrossRef] [PubMed]
50. Palterer, B.; Vitiello, G.; Carraresi, A.; Giudizi, M.G.; Cammelli, D.; Parronchi, P. Bench to bedside review of myositis autoantibodies. *Clin. Mol. Allergy* **2018**, *16*, 5. [CrossRef] [PubMed]
51. Lazzaroni, M.G.; Marasco, E.; Campochiaro, C.; DeVries-Bouwstra, J.; Gonzalez-Perez, M.I.; Rojas-Serrano, J.; Hachulla, E.; Zanatta, E.; Barsotti, S.; Furini, F.; et al. For the EUSTAR co-authors. The clinical phenotype of systemic sclerosiss patients with anti-Pm/scl antibodies: Results from the EUSTAR cohort. *Rheumatology* **2021**, *60*, 5028–5041. [CrossRef]
52. Hanke, K.; Bruckner, C.S.; Dahnrich, C.; Huscher, D.; Komorowski, L.; Meyer, W.; Janssen, A.; Backhaus, M.; Becker, M.; Kill, A.; et al. Antibodies against Pm/scl-75 and Pm/scl-100 are independent markers for different subsets of systemic sclerosis patients. *Arthritis Res. Ther.* **2009**, *11*, R22. [CrossRef] [PubMed]
53. Bilizna, C.; Henrion, D.; Beucher, A.; Lavigne, C.; Ghaali, A.; Levesque, H. Anti-Ki antibodies: Clinical, genetic and diagnostics insights. *Autoimmun. Rev.* **2010**, *9*, 691–694. [CrossRef] [PubMed]
54. Chan, E.K.L. Anti-Ro52 autoantibody is common in Systemic Autoimmune Rheumatic Diseases and correlating with worse outcome when associated with interstitial lung disease in Sysyemic Sclerosis and autoimmune Myositis. *Clin. Rev. Allergy Immunol.* **2022**, *63*, 178–193. [CrossRef] [PubMed]
55. Ferrara, C.A.; La Rocca, G.; Ielo, G.; Libra, A.; Sambataro, G. Towards early diagnosis of Mixed connective tissue disease: Updated perspectives. *Immunotargets Ther.* **2023**, *12*, 79–89. [CrossRef]

56. Salam, S.; Di Machkie, M.M.; Hanna, M.G.; Machado, P.M. Diagnostic and prognostic value of anti-cN1A antibodies in inclusion body myositis. *Clin. Exp. Rheumatol.* **2022**, *40*, 384–393. [CrossRef] [PubMed]
57. Li, C.; Liu, H.; Yang, L.; Liu, R.; Yin, G.; Xie, Q. Immune-mediated necrotizing myopathy: A comprehensive review of the pathogenesis, clinical features, and treatments. *J. Autoimmun.* **2024**, *148*, 103286. [CrossRef] [PubMed]
58. Ichimura, Y.; Matsushita, T.; Hamaguchi, Y.; Kaji, K.; Hasegawa, M.; Tanino, Y.; Inokoshi, Y.; Kawai, K.; Kanemura, T.; Habuchi, M.; et al. Anti-NXP2 autoantibodies in adult patients with idiopathic infammatory myopathies: Possible association with malignancy. *Ann. Rheum. Dis.* **2012**, *71*, 710–713. [CrossRef]
59. De Vooght, J.; Vulsteke, J.B.; De Haes, P.; Bossuyt, X.; Lories, R.; De Langhe, E. Anti-TIF1-γ autoantibodies: Warning lights of a tumour autoantigen. *Rheumatology* **2020**, *59*, 469–477. [CrossRef]
60. Liang, L.; Zhang, Y.M.; Chen, H.; Ye, L.F.; Li, S.S.; Lu, X.; Wang, G.C.; Peng, Q.L. Anti-Mi-2 antibodies characterize a distinct clinical subset of dermatomyositis with favourable prognosis. *Eur. J. Dermatol.* **2020**, *in press*. [CrossRef] [PubMed]
61. Lu, X.; Peng, Q.; Wang, G. anti-MDA5 antibody-positive dermatomyositis: Pathogenesis and clinical progress. *Nat. Rev. Rheumatol.* **2024**, *20*, 48–62. [CrossRef] [PubMed]
62. Albayda, J.; Mecoli, C.; Casciola-Rosen, L.; Danoff, S.K.; Lin, C.T.; Hines, D.; Gutierrez-Alamillo, L.; Paik, J.J.; Tiniakou, E.; Mammen, A.L.; et al. A North American Cohort of anti-SAE Dermatomyositis: Clinical phenotype, testing and review of cases. *ACR Open Rheumatol.* **2021**, *28*, 287–294. [CrossRef]
63. Shi, J.; Li, S.; Yang, H.; Zhang, Y.; Peng, Q.W.; Lu, X.; Wang, G. Clinical profiles and prognosis of patients with distinct antisynthetase autoantibodies. *J. Rheumatol.* **2017**, *44*, 1051–1057. [CrossRef] [PubMed]
64. Hamaguchi, Y.; Fujimoto, M.; Matsushita, T.; Kaji, K.; Koumura, K.; Hasegawa, M.; Kodera, M.; Muroi, E.; Fujikawa, K.; Seishima, M.; et al. Common and distinct clinical features in adult patients with anti-aminoacyl-tRNA synthetase antibodies: Heterogeneity within the syndrome. *PLoS ONE* **2013**, *8*, e60442. [CrossRef]
65. Moll, S.A.; Platenburg, M.G.J.P.; Platteel, A.C.M.; Vorselaars, A.D.M.; Bonas, J.; Roodenburg-Benschop, C.; Meek, B.; van Moorsel, C.H.M.; Grutters, J.C. Prevalence of novel myositis autoantibodies in a large cohort of patients with interstitial lung disease. *J. Clin. Med.* **2020**, *9*, 2944. [CrossRef] [PubMed]
66. Fischer, A.; Antoniou, K.M.; Brown, K.K.; Cadranel, J.; Corte, T.J.; Du Bois, R.M.; Lee, J.S.; Leslie, K.O.; Lynch, D.A.; Matteson, E.L.; et al. An official European Respiratory Society/American Thoracic Society research statement: Interstitial pneumonia with autoimmune features. *Eur. Respir. J.* **2015**, *46*, 976–981. [CrossRef] [PubMed]
67. Sambataro, G.; Sambataro, D.; Spicazza, L.; Meloni, F.; Lorini, G.; Malatino, L.; Colaci, M.; Sebastiani, G.; Iuliano, A.; Canofari, C.; et al. Progression and prognosis of interstitial pneumonia with autoimmune features: A longitudinal, prosective, multi-centre study. *Clin. Exp. Reumatol.* **2023**, *41*, 1140–1148. [CrossRef]
68. Sambataro, D.; Sambataro, G.; Pignataro, F.; Zanframundo, G.; Codullo, V.; Fagone, E.; Martorana, E.; Ferro, F.; Orlandi, M.; Del Papa, N.; et al. Patients with interstitial lung disease secondary to autoimmune diseases: How to recognize them? *Diagnostics* **2020**, *10*, 208. [CrossRef]
69. Loganathan, A.; Zanframundo, G.; Yoshida, A.; Faghihi-Kashani, S.; Bauer Ventura, I.; Dourado, E.; Bozan, F.; Sambataro, G.; Yamano, Y.; Bae, S.S.; et al. Agreement between local and central anti-synthetase antibodies detection: Results from the Classificatin Criteria of Anti-synthetase Syndrome project biobank. *Clin. Exp. Rheumatol.* **2024**, *42*, 277–287. [CrossRef]
70. To, F.; Ventin-Rodriguez, C.; Elkhalifa, S.; Lilleker, J.B.; Chinoy, H. Line blot immunoassays in idiopathic inflammatory myopathies: Retrospective review of diagnostic accuracy and factor predicting true positive results. *BMC Rheumatol.* **2020**, *4*, 28. [CrossRef]
71. Espinosa-Ortega, F.; Holmqvist, M.; Alexanderson, H.; Storfors, H.; Mimori, T.; Lundberg, I.E.; Ronnelid, J. Comparison of autoantibody specificities tested by a line blot assay and immunoprecipitation-based algorithm in patients with idiopathic inflammatory myopathies. *Ann. Rheum. Dis.* **2019**, *78*, 858–860. [CrossRef] [PubMed]
72. Tansley, S.L.; Li, D.; Betteridge, Z.E.; McHugh, N.J. The reliability of immunoassays to detect autoantibodies in patients with myositis is dependent on autoantibody specificity. *Rheumatology* **2020**, *59*, 2109–2114. [CrossRef]
73. Lackner, A.; Tiefenthaler, V.; Mirzayeva, J.; Posch, F.; Rossmann, C.; Kastrati, K.; Radner, H.; Demel, U.; Gretler, J.; Stotz, M.; et al. The use and diagnostic value of testing myositis-specific and myositis associated autoantibodies by line immune-assay: A retrospective study. *Ther. Adv. Musculoskelet. Dis.* **2020**, *12*, 1759720X20975907. [CrossRef]
74. Infantino, M.; Tampoia, M.; Fabris, M.; Alessio, M.G.; Previtali, G.; Pesce, G.; Deleonardi, G.; Porcelli, B.; Musso, M.; Grossi, V.; et al. Combining immunofluorescence with immunoblot assay improves the specificity of autoantibody testing for myositis. *Rheumatology* **2019**, *58*, 1239–1244. [CrossRef]
75. Beaton, T.J.; Gillis, D.; Prain, K.; Morwood, K.; Anderson, J.; Goddard, J.; Baird, T. Performance of myositis-specific antibodies detected on myositis line immunoassay to diagnose and sub-classify patients with suspected idiopathic inflammatory myopathy, a retrospective records-based review. *Int. J. Rheum. Dis.* **2021**, *24*, 1167–1175. [CrossRef]
76. Betteridge, Z.; Tansley, S.; Shaddick, G.; Chinoy, H.; Cooper, R.G.; New, R.P.; Lilleker, J.B.; Vencovsky, J.; Chazarain, L.; Danko, K.; et al. Frequency, mutual exclusivity and clinical associations of myositis autoantibodies in a combined European cohort of idiopathic inflammatory myopathy patients. *J. Autoimmun.* **2019**, *101*, 48–55. [CrossRef]

77. Huang, H.L.; Lin, W.C.; Tsai, W.L.; Weng, C.T.; Weng, M.Y.; Wu, C.H.; Sun, Y.T. Coexistence of Multiple Myositis-Specific Antibodies in patients with Idiopathic Inflammatory Myopathies. *J. Clin. Med.* **2022**, *11*, 6972. [CrossRef]
78. Loarce-Martos, J.; Calvo Sanz, L.; Garrote-Corral, S.; Ballester Gonzalez, R.; Pariente Rodriguez, R.; Rita, C.G.; Garcia-Soidan, A.; Bachiller-Corral, J.; Roy Arino, G. Myositis autoantibodies detected by line blot immunoassay: Clinical associations and correlation with antibody signal intensity. *Rheumatol. Int.* **2023**, *43*, 1101–1109. [CrossRef] [PubMed]
79. Wu, C.; Wang, Q.; He, L.; Yang, E.; Zeng, X. Hospitalization mortality and associated risk factors in patients with polymyositis and dermatomyositis: A retrospective case-control study. *PLoS ONE* **2018**, *13*, e0192491. [CrossRef] [PubMed]
80. Vojinovic, T.; Cavazzana, I.; Ceruti, P.; Fredi, M.; Modina, D.; Berlendis, M.; Franceschini, F. Predictive Features and Clinical Presentation of Interstitial Lung Disease in Inflammatory Myositis. *Clin. Rev. Allergy Immunol.* **2021**, *60*, 87–94. [CrossRef] [PubMed]
81. Chen, F.; Wang, J.; Zhang, P.; Zuo, Y.; Ye, L.; Wang, G.; Shu, X. Interstitial Lung Disease in Dermatomyositis without myositis-specific and myosiytis associated autoantibodies: Study of a series of 72 patients. *Front. Immunol.* **2022**, *13*, 879266.
82. Braun, N.M.; Arora, N.S.; Rochester, D.F. respiratory muscle and pulmonary function in polymyositis and other proximal myopathies. *Thorax* **1983**, *38*, 616–623. [CrossRef] [PubMed]
83. Jeganathan, N.; Sathananthan, M. Connective Tissue Disease-Related Interstitial Lung Disease: Prevalence, Patterns, Predictors, Prognosis, and Treatment. *Lung* **2020**, *198*, 735–759. [CrossRef]
84. Solomon, J.; Swigris, J.J.; Brown, K.K. Myositis-related interstitial lung disease and antisynthetase syndrome. *J. Bras. Pneumol.* **2011**, *37*, 100–109. [CrossRef]
85. Mehta, P.; Aggarwal, R.; Porter, J.C.; Gunawardena, H. Management of interstitial lung disease (ILD) in myositis syndromes: A practical guide for clinicians. *Best. Pract. Res. Clin. Rheumatol.* **2022**, *36*, 101769. [CrossRef] [PubMed]
86. Palmucci, S.; Di Mari, A.; Cancemi, G.; Pennisi, I.; Mauro, L.A.; Sambataro, G.; Sambataro, D.; Galioto, F.; Fazio, G.; Ferlito, A.; et al. Clinical and radiological features of interstitial lung disease associated with polymyositis and dermatomyositis. *Medicina* **2022**, *58*, 1757. [CrossRef]
87. Mehta, P.; Machado, P.M.; Gupta, L. Understanding and managing anti-MDA5 dermatomyositis, including potential CONVID-19 mimicry. *Rheumatol. Int.* **2021**, *41*, 1021–1036. [CrossRef]
88. Wells, A.U.; Flaherty, K.R.; Brown, K.K.; Inoue, Y.; Devaraj, A.; Richeldi, L.; Moua, T.; Crestani, B.; Wuyts, W.A.; Stowasser, S.; et al. Nintedanib in patients with progressive fibrosing interstitial lung diseases—Subgroup analyses by interstitial lung disease diagnosis in the INBUILD trial: A randomised, double-blind, placebo-controlled, parallel-group trial. *Lancet Respir. Med.* **2020**, *8*, 453–460. [CrossRef] [PubMed]
89. Kwon, B.S.; Choe, J.; Chae, E.J.; Hwang, H.S.; Kim, Y.G.; Song, J.W. Progressive fibrosing interstitial lung disease: Prevalence and clinical outcome. *Resp. Res.* **2021**, *22*, 282. [CrossRef]
90. Fu, H.; Zheng, Z.; Zhang, Z.; Yang, Y.; Cui, J.; Wang, Z.; Xue, J.; Chi, S.; Cao, M.; Chen, J. Prediction of progressive pulmonary fibrosis in patients with anti-synthetase syndrome-associated interstitial lung disease. *Clin. Rheumatol.* **2023**, *42*, 1917–1929. [CrossRef]
91. Li, T.; Guo, L.; Chen, Z.; Gu, L.; Sun, F.; Tan, X.; Chen, S.; Wang, X.; Ye, S. Pirfenidone in patients with rapidly progressive interstitial lung disease associated with clinically amyopathic dermatomyositis. *Sci. Rep.* **2016**, *6*, 33226. [CrossRef]
92. Sambataro, G.; Vancheri, C.; Sambataro, D. Interstitial Pneumonia with Autoimmune Features (IPAF): Time to redefine the classification criteria. *Expert. Rev. Clin. Immunol.* **2022**, *19*, 131–133. [CrossRef]
93. Sambataro, G.; Sambataro, D.; Torrisi, S.E.; Vancheri, A.; Pavone, M.; Rosso, R.; Schisano, M.; Crimi, C.; Pignataro, F.; Fischer, A.; et al. State of the art in interstitial pneumonia with autoimmune features: A systematic review on retrospective studies and suggestions for further advances. *Eur. Respir. Rev.* **2018**, *27*, 170139. [CrossRef]
94. Graham, J.; Bauer Ventura, I.; Newton, C.A.; Lee, C.; Boctor, N.; Pugashetti, J.V.; Cutting, C.; Joerns, E.; Sandhu, H.; Chung, J.H.; et al. Myositis-specific antibodies identify a distinct interstitial pneumonia with autoimmune features phenotype. *Eur. Respir. J.* **2020**, *16*, 2001205. [CrossRef]
95. Libra, A.; Colaci, M.; Spicuzza, L.; Luca, G.; Fischetti, S.; Pashalidis, G.; Ferrara, C.A.; Ielo, G.; Sambataro, D.; La Rosa, G.; et al. The pattern and progression of "Usual" Interstitial Pneumonia with autoimmune features: Comparison with patients with classic interstitial pneumonia with autoimmune features and idiopathic Pulmonary fibrosis. *J. Clin. Med.* **2024**, *13*, 369. [CrossRef] [PubMed]
96. Yamano, Y.; Kataoka, K.; Takei, R.; Sasano, H.; Yokoyama, T.; Matsuda, T.; Kimura, T.; Mori, Y.; Furukawa, T.; Fukuoka, J.; et al. Interstitial Pneumonia with autoimmune features and histologic usual interstitial pneumonia treated with anti-fibrotic versus immunosuppressive therapy. *Respir. Investig.* **2023**, *61*, 297–305. [CrossRef] [PubMed]

diagnostics MDPI

Review

Vascular Endothelial Damage in COPD: Where Are We Now, Where Will We Go?

Gianluca Screm [1,†], Lucrezia Mondini [1,†], Francesco Salton [1], Paola Confalonieri [1], Liliana Trotta [1], Mariangela Barbieri [1], Antonio Romallo [1], Alessandra Galantino [1], Michael Hughes [2], Selene Lerda [3], Marco Confalonieri [1] and Barbara Ruaro [1,*]

1 Pulmonology Unit, Department of Medical Surgical and Health Sciences, University of Trieste, Hospital of Cattinara, 34149 Trieste, Italy
2 Division of Musculoskeletal and Dermatological Sciences, Faculty of Biology, Medicine and Health, The University of Manchester & Salford Royal NHS Foundation Trust, Manchester M6 8HD, UK
3 Graduate School, University of Milan, 20149 Milan, Italy
* Correspondence: barbara.ruaro@yahoo.it
† These authors contributed equally to this work.

Abstract: Background: Chronic obstructive pulmonary disease (COPD) has higher rates among the general population, so early identification and prevention is the goal. The mechanisms of COPD development have not been completely established, although it has been demonstrated that endothelial dysfunction plays an important role. However, to date, the measurement of endothelial dysfunction is still invasive or not fully established. Nailfold video capillaroscopy (NVC) is a safe, non-invasive diagnostic tool that can be used to easily evaluate the microcirculation and can show any possible endothelial dysfunctions early on. The aim of this review is to evaluate if nailfold microcirculation abnormalities can reflect altered pulmonary vasculature and can predict the risk of cardiovascular comorbidities in COPD patients. Methods: A systematic literature search concerning COPD was performed in electronic databases (PUBMED, UpToDate, Google Scholar, ResearchGate), supplemented with manual research. We searched in these databases for articles published until March 2024. The following search words were searched in the databases in all possible combinations: chronic obstructive pulmonary disease (COPD), endothelial damage, vascular impairment, functional evaluation, capillaroscopy, video capillaroscopy, nailfold video capillaroscopy. Only manuscripts written in English were considered for this review. Papers were included only if they were able to define a relationship between COPD and endothelium dysfunction. Results: The search selected 10 articles, and among these, only three previous reviews were available. Retinal vessel imaging, flow-mediated dilation (FMD), and skin autofluorescence (AF) are reported as the most valuable methods for assessing endothelial dysfunction in COPD patients. Conclusions: It has been assumed that decreased nitric oxide (NO) levels leads to microvascular damage in COPD patients. This finding allows us to assume NVC's potential effectiveness in COPD patients. However, this potential link is based on assumption; further investigations are needed to confirm this hypothesis.

Keywords: chronic obstructive pulmonary disease (COPD); emphysema; nailfold video capillaroscopy (NVC); endothelial dysfunction; nitric oxide (NO)

1. Introduction

Chronic obstructive pulmonary disease (COPD) is a global problem and currently is the third cause of death worldwide, leading to almost 6 million deaths annually around the globe [1–6]. While the management of this disease is a hugely important objective, at the same time, it is quite complicated and challenging because COPD is commonly exhibited in patients with multiple comorbidities [2–8]. COPD often occurs with some relevant cardiovascular comorbidities, such as arterial hypertension, heart failure, acute myocardial infarction, stroke, or peripheral arterial disease. These comorbidities largely contribute to increasing COPD's severity and accelerate the disease's progression, leading to progressive deterioration in quality of life, resulting in poorer clinical outcomes [3–10]. In-depth studies of this pathology and progress in our knowledge of the underlying pathogenetic mechanisms have led to evidence in the literature that support a strict correlation between chronic systemic inflammation, which characterizes COPD patients, and comorbidities that are commonly associated with COPD, especially cardiovascular comorbidities [11–20]. In particular, this generally increased inflammatory state plays an important role in promoting atherosclerotic mechanisms, which are obviously a main cause of cardiovascular events [20–32].

Except for a few unusual situations, COPD can undoubtedly be considered a medical condition that is specifically related to old age. Both prevalence and incidence of this disease are higher within age groups >50 years old, and the average age of patients with COPD is 70 years old [4–8].

COPD is a virtually untreatable disease; it is defined as chronic airflow obstruction, not fully manageable, and with an inevitable progression over time [5–9]. In this chronic lung disease, there is increased inflammation in the airways and pulmonary remodelling, often related to parenchymal destruction and primarily responsible for emphysema; in addition, mucous hyper-secretion and loss of elastic recoil are major events that are related to the development of airflow obstruction [6–10]. These mechanisms underlie the typical symptomatology of COPD, in particular dyspnoea, wheezing, fatigue, cough, and possibly sputum production [6–10].

It is stated that cigarette smoking is the main risk factor of COPD; in fact, the literature has proven that non-smokers develop this disease less frequently [7–10]. Nevertheless, smoking is not the only cause of this disease, and as a matter of fact, only 30% to 50% of smokers develop COPD, so there are other major risk factors that play an almost equally key role related to the disease [5,8]. For example, exposure to high levels of indoor and outdoor air pollution, tuberculosis infection, frequent childhood infections, or childhood asthma all contribute to COPD development [5,8].

On the other hand, there is also a small percentage of people affected by COPD due to genetic abnormalities leading to a deficiency of alpha-1 antitrypsin, which plays a major role in protecting the lungs against proteolytic damage [5–9].

Diagnosis of COPD should be made starting from an analysis of exposure history, considering both smoking and the other main risk factors previously indicated, then moving towards clinical examination and instrumental exams [2–6].

A firm diagnosis is mainly based on spirometry parameters which show airway obstruction, characterized by irreversibility, revealed after the administration of a bronchodilator.

Moreover, the two major spirometry parameters of interest are FEV1 (forced expiratory volume during the first second) and the FEV1/FVC ratio, both of which are decreased in COPD patients.

So, a positive patient history, clinical signs, and a suggestive pulmonary function test are enough for a firm diagnosis. According to GOLD classification, COPD severity is graded according to the predicted FEV1 value as follows: mild (FEV1 > 80%), moderate (FEV1 from 50% to 80%), severe (FEV1 from 30% to 50%) and very severe (FEV1 < 30%) [2–6,32–38].

One the most serious challenges in COPD management is to control, treat, and prevent exacerbations of the disease, defined as severe and acute worsening of the symptoms and/or patient wellbeing. Within the Asia–Pacific region, almost 50% of patients with COPD had exacerbations and almost 20% of them needed hospitalization due to severe symptoms [9].

2. Role of Endothelium in COPD

COPD is a disorder strictly associated with a general increased inflammation status. Almost all researchers have focused on the role of inflammatory cells in COPD and did not give proper attention to endothelial function [38–47].

Nowadays, it is well established that, in COPD, the pulmonary vasculature does not work properly due to endothelial dysfunction, which is associated with disease severity and clinical outcomes; in addition, it is probably also an important pathogenic factor [13–15].

The vascular endothelium is composed of a monolayer of endothelial cells that lies between the luminal surface of the vasculature and the vessel wall [16]. In the lung, the endothelium represents the interface between the blood and other lung compounds, such as the parenchyma and airways, so it has a critical position and must function properly to ensure the balance of the whole system. Like in the lung, the endothelium has other similar critical roles for the whole body [48–58].

Endothelial activity plays a key role in the development of COPD, as shown by Table 1.

Table 1. Endothelial mechanism in COPD. In the lung, the endothelium has a critical position and must function properly to ensure the balance of the whole system. The endothelial activity plays a key role in the development of the disease.

Transendothelial migration (TEM)	TEM involves the migration of neutrophils through the endothelial cell, bypassing the normal paracellular route which involves its junction [52]. This mechanism appears to be upregulated in patients with COPD, with a unique pathway [13,52]. Since NO levels have the purpose of regulating neutrophil–endothelial interactions, lower levels of NO probably play a role in stimulating the TEM route [53]. In addition, ICAM-1, which is actively involved in TEM, seems to be inversely related to lung function and proportionally related to the severity of emphysema on CT scans [13].
Endothelial apoptosis	Apoptosis is a process regulated by the cell in response to various stimuli or triggers, such as DNA damage or oxidative stress [54]. Initial clinical trials suggested pulmonary vascular endothelial cell apoptosis may plays a significant role in emphysema development [13,54,55].
Endothelial cell senescence	In patients with COPD, especially smokers, oxidative stress is obviously more elevated as a result of many altered mechanisms. Consequently, this results in an accelerated senescence, which is related to BPCO development and to increases in systemic inflammation [56]. Interestingly, it has been proven that corticosteroids appear to have beneficial role in protecting pulmonary endothelial cells from cell senescence [56].
VEGF	VEGF is released in response to hypoxia and plays the role of a growth factor for endothelial cells. VEGF expression prevents endothelial cell apoptosis and induces cell proliferation [13]. Levels of VEGF have been shown to be decreased in patients with COPD [13,28].

So, in COPD, the endothelium appears to not work as it is supposed to. Endothelial dysfunction is an abnormal action of the endothelium, mainly characterized by decreased vasodilatation, promotion of the inflammatory state, and loss of its anti-thrombotic functions [13,17,18]. There are three main mechanisms responsible for endothelial dysfunction: oxidative stress, which is highly increased in COPD; systemic inflammation, a major characteristic in the pathogenesis of COPD; and in particular the reduced availability of nitric oxide (NO) [13,17,18]. The latter is one of the most important vasodilator molecules, as it inhibits crucial events in the development of atherosclerosis, and abnormalities in its levels reflect mainly in the microcirculation [17]. A link seems to exist between oxidative stress and NO, since reactive oxygen species (ROS) act as enzymatic inhibitors of NO synthase, leading to decreased levels of NO [57]. Loss of NO-dependent vasodilatation may result in a reduced perfusion, as well as in the maintenance of inflammatory status [58].

Patients with severe COPD or who experience exacerbations of the disease demonstrate a greater rate of circulating inflammatory cells, with upregulation of various molecules, among which are cytokines, chemokines, and acute-phase proteins cells [20]. The literature states that some abnormalities within circulating cells may maintain the inflammatory state and contribute to the different mechanisms indicated in Table 1 [18]. Endothelial dysfunction plays an important role in COPD's severity, mainly contributing to the pathogenesis of atherosclerosis, an important comorbidity in COPD patients that leads to the worst disease outcomes. Furthermore, the degree of endothelial dysfunction seems to have a significant prognostic value in cardiovascular events [18].

Nevertheless, some works in the literature showed evidence of alterations in lung vascularity in patients in the early stages of COPD, suggesting that these alterations may take part in the initial development of the disease, not only as an alteration during more severe COPD stages or during exacerbations [19].

Despite the proven role of the endothelium in the pathogenesis of various diseases and the scientific community's focus on the endothelium, data on the peripheral changes that occur in COPD, peripheral abnormalities, and pulmonary function tests are still missing.

Endothelial dysfunction is present in both the coronary circulation and in other microvascular networks, such as retinal vessels, which may suggest that pulmonary endothelial dysfunction may have an effect on other organs [10,11,17]. In addition, endothelial dysfunction is not restricted to central circulation, but it also affects smaller vessels in the peripheral microcirculation [21]. Moreover, these findings suggest that microvascular assessment may be used as an alternative marker of pulmonary endothelial dysfunction, allowing us to detect cardiovascular risk factors that can worsen existing COPD. This possible link has been previously shown to be potentially associated with induced retinal vasodilation [11]. Over the last few years, new and non-invasive methods such as nailfold video capillaroscopy (NVC) have gained relevance in the assessment of microvascular activity in unconventional diseases. Up to now, there is a lack of literature regarding these themes, despite the potential of these methods in terms of their diagnostic and prognostic relevance.

3. Nailfold Video Capillaroscopy (NVC)

NVC is a non-invasive technique employed to evaluate nailfold microcirculation, allowing us to define characteristics and functionalities of the nailfold capillaries [22]. Nowadays, it plays a well-established role in the assessment of Raynaud's phenomenon (RP) and systemic sclerosis (SSc); starting from 2013, NVC assessment of scleroderma pattern was included in the classification criteria for SSc.

An initial clinical trial showed that patients who suffer from RP without NVC abnormalities, monitored with a strict NVC follow-up, may show new possible NVC alterations that can predict a transition to secondary Raynaud phenomenon (SRP) [51].

Moreover, the importance of NVC is becoming an increasingly valuable asset for other diseases such as antineutrophil cytoplasmic antibody (ANCA)-associated vasculitis (AAV) and non-rheumatic diseases, even though its role in this field is still not well assessed [23].

NVC is a safe technique that can be used to easily evaluate several microvascular parameters: qualitative assessment (global capillaries array and morphology), semiquantitative analysis (presence of giant capillaries, capillary architecture disorganization, microhaemorrhages, neoangiogenesis, and capillary loss), quantitative analysis (estimates of capillary density, avascular areas, diameter of enlarged capillaries), and dynamic parameters (blood flow velocity) [24,25].

New methods, including laser speckle contrast analysis (LASCA), able to measure peripheral blood perfusion are being tested and are showing good results [49].

Patients with SSc present microvascular changes mainly due to endothelial cell dysfunction, characterized by an imbalance between the decrease in vasodilatation levels and the upregulation of vasoconstrictor molecules [26,27]. This impairment of microvascular homeostasis leads to microvascular abnormalities such as increase in capillary wall permeability and progressive microvascular weakness, conditions that appear in NVC as microhaemorrhages and local oedema [26,27]. In SSc patients, the resulting hypoxia leads to an overproduction of VEGF that determines the presence of bizarre and ramified capillaries, reflecting a neoangiogenic process. Nevertheless, in these patients, the chronic hypoxia induces irreversible microangiopathy, seen in NVC as avascular areas [27].

Several studies have demonstrated NVC's usefulness in lung diseases such as idiopathic pulmonary fibrosis (IPF) [29]. The underlying mechanism of endothelial dysfunction of SSc patients shares certain characteristics with COPD, despite the different levels of

VEGF that are upregulated in SSc patients, in contrast to COPD, where they are underregulated [13,28].

The structural assessment of nailfold microvasculature could add important information allowing clinicians to better establish the cardiovascular risk and perhaps even COPD severity. Despite the proven association between retinal vessel diameter and cardiovascular risk, in-depth studies have not been conducted yet [10]. Studying microvascular reactivity is one of the most effective ways to assess endothelial function, as alterations in reactivity could be used as a surrogate marker of its function, but more data are required to confirm this theory [10–12,30].

However, the potential to evaluate the endothelial function of large and small vessels in the periphery with non-invasive methods is promising. Retinal imaging, finger-pulse plethysmography, and especially brachial artery flow-mediated dilation (FMD) have revealed significant results in COPD. The interest for new reliable methods of measuring endothelial functions is growing, even if the prognostic and diagnostic utility is still unclear. There are neither data nor known limitations regarding the possible utility of NVC in COPD, so the goal of this review is to assess any possible use of NVC in COPD patients based on the evidence gained from other diseases.

4. Assessment of Endothelial Dysfunction in COPD Patients

The research work retrieved 17 articles, of which 7 were excluded because they lacked data of interest. In detail, three studies assessed the microvasculature through high-resolution images of retinal blood vessels [10,12,31]; three studies evaluated endothelial functioning by both FMD and nitrate-mediated dilatation (NMD) [32,35,36]; one review evaluated articles with different assessment methods such as optical coherence tomography (OCT), retinal fundus imaging, retinal oximetry, and colour Doppler ultrasonography [37]; one study measured peripheral endothelial dysfunction by using the EndoPat-2000 [33]; one study measured retinal vessel calibres, urine albumin, and myocardial blood flow on MRI as markers of microvascular dysfunction [39]; and lastly, one review included several methods for detecting endothelial dysfunction such as FMD, peripheral arterial tonometry (PAT), flow-mediated skin fluorescence (FMSF), and forearm blood flow after bradykinin infusion (VOP) [34].

4.1. Article Features

The 10 selected articles are listed in the table below, with the aim of the table being to highlight their similarities and differences and especially to define the aim of each paper and the different methods used to assess possible endothelial dysfunction in COPD patients. The main clinical papers are listened in Table 2.

Table 2. List of the selected articles reporting the patient and study characteristics.

Study	Pop (*n*)	Age (Years)	Comorbidities (%)	Smoking History (%)	Assessment Methods	Aim of the Study
Vaes AW et al. [10]	30	64 ± 7	60% were hypertensive, 40% had (pre)diabetes, 77% had dyslipidaemia, 15% had markedly high levels of PCR.	43.3% were current smokers, 53.3% were ex-smokers.	High-resolution images of ocular fundus.	The primary aim was to assess the effects of exercise on retinal microvasculature in COPD patients.
Vaes AW et al. [12]	246	64.4 ± 8.5	68.3% were hypertensive, 24% had (pre)diabetes, 58.1% had dyslipidaemia, 14.2% had markedly high levels of PCR.	28% were current smokers, 65.9% were ex-smokers.	High-resolution images of ocular fundus.	The primary aim was to assess the relationship between retinal vessel widths and pulmonary function tests, disease outcomes, and cardiovascular risk.

Table 2. *Cont.*

Study	Pop (*n*)	Age (Years)	Comorbidities (%)	Smoking History (%)	Assessment Methods	Aim of the Study
Vaes AW et al. [31]	62	64.4 ± 8.4	62.9% were hypertensive, 24.2% had (pre)diabetes, 53.2% had dyslipidaemia, 9.8% had markedly high levels of PCR.	48.4% were current smokers, 51.6% were ex-smokers	Retinal vessel images were used to assess microvascular health, and skin accumulation of AGEs was assessed by skin AF.	Primary aim was to determine the association between skin AF and microvascular health in COPD patients.
Moro et al. [32]	44	76.7	72.7% were hypertensive, 22.7% had diabetes, and higher prevalence of cardiovascular disease than healthy controls.	29.5% were current smokers, 40.9% were ex-smokers.	Evaluated the forearm blood flow induced by both FMD and NMD.	The primary aim was to evaluate the relationship between severity of bronchial obstruction and endothelial dysfunction.
Vaes AW et al. [33]	40	62.8 ± 7.3	62.5% were hypertensive, 32.5% had (pre)diabetes, 85% had dyslipidaemia, 25% had markedly high levels of PCR.	37.5% were current smokers, 62.5% were ex-smokers.	Microvascular endothelial dysfunction was measured with novel EndoPAT.	The primary aim was to expand the knowledge about the relationship between peripheral endothelial function and COPD patients.
Theodorakopoulou MP et al. [34]	–	–	–	–	Review.	The aim was to define the differences in endothelial dysfunction between COPD patients and control cohort.
Rodriguez-Miguelez P et al. [35]	17	56 ± 2	–	–	Brachial artery FMD was used for the assessment of endothelial function; this exam was combined with arterial tonometry.	The primary aim was to investigate the reproducibility of FMD and arterial stiffness in COPD patients.
Ambrosino et al. [36]	–	–	0–86.7% were hypertensive, 0–43.3% had diabetes, 0–56.7% had dyslipidaemia.	70.5–100% had positive smoking history.	Reviewed the data concerning FMD and NMD in COPD patients.	The aim was to bring together and simplify the role of FMD and NMD as markers of endothelial dysfunction and cardiovascular risk assessment.
Vaes AW et al. [37]	–	–	–	–	Reviewed 10 articles that used 4 different assessment methods: OCT, retinal fundus imaging, retinal oximetry, colour Doppler ultrasonography.	The primary aim was to bring together as much information as possible about retinal vessel imaging and COPD patients.
Harris B et al. [39]	3397	64.5 ± 19	Patients without cardiovascular diseases were recruited; 41.6% were hypertensive, 10.4% had diabetes.	14% were current smokers, 33% were ex-smokers.	Microvascular function was assessed by high-resolution images of ocular fundus, measurement of urine albumin and creatin, and myocardial perfusion evaluation.	The primary aim was to define whether systemic microvascular changes are related to lung function and density.

4.2. Association between Microvascular Impairment and COPD

Spirometry was performed to determine the main pulmonary function parameters of interest, and the conclusions are listed in Table 3.

Table 3. Spirometry parameters and main findings.

Articles	FEV1 (% Predicted)	FEV1/FVC Ratio	COPD Severity	Main Findings
Vaes AW et al. [10]	44.6 ± 17.4	36.2 (14.1)	3.3% GOLD I, 33.3% GOLD II, 43.3% GOLD III, 20% GOLD IV	Exercise did not lead to any significant changes in retinal blood vessel diameters. The article suggested that this might be due to inappropriate endothelium vasodilatory response, or inadequate exercise intensity during the test. The study cohort was small in terms of GOLD stage I patients.
Vaes AW et al. [12]	47.3 ± 19.9	35.9 ± 13.0	7.7% GOLD I, 28% GOLD II, 46.7% GOLD III, 17.5% GOLD IV	59% patients showed retinal abnormalities (most common sign was vessel tortuosity). These alterations seem to be related with cardiovascular risk, and above all with hypertension and levels of systemic inflammation. A relationship with lung function parameters was not detected, which is supposedly due to the already highly compromised lung function.
Vaes AW et al. [31]	45.0 ± 20.7	34.8 ± 14.5	8.1% GOLD I, 22.6% GOLD II, 48.4% GOLD III, 21% GOLD IV	Demonstrated an independent association between skin AF and retinal vessel diameter as well as with lower pulmonary functional tests, potentially adding a new tool for the assessment of endothelial function.
Moro et al. [32]	[1.43 L/min]	-	-	The study shows evidence about a predictable inverse relationship between endothelial vasodilatation and bronchial obstruction in COPD. The latter is strictly associated with bronchial inflammation, which in turn is associated with systemic inflammation.
Vaes AW et al. [33]	45.8 ± 17.5	35.8 ± 13.3	7.5% GOLD I, 25% GOLD II, 52.5% GOLD III, 15% GOLD IV	Patients with peripheral endothelial dysfunction demonstrated a lower maximal aerobic capacity, the latter being evaluated by VO2, which is measured using CPET. Nevertheless, using EndoPAT, an association between endothelial function and systemic inflammation was not found.
Theodorakopoulou MP et al. [34]	Values are highly variable depending on the article chosen.			Looking at almost all available functional methods, a significant endothelial impairment was found in COPD patients compared to healthy controls.
Rodriguez-Miguelez P et al. [35]	51.5 ± 3.4	54.8 ± 3.8	23% GOLD I, 53% GOLD II, 24% GOLD III	This study proved for the first time the reproducibility of both FMD in the evaluation of endothelial function and PWV to assess arterial stiffness in COPD patients.
Ambrosino et al. [36]	Vary from 41 to 61.9	Vary from 43 to 59	10% GOLD I, 43.3% GOLD II, 26.7% GOLD III, 20% GOLD IV	It has been shown that patients with a more severe disease, expressed by GOLD classification, had wider gaps in FMD analysis between COPD and healthy controls. Moreover, it has been proven that FMD is strictly associated with endothelial disfunction and cardiovascular risk.
Vaes AW et al. [37]	Values are highly variable depending on the article chosen.			Changes in retinal microvasculature showed higher rates in COPD patients compared to healthy controls.
Harris B et al. [39]	95.7	74.5	-	This article introduced new systemic markers for endothelial damage, all of which are associated with lung function, suggesting that COPD patients show diffuse microvascular alterations.

As stated in Table 3, retinal vessel diameters and, more generally, the functional and structural analysis of retinal vasculature are considered good markers of changes in endothelial functionality, as well as in cardiovascular risk assessment, since an association exists between hypertension and systemic inflammation [12,37]. It is known that COPD patients have significantly greater rates of wider retinal vessels compared to healthy controls [39]. Moreover, Harris B et al. described a potential link between retinal vessel diameters and decreased lung function (assessed by FEV1), and they defined a possible association between retinal vascular calibre and brachial artery FMD [39]; these findings were not confirmed by other studies considered in our review [10,12]. FMD is a widely used method aimed to evaluate endothelial function which seems to be correlated with decreased performance in pulmonary function tests and COPD severity [36,39].

COPD patients have a linear relationship between aerobic capacity and disease mortality. In addition, some articles showed that both retinal vessel evaluation and FMD could have a role in the assessment of endothelial response to exercise [10,33]. In particular, Vaes et al. did not detect any significant changes in vessel diameters during exercise in COPD patients, potentially due to an inadequate endothelium-mediated vasodilatory response [10]. On the other hand, endothelial function was shown to be strictly related to aerobic exercise, since it was negatively associated with VO2 values [33]. Overall, a more impaired FMD is correlated with worsening bronchial obstruction in COPD [32,34,36].

Other methods have been evaluated by Vaes et al. in 2020 for the assessment of endothelial dysfunction in COPD patients. Skin AF showed associations with retinal vessel diameters and reductions in the predicted FEV1% and FEV1/FVC. Moreover, this study also suggested that skin AF is also related to cardiovascular risk [31].

There is not enough evidence towards EndoPAT as a useful tool in COPD patients. Nevertheless, while some studies did not detect an adequate association between EndoPAT and FMD [17,33], it is proven to provide a potential value for predicting cardiovascular events and early atherosclerosis [17]. Moreover, this method can easily detect peripheral finger endothelial dysfunction, which is related to metabolic and cardiovascular risk factors, playing a role as a potential tool in predicting non-obstructive coronary atherosclerosis [17].

Asset endothelial function through the evaluation of peripheral fingers could have potential for cardiovascular risk assessment in COPD patients, considering that cardiovascular comorbidities are, aside from exacerbations, the first cause of poorer outcomes in COPD patients. Nailfold video capillaroscopy has never been used in this disease, and no evaluation of the relationship between capillaroscopic alteration and cardiovascular risk, exacerbations risk, or COPD severity has ever been conducted. Nevertheless, NVC is starting to see potential use in chronic pulmonary diseases such as idiopathic lung fibrosis, where the main alterations found were higher rates of neoangiogenesis and lower capillary density [29].

5. Potential Role of NVC in COPD

The link between nailfold microvascular disarrangement, which is seen in NVC (Figure 1), and endothelial dysfunction, primarily assessed via FMD, is yet to be clarified in COPD. Few studies have evaluated this possible relationship [40–45].

Peripheral microvascular endothelial dysfunction has shown to play a role in the development of RP, where decreasing levels of NO lead to vasospasms reflected by nailfold microcirculation, suggesting a close relationship between endothelial dysfunction (defined as a decrease in NO availability) and capillaroscopic abnormalities that are usually present in SRP patients [40]. In patients with secondary RP, the reduction in FMD is reported to be associated with nailfold microcirculation impairment. Moreover, an inverse association has been found between FMD and microangiopathy evolution score, corroborating the theory explained before; so, NVC has been proposed has a useful tool for the evaluation of endothelial dysfunction [44]. The same inverse relation was detected by Rollando et al. in 2010 in a study which enrolled SSc patients within asymptomatic cardiovascular disease and found a reduction in FMD values in early NVC microangiopathy patterns and lower FMD values according to late NVC microangiopathy patterns [41]. FMD is a marker for vascular function and NO release, both of which are demonstrated to be significantly reduced in COPD and SSc patients [13,28,41] and which seem to correlate with peripheral microvascular injury [40–43,45]. FMD, together with NVC, may play an important role in the assessment of vascular damage in SSc patients and may also predict future vascular complications, allowing clinicians to better stratify the cardiovascular risk in those patients [43,45].

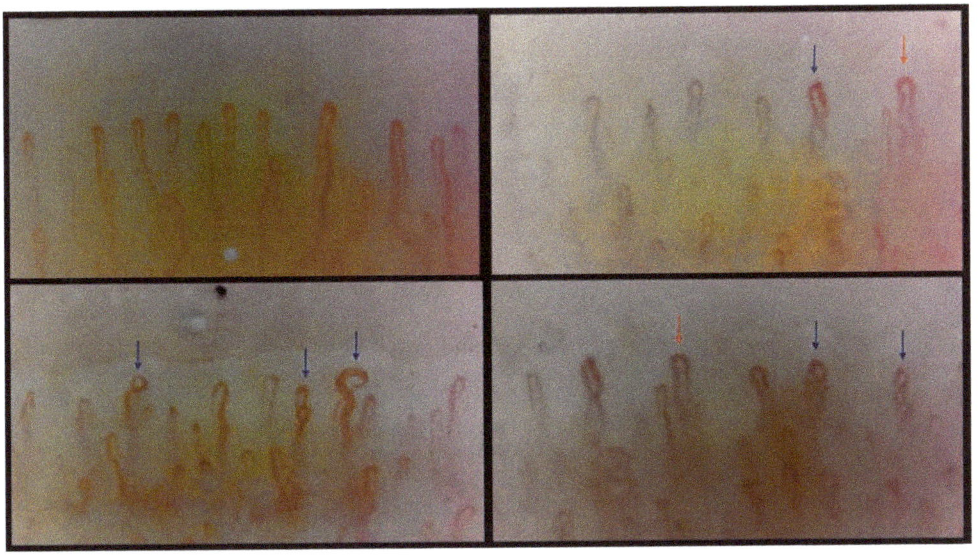

Figure 1. Capillaroscopic images recorded at our centre in a 78-year-old patient with moderate COPD, emphysematous phenotype: capillary density is preserved and capillaries with tortuous and winding morphology (red arrow) and capillaries with multiple and single cross-overs (blue arrow) are associated with normoconformed capillaries.

Controversially, a possible link between NVC and retinal imaging has not been found yet, despite few works explaining that retinal abnormalities may reflect vascular damage in SSc patients [46,47]. In 2020, Jakhar et al. found that retinal abnormalities were mostly associated with greater NVC alterations. However, no statistically significant data were found, and clinical relevance of this possible relationship must be analysed further to be confirmed [48].

So, taking everything into account, the association between NVC and FMD has been established in patients with SSc and SRP. The main cause of microvascular damage in those patients has been shown to be lack of NO availability, a main marker of endothelial damage in COPD patients [13,28,41]. All of these findings lead to the speculation that NVC may play a role in COPD patients, particularly together with FMD or other methods that assess endothelial damage. Moreover, clinical applications of FMD are difficult to perform due to the strict protocols involved, the expensive equipment, and the need for highly qualified operators and a controlled environment [34]. Nevertheless, one study has shown that FMD provides evidence for the reproducibility of endothelial function assessment in COPD patients [35]. However, this evidence is not supported by other studies and needs further investigations. On the other hand, NVC is nowadays a non-invasive, easily reproducible, non-operator-dependent tool, and thanks to initial evidence of its possible overlap with FMD values, its possible role in evaluating and studying COPD patients should be assessed.

NVC's enhanced ease-of-use are also due to new advances in digital imaging technology which have led to the development of a validated, fully automated algorithm, AUTOCAPI, with the aim of making NVC more easily reproducible and accurate [50].

We are aware that all these hypotheses strike us more as interpretations and suppositions than as scientifically well-founded evidence. To date, systematic studies characterizing the nailfold microvasculature in patients with COPD are missing, and no relevant articles were found concerning NVC in those patients.

6. Conclusions

In conclusion, this review pointed out that the COPD is not just a lung disease, but is strictly connected with systemic involvement. We can therefore consider COPD as a systemic disease. This could seriously open the door for new, possibly effective prospective tools, such as nailfold video capillaroscopy, that aim to evaluate different parameters concerning the microcirculation and can give more meaningful information about systemic endothelial function. This could add important medical evidence to help better understand the role of endothelial dysfunction as a primary factor leading to COPD development and progression, and at the same time, encourage the development of cardiovascular comorbidities and/or COPD exacerbations, which are the main factors of the worst patient outcomes.

7. Future Developments

More effective early diagnostic methods are needed for COPD, with the perspective that such tools will lead to early diagnosis and be able to predict the risk of disease exacerbations or cardiovascular comorbidities.

A further area of development is to consider NVC as a possible option available to better stratify COPD patients, since NVC is providing to be a useful tool for many diseases. Endothelial damage is proven to be a significant event during COPD, and therefore, greater understanding of its assessment can reveal novel targets in the management and prevention of this disease.

Author Contributions: Conceptualization, B.R.; methodology, B.R., G.S., L.M., F.S., P.C., L.T., M.B., A.R., A.G. and M.H., writing—original draft preparation, B.R., M.H. and S.L.; writing—review and editing, B.R. and M.C.; supervision, B.R. and M.C. All authors have read and agreed to the published version of the manuscript.

Funding: This research received no external funding.

Data Availability Statement: Data sharing not applicable.

Acknowledgments: The authors thank Chiesi S.P.A., Lombardi Adriano (from the University of Niccolò Cusano) for the support with the graphic acquisition and reproduction and Annalisa Bertolini for the support in supervising the writing of this manuscript.

Conflicts of Interest: The authors declare no conflicts of interest.

Abbreviations

COPD: chronic obstructive pulmonary disease; NVC: nailfold video capillaroscopy; FMD: flow-mediated dilation; FEV1: forced expiratory volume during the first second; TEM: transendothelial migration; NO: nitric oxide; RP: Raynaud's phenomenon; SRP: secondary Raynaud's phenomenon; SSc: systemic sclerosis; VEGF: vascular endothelial growth factor; IPF: idiopathic pulmonary fibrosis; NMD: nitrate-mediated dilatation; OCT: optical coherence tomography; FMSF: flow-mediated skin fluorescence; VOP: forearm blood flow after bradykinin infusion; PAT: peripheral arterial tonometry; CPET: cardiopulmonary exercise test; PWV: pulse wave velocity; AF: skin autofluorescence; AGEs: advanced glycation end-products.

References

1. Celli, B.; Fabbri, L.; Criner, G.; Martinez, F.J.; Mannino, D.; Vogelmeier, C.; Montes de Oca, M.; Papi, A.; Sin, D.D.; Han, M.K.; et al. Definition and Nomenclature of Chronic Obstructive Pulmonary Disease: Time for Its Revision. *Am. J. Respir. Crit. Care Med.* **2022**, *206*, 1317–1325. [CrossRef] [PubMed]
2. Confalonieri, M.; Braga, L.; Salton, F.; Ruaro, B.; Confalonieri, P. Chronic Obstructive Pulmonary Disease Definition: Is It Time to Incorporate the Concept of Failure of Lung Regeneration? *Am. J. Respir. Crit. Care Med.* **2023**, *207*, 366–367. [CrossRef] [PubMed]
3. Chen, W.; Thomas, J.; Sadatsafavi, M.; FitzGerald, J.M. Risk of cardiovascular comorbidity in patients with chronic obstructive pulmonary disease: A systematic review and meta-analysis. *Lancet Respir. Med.* **2015**, *3*, 631–639. [CrossRef] [PubMed]
4. Cavaillès, A.; Brinchault-Rabin, G.; Dixmier, A.; Goupil, F.; Gut-Gobert, C.; Marchand-Adam, S.; Meurice, J.C.; Morel, H.; Person-Tacnet, C.; Leroyer, C.; et al. Comorbidities of COPD. *Eur. Respir. Rev.* **2013**, *22*, 454–475. [CrossRef]
5. Agustí, A.; Celli, B.R.; Criner, G.J.; Halpin, D.; Anzueto, A.; Barnes, P.; Bourbeau, J.; Han, M.K.; Martinez, F.J.; Montes de Oca, M.; et al. Global Initiative for Chronic Obstructive Lung Disease 2023 Report: GOLD Executive Summary. *Eur. Respir. J.* **2023**, *61*, 2300239. [CrossRef] [PubMed]
6. Petty, T.L. The history of COPD. *Int. J. Chron. Obstruct. Pulmon. Dis.* **2006**, *1*, 3–14. [CrossRef] [PubMed]
7. Forey, B.A.; Thornton, A.J.; Lee, P.N. Systematic review with meta-analysis of the epidemiological evidence relating smoking to COPD, chronic bronchitis and emphysema. *BMC Pulm. Med.* **2011**, *11*, 36. [CrossRef]
8. Bagdonas, E.; Raudoniute, J.; Bruzauskaite, I.; Aldonyte, R. Novel aspects of pathogenesis and regeneration mechanisms in COPD. *Int. J. Chron. Obstruct. Pulmon. Dis.* **2015**, *10*, 995–1013.
9. Lim, S.; Lam, D.C.; Muttalif, A.R.; Yunus, F.; Wongtim, S.; Lan le, T.T.; Shetty, V.; Chu, R.; Zheng, J.; Perng, D.W.; et al. Impact of chronic obstructive pulmonary disease (COPD) in the Asia-Pacific region: The EPIC Asia population-based survey. *Asia Pac. Fam. Med.* **2015**, *14*, 4. [CrossRef]
10. Vaes, A.W.; Spruit, M.A.; Van Keer, K.; Barbosa-Breda, J.; Wouters, E.F.M.; Franssen, F.M.E.; Theunis, J.; De Boever, P. Structural analysis of retinal blood vessels in patients with COPD during a pulmonary rehabilitation program. *Sci. Rep.* **2020**, *10*, 31. [CrossRef]
11. Lim, M.; Sasongko, M.B.; Ikram, M.K.; Lamoureux, E.; Wang, J.J.; Wong, T.Y.; Cheung, C.Y. Systemic associations of dynamic retinal vessel analysis: A review of current literature. *Microcirculation* **2013**, *20*, 257–268. [CrossRef] [PubMed]
12. Vaes, A.W.; Spruit, M.A.; Goswami, N.; Theunis, J.; Franssen, F.M.E.; De Boever, P. Analysis of retinal blood vessel diameters in patients with COPD undergoing a pulmonary rehabilitation program. *Microvasc. Res.* **2021**, *139*, 104238. [CrossRef]
13. Green, C.E.; Turner, A.M. The role of the endothelium in asthma and chronic obstructive pulmonary disease (COPD). *Respir. Res.* **2017**, *18*, 20. [CrossRef] [PubMed]
14. Skurikhin, E.; Pershina, O.; Zhukova, M.; Widera, D.; Pan, E.; Pakhomova, A.; Krupin, V.; Ermakova, N.; Skurikhina, V.; Sandrikina, L.; et al. Spiperone Stimulates Regeneration in Pulmonary Endothelium Damaged by Cigarette Smoke and Lipopolysaccharide. *Int. J. Chron. Obstruct Pulmon Dis.* **2021**, *16*, 3575–3591. [CrossRef] [PubMed]
15. Karnati, S.; Seimetz, M.; Kleefeldt, F.; Sonawane, A.; Madhusudhan, T.; Bachhuka, A.; Kosanovic, D.; Weissmann, N.; Krüger, K.; Ergün, S. Chronic Obstructive Pulmonary Disease and the Cardiovascular System: Vascular Repair and Regeneration as a Therapeutic Target. *Front. Cardiovasc. Med.* **2021**, *8*, 649512. [CrossRef] [PubMed]
16. Goldenberg, N.M.; Kuebler, W.M. Endothelial cell regulation of pulmonary vascular tone, inflammation, and coagulation. *Compr. Physiol.* **2015**, *5*, 531–559.
17. Flammer, A.J.; Anderson, T.; Celermajer, D.S.; Creager, M.A.; Deanfield, J.; Ganz, P.; Hamburg, N.M.; Lüscher, T.F.; Shechter, M.; Taddei, S.; et al. The assessment of endothelial function: From research into clinical practice. *Circulation* **2012**, *126*, 753–767. [CrossRef] [PubMed]
18. Endemann, D.H.; Schiffrin, E.L. Endothelial dysfunction. *J. Am. Soc. Nephrol.* **2004**, *15*, 1983–1992. [CrossRef]
19. Peinado, V.I.; Barbera, J.A.; Ramirez, J.; Gomez, F.P.; Roca, J.; Jover, L.; Gimferrer, J.M.; Rodriguez-Roisin, R. Endothelial dysfunction in pulmonary arteries of patients with mild COPD. *Am. J. Physiol.* **1998**, *274 Pt. 1*, L908–L913. [CrossRef]
20. Barnes, P.J.; Celli, B.R. Systemic manifestations and comorbidities of COPD. *Eur. Respir. J.* **2009**, *33*, 1165–1185. [CrossRef]
21. Anderson, T.J.; Gerhard, M.D.; Meredith, I.T.; Charbonneau, F.; Delagrange, D.; Creager, M.A.; Selwyn, A.P.; Ganz, P. Systemic nature of endothelial dysfunction in atherosclerosis. *Am. J. Cardiol.* **1995**, *75*, 71B–74B. [CrossRef] [PubMed]
22. Tavakol, M.E.; Fatemi, A.; Karbalaie, A. Nailfold Capillaroscopy in Rheumatic Diseases: Which Parameters Should Be Evaluated? *Biomed. Res. Int.* **2015**, *2015*, 974530.
23. Screm, G.; Mondini, L.; Confalonieri, P.; Salton, F.; Trotta, L.; Barbieri, M.; Mari, M.; Reccardini, N.; Della Porta, R.; Kodric, M.; et al. Nailfold Capillaroscopy Analysis Can Add a New Perspective to Biomarker Research in Antineutrophil Cytoplasmic Antibody-Associated Vasculitis. *Diagnostics* **2024**, *14*, 254. [CrossRef] [PubMed]
24. Cutolo, M.; Sulli, A.; Smith, V. How to perform and interpret capillaroscopy. *Best Pract. Res. Clin. Rheumatol.* **2013**, *27*, 237–248. [CrossRef] [PubMed]
25. Bernero, E.; Sulli, A.; Ferrari, G.; Ravera, F.; Pizzorni, C.; Ruaro, B.; Zampogna, G.; Alessandri, E.; Cutolo, M. Prospective capillaroscopy-based study on transition from primary to secondary Raynaud's phenomenon: Preliminary results. *Reumatismo* **2013**, *65*, 186–191. [CrossRef] [PubMed]

26. Bruni, C.; Frech, T.; Manetti, M.; Rossi, F.W.; Furst, D.E.; De Paulis, A.; Rivellese, F.; Guiducci, S.; Matucci-Cerinic, M.; Bellando-Randone, S. Vascular Leaking, a Pivotal and Early Pathogenetic Event in Systemic Sclerosis: Should the door be closed? *Front. Immunol.* **2018**, *9*, 2045–2048. [CrossRef]
27. Mansueto, N.; Rotondo, C.; Corrado, A.; Cantatore, F.P. Nailfold capillaroscopy: A comprehensive review on common findings and clinical usefulness in non-rheumatic disease. *J. Med. Investig.* **2021**, *68*, 6–14. [CrossRef]
28. Guo, X.; Lin, Y.; Lin, Y.; Zhong, Y.; Yu, H.; Huang, Y.; Yang, J.; Cai, Y.; Liu, F.; Li, Y.; et al. PM2.5 induces pulmonary microvascular injury in COPD via METTL16-mediated m6A modification. *Environ. Pollut.* **2022**, *303*, 119115. [CrossRef] [PubMed]
29. Corrado, A.; Carpagnano, G.E.; Gaudio, A.; Foschino-Barbaro, M.P.; Cantatore, F.P. Nailfold capillaroscopic findings in systemic sclerosis related lung fibrosis and in idiopathic lung fibrosis. *Jt. Bone Spine* **2010**, *77*, 570–574. [CrossRef]
30. Heitmar, R.; Summers, R.J. Assessing vascular function using dynamic retinal diameter measurements: A new insight on the endothelium. *Thromb. Haemost.* **2012**, *107*, 1019–1102. [CrossRef]
31. Vaes, A.W.; Spruit, M.A.; Reynaert, N.L.; Franssen, F.M.E.; Wouters, E.F.M.; Theunis, J.; De Boever, P. Skin auto-fluorescence as a measure of advanced glycation end-products is associated with microvascular health in patients with COPD. *Microvasc. Res.* **2020**, *132*, 104053. [CrossRef] [PubMed]
32. Moro, L.; Pedone, C.; Scarlata, S.; Malafarina, V.; Fimognari, F.; Antonelli-Incalzi, R. Endothelial dysfunction in chronic obstructive pulmonary disease. *Angiology* **2008**, *59*, 357–364. [CrossRef] [PubMed]
33. Vaes, A.W.; Spruit, M.A.; Theunis, J.; Wouters, E.F.M.; De Boever, P. Peripheral endothelial function is positively associated with maximal aerobic capacity in patients with chronic obstructive pulmonary disease. *Respir. Med.* **2018**, *142*, 41–47. [CrossRef] [PubMed]
34. Theodorakopoulou, M.P.; Alexandrou, M.E.; Bakaloudi, D.R.; Pitsiou, G.; Stanopoulos, I.; Kontakiotis, T.; Boutou, A.K. Endothelial dysfunction in COPD: A systematic review and meta-analysis of studies using different functional assessment methods. *ERJ Open Res.* **2021**, *7*, 00983–2020. [CrossRef] [PubMed]
35. Rodriguez-Miguelez, P.; Seigler, N.; Bass, L.; Dillard, T.A.; Harris, R.A. Assessments of endothelial function and arterial stiffness are reproducible in patients with COPD. *Int. J. Chron. Obstruct Pulmon Dis.* **2015**, *10*, 1977–1986. [PubMed]
36. Ambrosino, P.; Lupoli, R.; Iervolino, S.; De Felice, A.; Pappone, N.; Storino, A.; Di Minno, M.N.D. Clinical assessment of endothelial function in patients with chronic obstructive pulmonary disease: A systematic review with meta-analysis. *Intern. Emerg. Med.* **2017**, *12*, 877–885. [CrossRef]
37. Vaes, A.W.; Spruit, M.A.; Theunis, J.; Goswami, N.; Vanfleteren, L.E.; Franssen, F.M.E.; Wouters, E.F.M.; De Boever, P. Looking into the eye of patients with chronic obstructive pulmonary disease: An opportunity for better microvascular profiling of these complex patients. *Acta Ophthalmol.* **2018**, *96*, 539–549. [CrossRef] [PubMed]
38. Laisure, M.; Covill, N.; Ostroff, M.L.; Ostroff, J.L. Summarizing the 2021 Updated GOLD Guidelines for COPD. *US Pharm.* **2021**, *46*, 30–35.
39. Harris, B.; Klein, R.; Jerosch-Herold, M.; Hoffman, E.A.; Ahmed, F.S.; Jacobs, D.R.; Klein, B.E., Jr.; Wong, T.Y.; Lima, J.A.; Cotch, M.F.; et al. The association of systemic microvascular changes with lung function and lung density: A cross-sectional study. *PLoS ONE* **2012**, *7*, e50224. [CrossRef]
40. Taher, R.; Sara, J.D.; Toya, T.; Shepherd, R.; Moder, K.; Lerman, L.O.; Lerman, A. Secondary Raynaud's phenomenon is associated with microvascular peripheral endothelial dysfunction. *Microvasc. Res.* **2020**, *132*, 104040. [CrossRef]
41. Rollando, D.; Bezante, G.P.; Sulli, A.; Balbi, M.; Panico, N.; Pizzorni, C.; Negrini, S.; Brunelli, C.; Barsotti, A.; Cutolo, M.; et al. Brachial artery endothelial-dependent flow-mediated dilation identifies early-stage endothelial dysfunction in systemic sclerosis and correlates with nailfold microvascular impairment. *J. Rheumatol.* **2010**, *37*, 1168–1173. [CrossRef] [PubMed]
42. Silva, I.; Teixeira, A.; Oliveira, J.; Almeida, I.; Almeida, R.; Águas, A.; Vasconcelos, C. Endothelial Dysfunction and Nailfold Videocapillaroscopy Pattern as Predictors of Digital Ulcers in Systemic Sclerosis: A Cohort Study and Review of the Literature. *Clin. Rev. Allerg. Immunol.* **2015**, *49*, 240–252. [CrossRef] [PubMed]
43. Silva, I.; Loureiro, T.; Teixeira, A.; Almeida, I.; Mansilha, A.; Vasconcelos, C.; Almeida, R. Digital ulcers in systemic sclerosis: Role of flow-mediated dilatation and capillaroscopy as risk assessment tools. *Eur. J. Dermatol.* **2015**, *25*, 444–451. [CrossRef] [PubMed]
44. Le, J.H.; Cho, K.I. Association between endothelial function and microvascular changes in patients with secondary Raynaud's phenomenon. *Clin Rheumatol.* **2014**, *33*, 1627–1633. [CrossRef] [PubMed]
45. Corrado, A.; Mansueto, N.; Correale, M.; Rella, V.; Tricarico, L.; Altomare, A.; Brunetti, N.D.; Cantatore, F.P.; Rotondo, C. Flow Mediated Dilation in Systemic Sclerosis: Association with clinical findings, capillaroscopic patterns and endothelial circulating markers. *Vascul. Pharmacol.* **2024**, *154*, 107252. [CrossRef] [PubMed]
46. Ushiyama, O.; Ushiyama, K.; Yamada, T.; Koarada, S.; Tada, Y.; Suzuki, N.; Ohta, A.; Nagasawa, K. Retinal findings in systemic sclerosis: A comparison with nailfold capillaroscopic patterns. *Ann. Rheum. Dis.* **2003**, *62*, 204–207. [CrossRef] [PubMed]
47. Shenavandeh, S.; Afarid, M.; Hasanaghaei, T.; Nazarinia, M.A. Prevalence of retinal changes in patients with systemic sclerosis: The association between retinal vascular changes and nailfold capillaroscopic findings. *Reumatologia* **2021**, *59*, 27–34. [CrossRef] [PubMed]
48. Jakhar, D.; Grover, C.; Singal, A.; Das, G.K. Nailfold Capillaroscopy and Retinal Findings in Patients with Systemic Sclerosis: Is There an Association? *Indian Dermatol. Online J.* **2020**, *11*, 382–386. [PubMed]

49. Cutolo, M.; Vanhaecke, A.; Ruaro, B.; Deschepper, E.; Ickinger, C.; Melsens, K.; Piette, Y.; Trombetta, A.C.; De Keyser, F.; Smith, V.; et al. Is laser speckle contrast analysis (LASCA) the new kid on the block in systemic sclerosis? A systematic literature review and pilot study to evaluate reliability of LASCA to measure peripheral blood perfusion in scleroderma patients. *Autoimmun. Rev.* **2018**, *17*, 775–780. [CrossRef]
50. Cutolo, M.; Trombetta, A.C.; Melsens, K.; Pizzorni, C.; Sulli, A.; Ruaro, B.; Paolino, S.; Deschepper, E.; Smith, V. Automated assessment of absolute nailfold capillary number on videocapillaroscopic images: Proof of principle and validation in systemic sclerosis. *Microcirculation* **2018**, *25*, e12447. [CrossRef]
51. Mondini, L.; Confalonieri, P.; Pozzan, R.; Ruggero, L.; Trotta, L.; Lerda, S.; Hughes, M.; Bellan, M.; Confalonieri, M.; Ruaro, B.; et al. Microvascular Alteration in COVID-19 Documented by Nailfold Capillaroscopy. *Diagnostics* **2023**, *13*, 1905. [CrossRef] [PubMed]
52. Gane, J.; Stockley, R. Mechanisms of neutrophil transmigration across the vascular endothelium in COPD. *Thorax* **2012**, *67*, 553–561. [CrossRef] [PubMed]
53. Burns, A.R.; Zheng, Z.; Soubra, S.H.; Chen, J.; Rumbaut, R.E. Transendothelial flow inhibits neutrophil transmigration through a nitric oxide-dependent mechanism: Potential role for cleft shear stress. *Am. J. Physiol. Heart Circ. Physiol.* **2007**, *293*, H2904–H2910. [CrossRef] [PubMed]
54. Song, Q.; Chen, P.; Liu, X.-M. The role of cigarette smoke-induced pulmonary vascular endothelial cell apoptosis in COPD. *Respir. Res.* **2021**, *22*, 39. [CrossRef] [PubMed]
55. Salton, F.; Ruaro, B.; Confalonieri, P.; Confalonieri, M. Epithelial-Mesenchymal Transition: A Major Pathogenic Driver in Idiopathic Pulmonary Fibrosis? *Medicina* **2020**, *56*, 608. [CrossRef] [PubMed]
56. Paschalaki, K.; Rossios, C.; Pericleous, C.; MacLeod, M.; Rothery, S.; Donaldson, G.C.; A Wedzicha, J.; Gorgoulis, V.; Randi, A.M.; Barnes, P.J. Inhaled corticosteroids reduce senescence in endothelial progenitor cells from patients with COPD. *Thorax* **2022**, *77*, 616–620. [CrossRef] [PubMed]
57. Maricić, L.; Vceva, A.; Visević, R.; Vcev, A.; Milić, M.; Serić, V.; Fijacko, V. Assessment of endothelial dysfunction by measuring von Willebrand factor and exhaled nitric oxide in patients with chronic obstructive pulmonary disease. *Coll. Antropol.* **2013**, *37*, 1153–1160.
58. A Hatoum, O.; Binion, D.G.; Otterson, M.F.; Gutterman, D.D. Acquired microvascular dysfunction in inflammatory bowel disease: Loss of nitric oxide-mediated vasodilation. *Gastroenterology* **2003**, *125*, 58–69. [CrossRef]

diagnostics MDPI

Brief Report

Oscillometry Assesses Small Airway Disease and Reveals Peripheral Lung Pathology in Early Pulmonary Fibrosis: A Cross-Sectional Study

Athena Gogali [1,*,†], Georgia Gkrepi [1,†], Christos Kyriakopoulos [1], Konstantinos Tatsis [1], Konstantinos Katsoulis [2], Chara Tselepi [1] and Konstantinos Kostikas [1]

[1] Respiratory Medicine Department, University of Ioannina, 45110 Ioannina, Greece; geogkrepi@gmail.com (G.G.); ckyriako@yahoo.gr (C.K.); konstantatsis@gmail.com (K.T.); haratselepi@gmail.com (C.T.); ktkostikas@gmail.com (K.K.)
[2] Respiratory Medicine Department, 424 Army General Hospital, 56429 Thessaloniki, Greece; kostascmax@gmail.com
* Correspondence: athenagogali@yahoo.com
† These authors contributed equally to this work.

Abstract: Background/Objectives: Small airway disease/dysfunction (SAD) is crucial in obstructive airway diseases but is less investigated in interstitial lung disease (ILD). There are only a few physiological studies investigating SAD in the context of pulmonary fibrosis. Oscillometry is a simple technique that assesses SAD with minimal patient effort. In this study, we investigated the role of oscillometry in patients with mild pulmonary fibrosis without evident obstructive disorder, focusing on small airways. **Methods:** Oscillometry and pulmonary function test (PFT) data of consecutive patients newly diagnosed with pulmonary fibrosis of unknown etiology in a university hospital ILD clinic were collected and analyzed. **Results:** Data from 34 patients with mild pulmonary fibrosis were collected in 6 months. Disease severity, as evaluated by FVC, presented strong correlations with the oscillometry parameters: resistance (R5: $r = -0.588$, $p < 0.001$), reactance (X5: $r = 0.671$, $p < 0.001$), resonant frequency (Fres: $r = -0.562$, $p = 0.001$), and the area of reactance (AX: $r = -0.515$, $p = 0.002$). The oscillometry parameter R5-19-expressing was abnormal in 27% of patients, correlated with FEF25-75% ($r = -0.370$, $p = 0.021$) and was a predictor of a FEF25-75% < 60% pred. with AUC 0.738 (95%Cl 0.519–0.956). R5-19 correlated with FVC ($r = -0.481$, $p = 0.004$) and was the only SAD parameter that correlated with the composite physiologic index (CPI, $r = 0.338$, $p = 0.04$), while FEF 25-75% and RV/TLC% did not. **Conclusions:** Oscillometry is an easy to perform technique that may reveal early mechanical alterations caused by pulmonary fibrosis. Peripheral resistance, as expressed by R5-19, which identifies small airway dysfunction as a marker of peripheral lung pathology, may be complementary to pulmonary function testing and may also have prognostic implications for ILD patients.

Keywords: interstitial lung disease; small airway dysfunction; oscillometry; R5-19

1. Introduction

Small airways, airways with a diameter of 2 mm or less, are often called the "silent zone" of the lung, mainly because they are affected early in the course of disease before any symptoms or spirometric changes occur [1]. Although small airway disease/dysfunction (SAD) is a fundamental element of chronic obstructive pulmonary disease (COPD) and asthma, this factor is less investigated in interstitial lung disease (ILD), where findings are mainly focused on interstitial and alveolar abnormalities [2]. Recent studies combining pathology and HRCT observations have demonstrated that the perception that idiopathic pulmonary fibrosis (IPF) spares the airways is not quite right, as regions of minimal fibrosis are associated with a significant reduction in the number of terminal bronchioles, while the remainder exhibit wall thickening and lumen distortion [3]. There are only a few physiological studies that have investigated SAD in the context of pulmonary fibrosis. A small number of previous studies have shown that spirometry-defined SAD is a feature of advanced fibrosis and correlates with decreased survival, especially when it is combined with obstructive disease [4,5].

Oscillometry is a simple technique that assesses small airway dysfunction (SAD) with minimal patient effort [6]. Forced oscillation technique (FOT) assesses the mechanical properties of the lung and evaluates pulmonary obstruction by measuring the respiratory system's response to small-pressure stimuli generated by a loudspeaker during normal breathing [7]. Oscillometry provides additional valuable information about lung periphery, the first area affected in pulmonary fibrosis, including the compliance of the lung parenchyma and the small airways.

Therefore, in this pilot study, we investigated whether oscillometry provides complementary data to pulmonary function tests (PFTs) in patients with mild pulmonary fibrosis without evident obstructive disorder, focusing on small airways.

2. Materials and Methods

In this prospective observational pilot study, we evaluated consecutive patients with newly diagnosed idiopathic fibrosing ILD in our tertiary ILD clinic, which is a referral university hospital clinic, in a 6-month period. All participants had a predominantly fibrotic pattern on high-resolution computerized tomography (HRCT) of less than 20% extent. HRCT scans were scored according to the easily applicable limited/extensive staging system reported by Goh et al. [8]. Two experienced clinicians (A.G. and K.K.) meticulously evaluated the study population and excluded individuals with ILDs of known etiology (e.g., in the context of collagen vascular disease or hypersensitivity pneumonitis) and also patients with asthma, COPD, or radiological signs of airway obstruction such as mosaic attenuation, air trapping, emphysema, and centrilobular nodules. This study was performed in accordance with the recommendations of the Declaration of Helsinki, the International Conference of Harmonisation—Good Clinical Practice (ICH-GCP) Guidelines, the EU-Directive 2001/20, and all national requirements and was approved by the Scientific Council of our hospital. Written informed consent was obtained from all participants prior to their inclusion in this study.

All patients underwent FOT (RESMON pro Full-V3, MCG Diagnostics, Saint Paul, MN, USA) and full pulmonary function tests (PFT, Vyntus Body Plethysmograph, Vyaire Inc., Mettawa, IL, USA). The forced expiratory flow rate between 25 and 75% of forced vital capacity (FVC) (FEF25-75%) and the ratio of residual volume (RV) to total lung capacity (TLC) (RV/TLC) were assessed as SAD indices. The oscillometry parameters resistance (R5), reactance (X5), resonant frequency (Fres), and the area of reactance (AX) were evaluated, while resistance heterogeneity measured between 5 and 19 Hz (R5-19) assessed SAD. Specifically, R5-19 was considered the primary oscillometry parameter indicative of SAD. We also calculated the composite physiologic index (CPI), an index that reflects the extent of pulmonary fibrosis better than individual pulmonary function tests and is associated with prognosis including mortality [9].

Descriptive statistics were presented as mean (SD), and correlations were performed with Spearman's rank correlation coefficient. Correlations between FOT and PFT were investigated. Statistical analysis was performed with SPSS (Version 22; IBM, Chicago, IL, USA) and GraphPad Prism v7 software (GraphPad, Inc., San Diego, CA, USA). p-values < 0.05 were considered statistically significant.

3. Results

Data from a total of 34 patients newly diagnosed with lung fibrosis of unknown etiology were collected in the six months of the study; 19 (56%) were men, with a mean (standard deviation, SD) age of 71.4 (6.7) years and a mean BMI of 28.8 (4.8) kg/m^2, while half of them (n = 17) were current or ex-smokers with a mean history of 19.1 (28.2) pack-years. The majority received an IPF diagnosis (29 patients) according to the Official ATS/ERS/JRS/ALAT Clinical Practice Guideline [10], 3 had nonspecific interstitial pneumonia (NSIP, this pattern was proven histologically), and 2 were characterized as having unclassifiable lung fibrosis.

Overall, our patients exhibited mild ILD at diagnosis with an FVC 89.8 (20.7)% pred. and diffusing capacity (DLCO) 66.5 (23.6)% pred. Additional spirometric parameters included forced expiratory volume in 1 s (FEV1), 2.20 (0.62) L [92.7 (21.2)% pred.]; FEV1/FVC, 81.6 (5.4)%; DLCO/VA, 93.9 (26.7)%; and TLC, 73.0 (14.5)% pred. The SAD values were FEF25-75%, 85.6 (30.1)% pred.; FEF25-75%, 2.31 (0.73) L; RV/TLC%, 88.1 (14.8)% pred.; and RV/TLC, 38.0 (5.5)%. The oscillometry measurements were R5 act., 3.00 (1.10) $cmH_2O/L/s$; R5, 93.12 (37.78)% pred.; X5 act., −1.84 (0.95) cmH2O/L/s; X5, 139.28 (84.8)% pred.; Fres act, 14.97 (5.13) Hz; Fres, 117.4 (53.7)% pred.; AX act., 9.22 (9.03) cmH_2O/L; AX, 230.6 (211.9)% pred.; R19 act., 4.98 (13.59) $cmH_2O/L/s$; R19, 85.7 (21.3)% pred.; R5-19, act. 0.38 (0.28) $cmH_2O/L/s$; and R5-19, 94.9 (84.2)% pred. A total of 27% of the patients had abnormal values of R5-19, using a threshold of 0.7 $cmH_2O/L/s$, as proposed previously [11].

The correlations between oscillometry parameters and PFT parameters are shown in Table 1. Disease severity, as evaluated by FVC, presented strong correlations with the oscillometry parameters: resistance (R5: $r = -0.588$, $p < 0.001$), reactance (X5: $r = 0.671$, $p < 0.001$), resonant frequency (Fres: $r = -0.562$, $p = 0.001$), and the area of reactance (AX: $r = -0.515$, $p = 0.002$). The oscillometry parameters also correlated with FEV1 but not FEV1/FVC%. The SAD FOT parameter R5-19 was negatively correlated with FEF25-75% ($r = -0.370$, $p = 0.021$; Figure 1A) and had an acceptable performance in the prediction of small airway disease, with FEF25-75% < 60% as the "gold standard" in a ROC analysis (AUC 0.738, 95%Cl 0.519–0.956) (Figure 1B). R5-19 did not correlate with smoking history but correlated with disease severity assessed by FVC ($r = -0.481$, $p = 0.004$). Furthermore, R5-19 was the only SAD parameter that correlated with the composite physiologic index (CPI, $r = 0.338$, $p = 0.04$) (Figure 1C), while FEF25-75% ($r = -0.089$, $p = 0.619$) and RV/TLC% ($r = -0.111$, $p = 0.539$) did not.

Table 1. Correlations of oscillometry parameters with pulmonary function test parameters.

	R5 Act.		**X5 Act.**		**Fres Act.**		**AX act.**		**R5-19 Act.**	
Variables	r	*p*	r	*p*	r	*p*	r	*p*	r	*p*
FEV_1, % pred.	−0.290	0.096	0.374	0.064	−0.396	**0.020**	−0.364	**0.034**	−0.240	0.172
FEV_1, L	−0.601	**<0.001**	0.666	0.224	−0.552	**0.001**	−0.500	**0.003**	−0.483	**0.004**
FVC, % pred.	−0.287	0.099	0.397	**0.047**	−0.454	**0.007**	−0.515	**0.002**	−0.254	0.148
FVC, L	−0.588	**<0.001**	0.671	**<0.001**	−0.562	**0.001**	−0.515	**0.002**	−0.481	**0.004**
FEV_1/FVC, %	0.218	0.216	−0.289	0.549	0.243	0.166	0.245	0.163	0.174	0.324
FEF25-75, % pred.	−0.366	**0.033**	0.337	0.736	−0.228	0.195	−0.166	0.347	−0.370	**0.021**
DLCO, % pred.	−0.134	0.458	0.208	0.405	−0.251	0.159	−0.184	0.304	−0.234	0.190
DLCO/VA, % pred.	0.084	0.680	0.069	0.767	0.049	0.785	0.081	0.654	−0.085	0.639
RV/TLC, % pred.	0.074	0.680	−0.106	0.750	0.036	0.842	0.081	0.653	−0.017	0.924
RV/TLC, %	0.076	0.690	−0.145	0.973	0.238	0.205	0.159	0.401	0.131	0.468
TLC, % pred.	−0.324	0.066	0.448	0.100	−0.479	**0.005**	−0.405	**0.019**	−0.311	0.078

Bold numbers indicate statistically significant values.

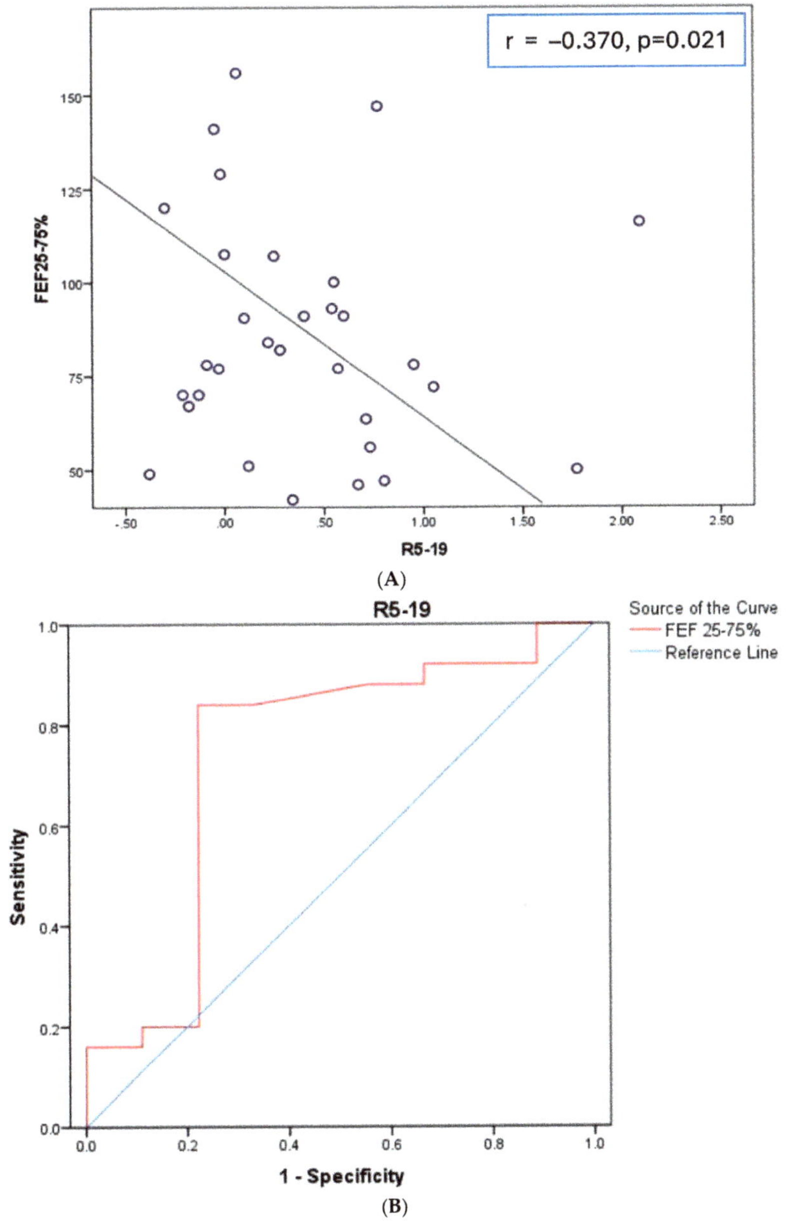

Figure 1. *Cont.*

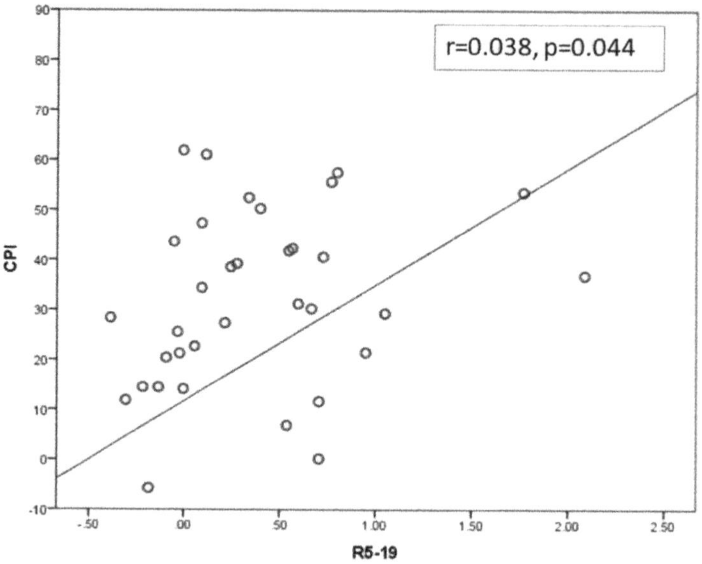

(C)

Figure 1. (**A**) Correlation of R5-19 with FEF25-75%; (**B**) receiver operating characteristic (ROC) analysis evaluating the performance of R5-19 in the identification of small airway disease as expressed by FEF25-75% < 60% pred.; (**C**) correlation of R5-19 with the composite physiologic index (CPI).

4. Discussion

In the present study, the oscillometric parameters correlated with ILD disease severity as assessed by FVC, and R5-19 had a good performance in the prediction of spirometric SAD as expressed by FEF25-75%. More importantly, R5-19 was the only SAD parameter related to CPI, a marker of disease prognosis. This study provides evidence for the potential use of oscillometry in the evaluation of disease severity and lung periphery in interstitial lung disease.

Although small airway disease is a major characteristic of COPD and this feature is less investigated in ILD, recent studies based on combined pathology and HRCT observations have demonstrated that airways are often not spared in pulmonary fibrosis. Small airway disease seems to represent an early pathologic feature of idiopathic pulmonary fibrosis, as regions of minimal fibrosis are associated with a significant reduction in the number of terminal bronchioles, while the remaining exhibit major alterations [12]. Furthermore, abnormal small airways exhibited increased expression of matrix metalloproteinases 7 and 9 in the bronchial epithelium [13]. These findings highlight the potential pathogenetic role in lung remodeling of small airways in ILD and raise treatment implications.

SAD may not be easy to assess, and many methods have been developed, including spirometry, body plethysmography, inert gas washout techniques, and imaging modalities with various concerns in terms of availability, complexity, and reproducibility [14]. In our study, oscillometry was evaluated as the main modality to assess SAD in patients with early pulmonary fibrosis. Oscillometry is a novel, easy technique evaluating proximal and distal airway function without performing a forced expiratory maneuver. This has a particular utility in children, the elderly, and patients with severe respiratory disease or specific contraindications, as it requires minimal effort [15]. Oscillometry was proven to be more sensitive than spirometry in detecting SAD in populations exposed to the World Trade Center collapse on 11 September 2001; spirometry, including measures of midexpiratory airflow rates, remained within normal limits, while small airway abnormality was correlated with severity and frequency of wheeze and was independently associated with the presence of systemic inflammation confirmed on histologic evaluation [16]. The value of oscillometry has been more extensively investigated in obstructive diseases. Previous studies have shown that oscillometry may be a more sensitive marker of SAD than FEF25-75% in asthma and COPD with preserved pulmonary function, as patients with oscillometry-defined SAD had a significantly higher incidence of respiratory symptoms compared to those with SAD defined by spirometry [17]. Furthermore, oscillometry data contribute to more accurate characterization of asthma phenotype [18].

Our results indicate that oscillometric parameters correlate with ILD severity, as assessed by FVC, even in mild ILD, consistent with previous reports in more advanced fibrotic disease. Using the method described by Horita and colleagues [19] to identify potential differences in the r values in different correlations between oscillometric and pulmonary function testing parameters, we were not able to identify differences that exceeded the minimal clinically important difference. In a cohort of eighty Japanese patients with IPF that underwent high-resolution computed tomography (HRCT) and aimed to assess the utility of oscillometry as a potential marker for traction bronchiectasis and airflow obstruction, spirometry and oscillometry were investigated in relation to fibrosis-related HRCT findings [20]. FVC correlated negatively with oscillometric parameters and HRCT scores, while respiratory reactance correlated positively with all fibrosis-related HRCT scores [19]. Zhiang et al. showed that spirometry-defined SAD was associated with significantly increased risk of mortality in patients with IPF (HR 1.73, 95% CI 1.02–2.92) [5]. In a retrospective analysis, Yin and colleagues demonstrated that one-third of IPF patients had SAD, assessed by spirometric indices; these patients had a significantly shorter survival compared to non-SAD patients. Histopathological examination in those undergoing lung transplantation presented various degrees of airway distortion and obliteration in triple the number of patients than were captured by spirometry [4]. Finally, Mikamo et al. found that oscillometry was more sensitive than spirometry in the detection of HRCT-defined SAD [21]. Ninety patients with different ILDs were evaluated according to the HRCT findings (mosaic attenuation, air trapping, and centrilobular micronodules) as having or not having SAD. Although pulmonary function parameters (%FVC, %FEV1, FEV1/FVC, FEF 25-75, IC, TLC, FRC, RV/TLC, and %DLCO) did not differ between the two groups, oscillometry parameters were significantly associated with the radiological presence of SAD [21]. The absence of correlations between oscillometric parameters and DLCO% or DLCO/VA in our pilot study likely suggests that oscillometry, focusing on small airway pathology, evaluates different pathophysiological aspects than diffusing capacity in this population of patients with mild ILD and may be complementary in the holistic evaluation of this population.

Our study showed that in patients without radiological evidence of obstruction, R5-19 captured SAD abnormality in one-third, correlated with disease severity and the prognostic score CPI. SAD may play a role in the worsening of respiratory symptoms and contribute to mortality risk in more advanced ILD disease, but importantly, early involvement of small airways in the disease course may imply its potential role in pathogenesis of pulmonary fibrosis [4,5]. Despite the small number of participants and the cross-sectional design of this study, our data are derived from a well-characterized population of patients with early pulmonary fibrosis. We excluded patients with pulmonary fibrosis in the context of collagen vascular disease or hypersensitivity pneumonitis, where SAD may be a cardinal pathological and radiological feature caused by different mechanisms from idiopathic fibrotic diseases. In contrast to IPF, small airway dysfunction is well recognized in rheumatoid arthritis and Sjogren's syndrome, where obliterative and follicular bronchiolitis are common findings; similarly, in hypersensitivity pneumonitis (HP), granuloma formation and lymphocytic infiltrates lead to small airway obstruction [22]. Our results show that small airways may be abnormal in early pulmonary fibrosis as assessed by R5-19, irrespective of smoking habit, and this finding correlates with FVC and the CPI. Further research focused on small airways, and their epithelium is likely warranted in the context of pulmonary fibrosis, which may stimulate exploration of new treatable traits. Small airway-targeted treatment via aerosolized drug delivery onto the airway epithelium could be a therapeutic option in the early stage of pulmonary fibrosis [23].

The small sample size is a potential limitation of our pilot study, yet the identification of patients with mild ILD and the meticulous evaluation of these patients, including pulmonary function testing and forced oscillometry on the same day at the time of the initial diagnosis supports the validity of our observations. The cross-sectional design of this study represents an additional limitation, and future prospective cohort studies in larger populations are certainly needed to identify the role of oscillometry in the evaluation and management of patients with ILD.

In conclusion, this is the first physiological study to our knowledge, investigating patients with early pulmonary fibrosis, with no evidence of comorbid obstructive disease, focusing on small airways. Our data highlight that oscillometry is an easy-to-use technique revealing early mechanical alterations caused by pulmonary fibrosis and captures SAD, a sensitive marker of peripheral lung pathology, which may have prognostic implications for ILD patients.

Author Contributions: Conceptualization, A.G. and K.K. (Konstantinos Kostikas); methodology, A.G.; software, K.K. (Konstantinos Kostikas); validation, A.G., G.G., C.K., K.T. and K.K. (Konstantinos Kostikas); formal analysis, C.K.; writing—original draft preparation, A.G., G.G. and K.K. (Konstantinos Katsoulis); writing—review and editing, all authors; visualization, C.K.; supervision, K.K. (Konstantinos Kostikas); project administration, C.T.; funding acquisition, n/a. All authors have read and agreed to the published version of the manuscript.

Funding: This research received no external funding.

Institutional Review Board Statement: This study was performed in accordance with the recommendations of the Declaration of Helsinki, the International Conference of Harmonization—Good Clinical Practice (ICH-GCP) Guidelines, the EU-Directive 2001/20, and all national requirements and was approved by the Scientific Council of our hospital (approval number 21/4 April 2021).

Informed Consent Statement: Written informed consent was obtained from all participants prior to their inclusion in this study.

Data Availability Statement: The data are available from the authors upon request.

Conflicts of Interest: Athena Gogali has received honoraria for presentations and consultancy fees from AstraZeneca, Boehringer Ingelheim, Chiesi, ELPEN, GSK, and Menarini. Konstantinos Kostikas has received honoraria for presentations and/or consultancy fees from AstraZeneca, Boehringer Ingelheim, Chiesi, ELPEN, GSK, Guidotti, Menarini, Pfizer, Sanofi, and Specialty Therapeutics. Georgia Gkrepi, Christos Kyriakopoulos, Christos Kyriakopoulos, Konstantinos Tatsis, Konstantinos Katsoulis, and Chara Tselepi have no conflicts of interest to declare.

References

1. Shaw, R.J.; Djukanovic, R.; Tashkin, D.P.; Millar, A.B.; Du Bois, R.M.; Corris, P.A. The role of small airways in lung disease. *Respir. Med.* **2002**, *96*, 67–80. [CrossRef] [PubMed]
2. Nicholson, A.G.; Fulford, L.G.; Colby, T.V.; du Bois, R.M.; Hansell, D.M.; Wells, A.U. The relationship between individual histologic features and disease progression in idiopathic pulmonary fibrosis. *Am. J. Respir. Crit. Care Med.* **2002**, *166*, 173–177. [CrossRef] [PubMed]
3. Ikezoe, K.; Hackett, T.L.; Peterson, S.; Prins, D.; Hague, C.J.; Murphy, D.; LeDoux, S.; Chu, F.; Xu, F.; Cooper, J.D.; et al. Small Airway Reduction and Fibrosis Is an Early Pathologic Feature of Idiopathic Pulmonary Fibrosis. *Am. J. Respir. Crit. Care Med.* **2021**, *204*, 1048–1059. [CrossRef]
4. Yin, C.; Xie, H.; He, X.; Zhang, Y.; Zhang, A.; Li, H. Small airway dysfunction in idiopathic pulmonary fibrosis. *Front. Pharmacol.* **2022**, *13*, 1025814. [CrossRef]
5. Zhang, X.; Xie, B.; Ban, C.; Ren, Y.; Ye, Q.; Zhu, M.; Liu, Y.; Zhang, S.; Geng, J.; Jiang, D.; et al. Small airway dysfunction in Chinese patients with idiopathic pulmonary fibrosis. *BMC Pulm. Med.* **2022**, *22*, 297. [CrossRef]
6. Kaminsky, D.A.; Simpson, S.J.; Berger, K.I.; Calverley, P.; de Melo, P.L.; Dandurand, R.; Dellacà, R.L.; Farah, C.S.; Farré, R.; Hall, G.L.; et al. Clinical significance and applications of oscillometry. *Eur. Respir. Rev.* **2022**, *31*, 210208. [CrossRef]
7. King, G.G.; Bates, J.; Berger, K.I.; Calverley, P.; De Melo, P.L.; Dellacà, R.L.; Farre, R.; Hall, G.; Ioan, I.; Irvin, C.G.; et al. Technical standards for respiratory oscillometry. *Eur. Respir. J.* **2020**, *55*, 1900753. [CrossRef] [PubMed]

8. Goh, N.S.L.; Desai, S.R.; Veeraraghavan, S.; Hansell, D.M.; Copley, S.J.; Maher, T.M.; Corte, T.J.; Sander, C.R.; Ratoff, J.; Devaraj, A.; et al. Interstitial lung disease in systemic sclerosis. *Am. J. Respir. Crit. Care Med.* **2008**, *177*, 1248–1254. [CrossRef]
9. Wells, A.U.; Desai, S.R.; Rubens, M.B.; Goh, N.S.; Cramer, D.; Nicholson, A.G.; Colby, T.V.; Du Bois, R.M.; Hansell, D.M. Idiopathic pulmonary fibrosis: A composite physiologic index derived from disease extent observed by computed tomography. *Am. J. Respir. Crit. Care Med.* **2003**, *167*, 962–969. [CrossRef] [PubMed]
10. Raghu, G.; Remy-Jardin, M.; Richeldi, L.; Thomson, C.C.; Inoue, Y.; Johkoh, T.; Kreuter, M.; Lynch, D.A.; Maher, T.M.; Martinez, F.J.; et al. Idiopathic pulmonary fibrosis (an update) and Progressive pulmonary fibrosis in adults: An official ATS/ERS/JRS/ALAT clinical practice guideline. *Am. J. Respir. Crit. Care Med.* **2022**, *205*, e18–e47. [CrossRef] [PubMed]
11. Crisafulli, E.; Pisi, R.; Aiello, M.; Vigna, M.; Tzani, P.; Torres, A.; Bertorelli, G.; Chetta, A. Prevalence of small-airway dysfunction among COPD patients with different GOLD stages and its role in the impact of disease. *Respiration* **2017**, *93*, 32–41. [CrossRef]
12. Verleden, S.E.; Tanabe, N.; McDonough, J.E.; Vasilescu, D.M.; Xu, F.; Wuyts, W.A.; Piloni, D.; De Sadeleer, L.; Willems, S.; Mai, C.; et al. Small airways pathology in idiopathic pulmonary fibrosis: A retrospective cohort study. *Lancet Respir. Med.* **2020**, *8*, 573–584. [CrossRef]
13. Figueira de Mello, G.C.; Ribeiro Carvalho, C.R.; Adib Kairalla, R.; Saldiva, P.H.N.; Fernezlian, S.; Silva, L.F.F.; Dolhnikoff, M.; Mauad, T. Small airway remodeling in idiopathic interstitial pneumonias: A pathological study. *Respiration* **2010**, *79*, 322–332. [CrossRef]
14. Katsoulis, K.; Kostikas, K.; Kontakiotis, K. Techniques for assessing small airways function: Possible applications in asthma and COPD. *Respir. Med.* **2016**, *119*, e2–e9. [CrossRef]
15. Galant, S.P.; Komarow, H.D.; Shin, H.W. The case for impulse oscillometry in the management of asthma in children and adults Ann Allergy Asthma Immunol. *Ann. Allergy Asthma Immunol. Ann. Allergy Asthma Immunol.* **2017**, *118*, 664–671. [CrossRef] [PubMed]
16. Oppenheimer, B.W.; Goldring, R.M.; Herberg, M.E.; Hofer, I.S.; Reyfman, P.A.; Liautaud, S.; Rom, W.N.; Reibman, J.; Berger, K.I. Distal airway function in symptomatic subjects with normal spirometry following World Trade Center dust exposure. *Chest* **2007**, *132*, 1275–1282. [CrossRef]
17. Chiu, H.-Y.; Hsiao, Y.-H.; Su, K.-C.; Lee, Y.-C.; Ko, H.-K.; Perng, D.-W. Small airway dysfunction by impulse oscillometry in symptomatic patients with preserved pulmonary function. *J. Allergy Clin. Immunol. Pract.* **2020**, *8*, 229–235.e3. [CrossRef]
18. Chan, R.; Lipworth, B. Interactions between spirometry and oscillometry in patients with moderate to severe asthma. *Eur. Respir. J.* **2022**, *60*, 2200543. [CrossRef]
19. Horita, N.; Yamamoto, S.; Mizuki, Y.; Kawagoe, T.; Mihara, T.; Yamashiro, T. Minimal clinically important difference (MCID) of effect sizes other than mean difference. *J. Clin. Quest.* **2024**. [CrossRef]
20. Yamamoto, Y.; Miki, K.; Tsujino, K.; Kuge, T.; Okabe, F.; Kawasaki, T.; Matsuki, T.; Kagawa, H.; Miki, M.; Kida, H. Oscillometry and computed tomography findings in patients with idiopathic pulmonary fibrosis. *ERJ Open Res.* **2020**, *6*, 00391–02020. [CrossRef] [PubMed]
21. Mikamo, M.; Fujisawa, T.; Oyama, Y.; Kono, M.; Enomoto, N.; Nakamura, Y.; Inui, N.; Sumikawa, H.; Johkoh, T.; Suda, T. Clinical significance of forced oscillation technique for evaluation of small airway disease in interstitial lung diseases. *Lung* **2016**, *194*, 975–983. [CrossRef] [PubMed]
22. Barkas, G.I.; Daniil, Z.; Kotsiou, O.S. The role of small airway disease in pulmonary fibrotic diseases. *J. Pers. Med.* **2023**, *13*, 1600. [CrossRef] [PubMed]
23. Maher, T.M. Small airways in idiopathic pulmonary fibrosis: Quiet but not forgotten. *Am. J. Respir. Crit. Care Med.* **2021**, *204*, 1010–1011. [CrossRef] [PubMed]

MDPI AG
Grosspeteranlage 5
4052 Basel
Switzerland
Tel.: +41 61 683 77 34

Diagnostics Editorial Office
E-mail: diagnostics@mdpi.com
www.mdpi.com/journal/diagnostics

www.ingramcontent.com/pod-product-compliance
Lightning Source LLC
Chambersburg PA
CBHW061514210926
51850CB00008B/1911